T0362105

Fluid Dynamics of OIL PRODUCTION

Fluid Dynamics of OIL PRODUCTION

BAKYTZHAN ZHUMAGULOV

VALENTIN MONAKHOV

AMSTERDAM • BOSTON • HEIDELBERG • LONDON
NEW YORK • OXFORD • PARIS • SAN DIEGO
SAN FRANCISCO • SINGAPORE • SYDNEY • TOKYO
Gulf Professional Publishing is an imprint of Elsevier

Gulf Professional Publishing is an imprint of Elsevier
525 B Street, Suite 1900, San Diego, CA 92101-4495, USA
32 Jamestown Road, London NW1 7BY, UK
225 Wyman Street, Waltham, MA 02451, USA

British Library Cataloguing-in-Publication Data
A catalogue record for this book is available from the British Library

Library of Congress Cataloging-in-Publication Data
A catalog record for this book is available from the Library of Congress

ISBN: 978-0-12-416635-6

For information on all Gulf Professional Publishing visit our website at elsevierdirect.com

Printed and bound by CPI Group (UK) Ltd, Croydon, CR0 4YY
Transferred to digital print 2012

CONTENTS

PREFACE

SUBSURFACE FLUID DYNAMICS

At the present time, mathematical modelling is frequently applied to the mechanics of continuous media and in particular to subsurface fluid dynamics. The latter deals with such important theoretical and practical issues as water flow through dams, soil salinization, the spread of pollution by groundwater flows, oil production, groundwater flow into artesian wells and many others. The similarity of the physical processes involved in these phenomena means that their models also have many similarities, although the model equations all have their own special characteristics. It is in fact these special characteristics that make it very difficult to validate the models and solve the equations.

Filtration is defined as fluid flow through a porous medium. A medium is regarded as porous if it contains a large number of voids which are small by comparison with the typical dimensions of the medium. Porosity is defined quantitatively by the ratio of pore volume to bulk volume: $m = V_{por}/V_{total}$. Mathematical flow models are based on the law of conservation, the mechanics of continuous media, their effects, and other accepted equations. Primary equations include the continuity equation (taking porosity into account), the heat balance equation and equations of state. The main assumption of the flow theory is the replacement of Euler's or Navier-Stokes equations of motion with Darcy's Law.

The simplest two-phase flow model is the well-known Buckley-Leverett (BL) model (Chapter 1, Section 1.1), which assumes the equality of phase pressures, and therefore does not allow for the effect of capillary forces on fluid flow. The difficulties which arise in solving its equations (the potential ambiguity of the solution) are resolved by making the flow process mathematically ideal by assuming that the final function contains a point of inflexion. Convective processes are central to the BL model. To take additional effects into account, the mathematical flow model needs to be adjusted in various ways.

The introduction of capillary forces produces the Muskat-Leverett (ML) model (Chapter 1, Sections 1.1–1.4) which uses a Laplace equation to allow for these forces. Unidimensional transformation of the model produces a non-linear degenerate second-order differential equation. The solution of this equation has no point of inflexion, and the high-gradient region is confined to a limited area, which is perfectly justified in physical terms. Another advantage of this equation is that although it is parabolic, the model retains an important and physically natural property, in that it allows perturbations to propagate within a defined range of velocities (provided that its functional parameters have been correctly selected) (Chapter 2, Section 2.1).

The flow model is further refined (and therefore further complicated) if we allow for the interaction of velocity and temperature in oil-bearing formations, which means that an energy equation needs to be added to the model. Models of non-isothermal two-phase flow were studied by O.B. Bocharov, V.N. Monakhov, R. Yuing (MLT-model) [2; 15; 16; 44], E.B. Chekalyuk [140] and others. O.B. Bocharov and V.N. Monakhov [16] proposed and investigated an even more generalized MLT-model, which included variable (temperature-dependent) residual saturation.

Other generalized flow models include non-linear, multiphase and multicomponent flow models and others.

In our book, we concentrate on the effect of temperature on fluid flow processes as applied to modelling water-oil displacement and the production of fluid. The inclusion of non-isothermal flow makes it possible to approximate the real conditions, making the physical and therefore the mathematical model less abstract, and provides some corrections to the accepted hydrodynamic methods of calculating oil production.

Studies have shown that oil recovery factors can be significantly increased only by changing the physical and physico-chemical properties of the displaced phase, withthermal recovery being increasingly favoured. The importance of thermal recovery methods is largely due to the fact that they use easily available media—water and air. Another major advantage over most other methods (e.g. physico-chemical) is the potential for increasing recovery in a variety of physico-geological oil field conditions. Thermal recovery methods are based on the fact that the viscosity of oil decreases considerably when it is heated, so that their primary application is in high-viscosity oil fields. At the same time, thermal recovery involves virtually all known oil displacement mechanisms, together with a variety of phase transitions, so that it offers promise even in the case of low-viscosity oil fields which have long been operated under water injection. It should be noted that the injection of water at a temperature lower than formation temperature (e.g. sea water or injection during winter) reduces oil recovery. In particular, it may lead to wax precipitation directly in the porous medium.

It is well known that in water wet rocks capillary forces can play a very important role in the process of oil displacement. If a low-permeability section is surrounded by high–permeability rock, the water will flow around the oil contained in the section. If water wet formations are flooded, the oil can be frequently recovered only by the use of capillary forces. The existence of this mechanism has been confirmed both experimentally and by analysing fields consisting of inhomogeneous water wet rocks. Capillary saturation may also have a decisive influence on the mechanism of oil recovery in stratified beds. Therefore, we need to know how the non-isothermal process of oil displacement by capillary forces will affect the recovery of oil from such heterogeneous formations.

All these phenomena require thorough study, and the Muskat-Leverett thermal flow model provides an effective tool.

Many problems formulated using these models can be studied in a given sequence, forming a specific process cycle, such as for instance steam treatment (Chapter 2, Section 2.1), which may be described in a simplified form as consisting of the following steps (the corresponding mathematical statements are shown in brackets):

1. Steam (superheated water) injection at a specified temperature and flow rate (non-isothermal two-phase flow with convective forces predominating);
2. Soaking for a specified time without water injection (thermocapillary saturation)
3. Steam or water injection (possibly, at a different temperature and flow rate) (non-isothermal two-phase flow with convective forces predominating).

Therefore, if we know how to model these steps we can use them to study more complicated processes and make multivariate optimizing calculations.

For all the above models, we need to find specific solutions, including self-similar (analytical) solutions, and this problem is dealt with in Sections 2.1–2.5 of Chapter 2.

NUMERICAL MODELLING OF OIL PRODUCTION PROCESSES

The most common oil-field development systems are based on symmetrical well patterns. This means that rather than studying a whole field, we can study a single development unit, which usually consists of two wells. For example, for a five-spot water flood, the basic element is a rectangle with no flow boundaries, containing an injection and a production well in opposite corners.

Since calculations of the development of basic elements of symmetrical well patterns can be reduced to calculations of linear flow (for an in-line pattern) or a plane-radial flow (for an areal pattern), this simplifies flow model equations, making them one—dimensional.

The formulation of the initial and boundary conditions for the basic elements is also simplified: the production rate, pressure or saturation are specified for each well. Consolidated figures are then calculated for the production unit as a whole, followed by the calculation of the 2D process of two-phase flow in the basic element—this program can be attached to more detailed multi-parameter 1D programs, providing them with coefficients allowing for the fact that the processes are not unidimensional. However, the 2D basic element calculation is important not only because it supplements the 1D programs but also as a stand-alone petroleum engineer's tool, in which the multiple parameters of the 1D models can be incorporated, provided sufficient computing power is available. In addition, as field development proceeds, well patterns and well operation become asymmetrical, and this can only be allowed for by 2D calculations, performed by a program which calculates the process of oil production in a 2D basic element without assuming that the boundary conditions are symmetrical.

The calculation program produces oil saturation and pressure fields within the basic element and calculates the oil recovery factor and water cut as a function of the injected pore volumes of water. The information may be presented in graphical form and then printed out as data files for use in further analysis of the oil production process and/or printed out as numerical files.

The contents of the book. If we include submodels and combined models, Chapter 1 contains the description of over 30 different mathematical models of oil formations, provides analyses of a number of some generally accepted flow models and offers new models of some physical effects not covered elsewhere. In designing these models, we have attempted to achieve a good numerical implementation without increasing the number of their key parameters. As a rule, the proposed model design changes are accompanied by small "slippage" terms introduced into the equation by analogy with the computing "slippage" in finite-difference equations. It should be noted that other authors have also introduced some of the filtration model changes proposed in Chapter 1, but did not analyse the resultant models sufficiently.

S.N. Antontsev and V.N. Monakhov [4] proposed a general oil formation flow model containing a range of functional parameters. By making a careful selection, some of the models proposed in Chapter 1 can be derived from them.

Mathematical models can be subdivided into three main classes:

1. Single-phase Darcy models and contact models (Section 1.1);
2. Two-phase models (e.g. the Muskat-Leverett model – Sections 1.1–1.4);
3. Combined models (e.g. of two inhomogeneous liquids – Section 1.7).

In addition to the conventional Darcy and Muskat-Leverett models and the Muskat-Leverett thermal model (in the form proposed by O.B. Bocharov and V.N. Monakhov [15; 16]), Chapter 1 describes the Navier-Stokes and Zhukovsky models (Section 1.1), used by the authors to optimize oil production control and production forecasts.

The book also contains some unconventional models, such as the models describing the process of "foaming" in oil formations (Section 1.6), the combination of reservoir flow with liquid flow in wells (Section 1.5) and others.

Of the new and modified models (e.g. the reduced-pressure ML and MLT models) Chapter 1 discusses only the models developed by V.N. Monakhov and studied by him and his colleagues and students, S.N. Antontsev, O.B. Bocharov, A.A. Papin, R. Yuing, E.M. Turganbayev, V. N. Starovoitov, N.V. Khusnutdinov, A.E. Osokin, and others [4; 15; 16; 18; 20; 32; 44; 61; 69; 75; 91; 94; 101; 124; 134].

Chapter 2 presents a theoretical and numerical analysis of one-dimensional and self-similar (analytical) thermal two-phase flow patterns, while its Section 2.1 provides additional information based on the ordinary differential equation theory, which is also of independent interest.

The core of the chapter is formed by Sections 2.2 and 2.3, which present the results of V.N. Monakhov, O.B. Bocharov, A.E. Osokin, and T.V. Kantayeva's work [20; 69; 92]. These include the theorem of existence of self-similar (analytical) solutions of the MLT model for constant and variable residual saturation, the identification of a restricted range of velocities of propagation of perturbations, and the computer implementations of the numerical algorithms proposed by the authors and their substantiation.

Section 2.4 contains a theoretical analysis of the analytical solutions (B.T. Zhumagulov, V.N. Monakhov [58].

The existence and uniqueness of self-similar (analytical) solutions of the model of two-phase flow of non-linear-viscous liquids is demonstrated in 2.5 (E.G. Galkina, A.A. Papin [32]). Section 2.6 establishes the convergence of Rothe-type methods in a one-dimensional MLT model (A. E. Osokin [100]).

Section 2.7 is devoted to the substantiation of a new method of integrating ML and BL model solutions and to their numerical implementation (I.G. Telegin, [129]; B.T. Zhumagulov, Sh.S. Smagulov, V.N. Monakhov, N.V. Zubov [61]).

The existence and uniqueness of "im Kleinen" (small-scale) solutions of the first boundary-value problem, based on the initial data for the two interpenetrating viscous liquids flow model is demonstrated in Section 2.8 (A. A. Papin [101]).

Chapter 3 deals with numerical modelling of two-dimensional subsurface hydrodynamics processes with reference to Muskat-Leverett isothermal and temperature models as well as Navier-Stokes and Zhukovsky models.

In this chapter, Section 3.1 demonstrates the convergence and stability of effective finite–difference schemes [38] for Navier-Stokes velocity vs. pressure finite difference equations, while Section 3.2 uses velocity vs. flux function and the method of virtual regions to provide numerical calculations of reservoir flows in multiply connected regions [45] and geometrically complex regions (Section 3.3) [45]. In Section 3.4, similar numerical methods are applied to the Zhukovsky model. [49].

Section 3.5 provides a solution to a key problem of subsurface hydrodynamics—that of determining formation pressure from measured well pressure values [49]. We have performed a numerical calculation of formation heating, which forms one of the stages of steam treatment, based on the classical thermal convection model (Section 3.6) [63]. Section 3.9 [49] provides a numerical solution of water-oil displacement from inhomogeneous oil formations, based on the ML model, while Section 3.7 and Section 3.8 present the mathematical substantiation of the finite-difference equations used in Section 3.8 for more general models [18, 44]. And finally, Section 3.10 offers a hydrodynamic analysis of the results of numerical calculations of subsurface hydrodynamics problems based on different formation models [57].

Sections 3.1–3.6 and 3.9 present the results obtained by B.T. Zhumagulov and his colleagues, Sh.S. Smagulov, N.T. Danayev, B.G. Kuznetsov, G.T. Balakayeva, N.T. Temirbekov, K.Zh. Baigelov, K.M. Baimirov, K.B. Esikeyev [38; 47-52; 56; 59].

Results obtained jointly by V.N. Monakhov with R. Ewing [44], O. B. Bocharov [18] and B.T. Zhumagulov [57] appear in Sections 3.7, 3.8 and 3.10.

ACKNOWLEDGEMENTS

In 1996, B.T. Zhumagulov, N.V. Zubov, V.N. Monakhov, and Sh.S. Smagulov published a book entitled *New Computer Technologies in Oil Production* (Almaty, Gylym). It described a computer-aided oil and gas field development analysis system developed jointly by a team of Russian and Kazakh scientists, led by the authors.

Subsequently, the research continued independently at the Lavrentyev Institute of Hydrodynamics, where it was led by V.N. Monakhov, at the Kazakhstan Engineering Academy and the Al-Farabi Kazakh State University, where it was conducted by B.T. Zhumagulov, Sh.S. Smagulov and their colleagues.

These studies form the basis of this book, which presents both the authors' own and collaborative work, except for Sections 2.3, 2.5–2.8 of Chapter 2, which include the work of V.N. Monakhov's students, A. A. Panin, A. E. Osokin, T. V. Kantayeva, I.G. Telegin and E. G. Galkina.

There is no doubt of the important contribution to the content and the scientific value of the book made by our co-authors, Sh. S. Smagulov, S.N. Antontsev, N.T. Danayev, O.B. Bocharov, N.V. Khusnutdinov, B.G. Kuznetsov, and N.M. Temirbekov, to all of whom we wish to express our sincere gratitude. To our editor-in-chief, R.I. Nigmatulin, and our reviewer, Sh.S. Smagulov, our profound gratitude for the many years of fruitful support, which has in many ways determined the style and the conceptual framework of the book.

Thanks are due to N.M. Temirbekov, who read the manuscript and made many helpful suggestions.

The work was partly was financed by the Russian Fund for Fundamental Research (Project Code 99-01-00622), and the "Universities of Russia" science programme (Project 1788).

Fluid Dynamics of Oil Production was selected for publication from among scientific texts submitted in 2000, and is published by the Ministry of Education and Science of the Republic of Kazakhstan.

CHAPTER 1

Fluid Flow Models

1.1 INTRODUCTION

In this chapter we analyze a number of well-known mathematical models of homogeneous and non-homogeneous fluid flow in porous media, and propose some new models. As the existing models [89; 105; 143] are based on specific conceptions of these processes, the inclusion of each new effect requires a revision of their underlying assumptions, as well as a revision of the model.

The fact that there are many forms of Darcy's Law means that we need to ask ourselves how to select the form which will best describe each specific situation. While the work on this question has progressed in recent decades, it has involved virtually no review of the fundamentals of conventional models. Frequently, experimental data processed to fit the conventional models have been unstable (not easily reproducible), while published experimental results did not, as a rule, provide sufficient information to fit them to other models. Some effects are simply impossible to describe in terms of the existing models.

Basic mathematical analysis of the various forms of flow models may prove extremely useful for the modelling of phenomena. At the same time, new physical factors need to be taken into account, that is, the minor effects which stabilize the numerical calculations (i.e. the physical "slippage terms" in the equations). For instance, transition to linear models often leads to a loss of divergence in equations, and when it comes to numerical calculations, does not simplify the initial nonlinear model. Equally, striving to achieve a mathematically satisfactory model can lead to a lack of conformity with the physics of the phenomenon, as is the case with the divergent form of Darcy's Law for inhomogeneous media, proposed by Sheidegger [143].

At present, the widespread use of computers has led to the establishment of a well-defined "process flow diagram" for solving specific problems in the mechanics of continuous media, including multiphase fluid flow. The work flow progresses from the problem under consideration to a mathematical model, from there to a numerical algorithm, the implementation software and finally to the analysis of the results. While the individual components of the process are not isolated but interconnected, linking both forwards and backwards, the most important factor for success is likely to be the choice of an appropriate mathematical model.

There are several principal requirements applicable to phenomenological flow models:

1. Experiment reproducibility. The ability to define all parameters experimentally, without needing to involve additional "theories", and good reproducibility of the experiments.
2. A clear distinction between the underlying hypotheses, and a clear definition of the limits of their applicability, both in qualitative terms (what kinds of physical effects they can describe) and quantitatively.
3. Ability to incorporate simpler models into higher-level models, so that new physical factors can be taken into account.
4. Mathematical feasibility and correctness.

Needless to say, these are not rigid requirements and could even be seen as programmes of study of the models. Moreover, the features of phenomenological models can be determined in laboratory conditions, using higher-level models containing independently determined parameters. For instance, the Navier-Stokes model could be used to determine phase permeabilities in two-phase flow models and to check various properties (e.g. saturation). Below we comment on several examples of multiphase fluid flow. There is no point in calculating total oil recovery using models which specify the total flow rate for injection wells, and the flow rate of only the displaced phase for production wells. If the phases are incompressible, then the answer lies in correctly stating the well conditions.

With these examples, we hope to have provided some insight into the difficulties of choosing an appropriate model with which to describe the physical process of fluid flow in porous media as it occurs in reality.

1.2 SINGLE-PHASE AND TWO-PHASE FLUID FLOW MODELS

1.2.1 Darcy's Model and Contact Models

1.2.1.1 The Properties of Porous Media

The main property of a porous material, its porosity (effective or dynamic), is described by the ratio $m = V_n / V_0$, where V_n is the interconnected pore volume and V_0 is the bulk volume.

If the compressibility of the medium is taken into account, and it is assumed that the medium is elastic, i.e. it obeys Hooke's Law, then

$$m = m_0 + (p - p_0)\beta_m, \tag{1.1}$$

where p is the fluid pressure; p_0 is the average pressure of the porous medium.

As a rule in inhomogeneous medium m and m_0 are functions of the coordinates $x = (x_1, x_2, x_3)$.

The adsorption of molecules on the walls of porous materials results in high near-wall concentrations. Using the kinetic theory of gases, Langmuir calculated the mass w_a of adsorbed gas at a gas pressure p:

$$\omega_a = \frac{abp}{1 + ap}(a, b = const). \tag{1.2}$$

Changes in pore pressure can result in the escape of the adsorbed gas, i.e. in gas sorption. Taking these processes into account, we can calculate the total mass w of gas in a porous medium from the formula

$$\omega = mp(1 - m)\frac{abp}{1 + ap} = \omega(p, x). \tag{1.3}$$

The flow properties of porous media are described by a symmetrical flow tensor $K = \{k_{ij}\}$, whose terms k_{ij} have an areal dimension.

We shall define an *ideal porous medium* as consisting of a porous material which is incompressible, homogeneous and isotropic, and displays a linear resistance to the fluid flow.

For an ideal porous medium, the flow tensor is expressed by

$$K = kE,$$

where E is the identity matrix; $k = const$ is the permeability coefficient.

In the case of inhomogeneous media, the permeability coefficient is a function of the coordinates $k = k(x)$, and in the case of compressible media $k = k(p)$; if its resistance behaviour is nonlinear it is $k = k(\nabla p)$, whereas in the case of anisotropic media $K = \{k_{ij}\}$ is a symmetrical flow tensor.

1.2.1.2 Fluid Properties

Fluid properties are a flow velocity vector $\vec{v}$ whose value $v = |\vec{v}|$ coincides with specific fluid flow rate per unit of time through a porous area normal to $\vec{v}$; pressure p and density ρ. The relationship between flow velocity $\vec{v}$ and average fluid particlevelocity $\vec{u}$ is described by the formula

$$\vec{v} = m\vec{u} = m\frac{d\vec{x}}{dt}, \quad \vec{x} = (x_1, x_2, x_3).$$

1.2.1.3 Darcy's Model

Flow theory is based on Darcy's Law, which is a law derived experimentally for stationary flows and describing the resistance of porous media to fluids flowing through them [87, 89, 105]; Darcy's Law establishes the relationship between flow velocity $\vec{v}$; and pressure gradient p.

In the general case, Darcy's Model (D) is based on Darcy's Law combined with the equation of continuity and equation of state for fluids:

$$-\vec{v} = K\mu^{-1}(\nabla p + \rho\vec{g}); \quad \frac{\partial}{\partial t}\omega(p,x) + div\rho\vec{v} = 0; \quad \rho = \rho(p); \tag{D}$$

where μ is the dynamic fluid viscosity; $\rho\vec{g}$ is the gravity acceleration vector; the total mass of gas $\omega(\rho)$ is described by (1.3) (in the case of fluids $\omega = m\rho$).

By substituting $\vec{v} = \vec{v}(x, p, \nabla p)$, taken from the second equation, in the first equation (D), we obtain a single equation for pressure p:

$$\frac{\partial}{\partial t}\omega(p,x) + div\vec{v}(x, p, \nabla p) = 0, \quad \rho = \rho(p) \tag{1.4}$$

Equation (1.4) is usually a degenerate (simplified) parabolic equation.

In the specific case of an ideal porous medium and incompressible fluid ($\rho = const$), Darcy's model assumes the following simple form:

$$-\vec{v} = \frac{k}{\mu}(\nabla p + \rho\vec{g}), \quad div\vec{v} = 0. \tag{D0}$$

A mathematical theory of two-dimensional stationary flow in ideal porous media [in (D_0),$k = const$] was first put forward in P. Ya. Polubarinova-Kochina's ground-breaking work, described in her monograph [105], which also includes the research of her students.

V.N. Monakhov developed boundary value theory methods for elliptical equation systems and quasi-conformal mappings, which made it possible to investigate the mathematical correctness of fluid flow problems in specified regions, as well as free boundary regions in non-ideal porous media ([89, Chapter 8], [90]), when $K = k_{ij}$ is a tensor (D) and $k_{ij}(\vec{x}, p, |\vec{v}|)$.

1.2.1.4 The Forms of Darcy's Law in the Case of Homogeneous Fluid Flow in Inhomogeneous Porous Media

As noted by Sheidegger [143], if we allow for the inhomogeneity of the medium, we obtain the following two different generalized forms of Darcy's Law for the same type of fluid flow:

$$\vec{v} = k\mu^{-1}\nabla\varphi; \tag{D1}$$

$$\vec{v} = -\nabla(k\mu^{-1}\varphi), \quad k = k(x), \tag{D2}$$

These two forms are equivalent if the medium is homogeneous, and $k = const$ $(\varphi = p + g\rho h,\ g\nabla h = \vec{h})$. Form ($D_2$) of Darcy's Law is mathematically more convenient, since the introduction of a new potential $\Phi = \frac{k}{k_0}\varphi$, where $k_0 = const$- average permeability, reduces it to the form $\vec{v} = -k_0\mu^{-1}\nabla\Phi$, which corresponds to a homogeneous medium. However, in this case the inhomogeneity of the medium affects flow velocity $\vec{v}$ only through the values of $k(x)$ and ∇k at the boundary $\Gamma = \partial\Omega$ of flow region Ω. In particular, when $k(x)|_\Gamma = const$ and $\nabla k|_\Gamma = 0$, the distribution of flow velocity $\vec{v}$ in Ω does not depend on the inhomogeneity of the medium. Consequently, form (D_2) of Darcy's Law is physically unacceptable, while form (D_1) is not able to account properly for the anisotropy of permeability in stratified formations.

Darcy's Law can also take another form in inhomogeneous media:

$$\vec{v} = -k_0\mu^{-1}\nabla\varphi + \vec{q}(x), \tag{D3}$$

where $\vec{q}(x)$- mobility. Formally, (D_1) may be transformed into (D_3), if wesubstitute $\vec{q} = \varphi\nabla k\mu^{-1}$.

1.2.1.5 Contact Models of Flow

Let $\Omega = \Omega_1(t) \cup \Omega_2(t) \cup \Gamma$ be the region of inhomogeneous fluid flow, whose two components flow respectively through regions $\Omega_i(t)$ which have a common (contact) boundary $\Gamma(t) = \overline{\Omega}_1(t) \cap \overline{\Omega}_2(t)$. Regions $\Omega_i(t)$, satisfy equations (D0) which relate the fluid component characteristics $\vec{v}_i, p_i, \rho_i$ and μ_i to characteristics of the porous medium and m, k_i, at

$\Gamma(t) = \{\vec{x}, t | f(\vec{x}, t) = 0\}$ and at $t = 0$ with the following initial conditions and conditions of conjugation:

$$\left.\begin{array}{l} \vec{v}_i\,\vec{n} = v_n, \quad p_2 - p_1 = p_0, \quad mf_t + v_n f_n = 0, \quad (\vec{x}, t) \in \Gamma \\ f|_{t=0} = f_0(\vec{x}), \quad p_i|_{t=0} = p_i^0(\vec{x}), \quad \vec{x} \in \Omega_i(0), \quad i = 1, 2 \end{array}\right\} \tag{1.5}$$

where$f_0(\vec{x})$, $p_i^0(\vec{x})$ and $p_0(\vec{x}, t)$ (not equal to zero if capillary forces are taken into account) are the given and $v_n(\vec{x}, t)$ and $f(\vec{x}, t)$ are the target functions, $f_n = \vec{n}\ \nabla f_x$.

The problem described by (D_0) (1.5) was first formulated and solved for incompressible fluids ($\rho_i = const$) by Muskat (who made certain specific assumptions), and is named after him [105, 143]. The Muskat model is used to describe the movement of the oil—water contact (the outer oil limit) in oil formations and of the boundary between salt water and fresh water, and to solve other flow problems.

In the case of compressible fluids whose equation of state $\rho_i = \lambda_i(p_i + a_i)(\lambda_i, a_i - constant)$, the problem described by (D_0), (1.5) was studied by N.N. Verigin, and is frequently referred to as Verigin's problem. Several approximate methods of solving (D_0), (1.5) [110] have been proposed; A.M. Meiramov [88] was able to demonstrate that Verigin's problem was correct in the case of one-dimensional flows.

1.2.1.6 Nonhomogeneous Incompressible Fluid Flow Model (NF, Nonhomogeneous Fluid)

V.N. Monakhov proposed using the condition of incompressibility [91] as the equation of state in Darcy's model (D):

$$m\rho_t + \vec{v}\nabla\rho = 0,$$

This condition is true for fluids, providing a high degree of accuracy. It therefore follows from the equation of continuity in (D) that $div\vec{v} = 0$, and we can write (D) in the form

$$-\vec{v} = k(\rho, \vec{x})(\nabla p + \rho\vec{g}), \quad m\rho_t + \vec{v}\nabla\rho = 0; \quad div\vec{v} = 0. \tag{NF}$$

The law of logarithmic dependence of viscosity $\mu(\rho) = k_0(\vec{x})/k(\rho, \vec{x})$ on density ρ[113] is used to close equations (NF):

$$\ln(\mu/\mu_1) = \frac{\rho - \rho_2}{\rho_1 - \rho_2} \ln(\mu_2/\mu_1)$$

in which case

$$k = k_0\mu_1^{-1}e^{-\lambda S(\rho)};\quad \lambda = \ln \mu_2/\mu_1,\quad S = (\rho - \rho_2)(\rho_1 - \rho_2)^{-1}. \tag{1.6}$$

As in 1.5 above, we can use equations (NF) to describe the movement of the contact boundary $\Gamma(t)$ in a stratified fluid and, in particular, the movement of an (unknown) free boundary [105, 113], if we take into account the movement of its contacting air.

In his paper [91], V.N. Monakhov established that the stationary flow problem can be solved for the (NF) model.

Let us assume stratified fluid flow in $\Omega = (\Omega_1 \cup \Omega_2 \cup \Gamma)$, Γ being an unknown direction of flux and $\rho = \rho_i = const$ in Ω_i. In these conditions, at $\rho_2 \to 0$ the solution of the problem in $\Omega_1$is reduced to the solution of the classical problem of free-boundary flow [91].

V.N. Starovoitov [124] demonstrated that a similar three-dimensional non-stationary problem can also be solved for an (NF) model with natural physical conditions at unknown boundary Γ, relating to surface tension forces.

1.2.2 Navier-Stokes and Zhukovsky models

1.2.2.1 The Navier-Stokes (NS) model

This model assumes that a porous medium is a randomly ordered system for which the Gibbs concept of an assembly of identical systems is true, as is the ergodic theorem which states that time averaging may be replaced by assembly averaging and vice versa. It considers that the trajectory of an individual fluid particle obeys the micro-laws of viscous flow. Consequently, a micro-flow of fluid flowing at a velocity $\vec{u} = \frac{d\vec{x}}{dt}$ through pores (capillaries) can be described by Navier-Stokes equations for a viscous compressible fluids:

$$L\vec{u} \equiv \rho\frac{d\vec{u}}{dt} - \mu\Delta\vec{u} = -\nabla p - \vec{f},\quad \rho_t + div\rho\vec{u} = 0,\quad \rho = \rho(p), \tag{NS}$$

Where $\frac{d}{dt} = \frac{\partial}{\partial t} + (\vec{u}\cdot\nabla); \vec{f}$ – the vector of external forces.

There are several ways of deriving Darcy's law as approximations for the conservation of momentum in (NS). Two principal such methods are described below.

1.2.2.2 N.E. Zhukovsky's Hypothesis and Model (1896, see [105])

This hypothesis regards the porous medium and fluid as two components of an inhomogeneous fluid whose velocities are $\vec{v}_0 = 0$ (the stationary component) and $\vec{v} = m\vec{u}$ (flow velocity). It assumes that in each point $x \in R^3$ there exist both the porous medium and the flowing fluid. Therefore, the surface forces of resistance of the porous medium to the movement of the fluid may be regarded as bulk forces, and in accordance with N.E. Zhukovsky's hypothesis we may assume

$$\vec{f} = \rho\vec{g} + \lambda(\vec{v} - \vec{v}_0) = \rho\vec{g} + \lambda\vec{v}, \quad \lambda = \mu/k, \tag{1.7}$$

where λ is the resistance factor (for simplicity, the porous medium is assumed to be isotropic).

Let us now introduce Darcy's operator

$$D(\vec{v}, p) \equiv \frac{k}{\mu}(\nabla p + \rho\vec{g}) + \vec{v} \tag{1.8}$$

and incorporate (1.7), rewriting the (NS) equations in the form

$$L\vec{u} \equiv \rho\frac{d\vec{u}}{dt} - \mu\Delta\vec{u} = -\lambda D(\vec{v}, p), \quad \rho_t + div\rho\vec{u} = 0, \quad \rho = \rho(p) \qquad \text{(Zh)}$$

We shall name this system of equations Zhukovsky's model (Zh).

Zhukovsky's model [105] requires the following conditions to be met

$$(\rho|\vec{u}_t|, \quad \rho|(\vec{u}\cdot\nabla)\vec{u}|, \quad \mu|\nabla\vec{u}) \ll \lambda|\vec{v}|,$$

which makes the model similar to Darcy's model. A less stringent condition of their similarity is the inequality

$$|L\vec{u}| \ll \lambda|\vec{v}|.$$

An even less stringent condition is:

$$k\mu^{-1}|L\vec{u}| \ll |D(\vec{v}, p)| \approx \delta^{\alpha}, \quad \alpha \geq 1,$$

Where $\delta \ll 1$ is related to Reynolds number.

1.2.2.3 The Irmey Hypothesis of the Closeness of the NS and D Models

Let us now introduce the "static mean values" proposed by Irmey [24, p. 75], which are determined by the dimensionally correct relationship

$$\Delta\vec{u} = -\gamma\vec{u},$$

where γ depends on permeability K, viscosity μ and pore dimensions. Substituting this equation in (NS), with $\vec{f} = \rho\vec{g}$, and discarding the non-stationary and inertial terms $\vec{u}_t$and $(\vec{u} \cdot \nabla)\vec{u}$, as proposed by Zhukovsky, we arrive at Darcy's model (D).

1.2.2.4 The Oldroyd-Zhukovsky (OZh) Model

Let us consider viscoelastic fluid flow in a limited region Ω of Euclidean space R^3, consisting of a medium whose resistance is proportional to fluid velocity (Zhukovsky's hypothesis). Following Oldroyd [134] let us write out a system of equations describing the fluid flow:

$$\left.\begin{aligned} &Re[\vec{u}_t + (\vec{u} \cdot \nabla)\vec{u})] + \nabla p + \gamma\vec{u} = (1-\alpha) \cdot \Delta\vec{u} + \nabla \cdot S + \vec{f}, \\ &div\,\vec{u} = 0, \\ &S + We \cdot [S_t + (\vec{u} \cdot \nabla)S] = 2\alpha D \quad in \quad Q_T = \Omega \times [O, T], \end{aligned}\right\} \qquad \text{(OZh)}$$

where $\vec{u}$ is the fluid velocity; p is pressure; and S is the elastic part of the stress tensor, all the unknown values of time t and point x; $D = (\nabla \cdot \vec{u} + (\nabla \cdot \vec{u})^T)/2$- strain rate tensor; $Re = UL/\mu$ and $We = \lambda_1 U/L$ are respectively Reynolds and Weissenberg numbers; $\alpha = 1 - \lambda_2/\lambda_1$ is the numerical parameter; λ_1 is relaxation time; λ_2 is the delay time; $0 < \lambda_2 < \lambda_1$; U, L is the characteristic velocity and model size; $\gamma\vec{u}$ is the resistance force of the medium, $\gamma > 0$.

E. M. Turganbayev [134] demonstrated that the initial boundary value problem shown below can be solved for (OZh):

$$\left.\begin{aligned} &u(x,t) = 0, \quad x \in \Gamma, \quad t \in [0, T]; \\ &S(x, 0) = S_0(x), \\ &u(x, 0) = u_0(x), \quad x \in \Omega, \quad div\,u_0 = 0, \quad u_0| = 0. \end{aligned}\right\}$$

1.2.3 Two-phase Fluid Flow Models

1.2.3.1 The Muskat-Leverett Model (ML-model)

Mathematical models of the flow of two immiscible fluids (for example, water and oil) through a porous medium are much more complex than Darcy's model. Experiments have shown that in this particular case each of the fluids selects its own circuitous route which does not change.

As saturation s_i (the part of pore space occupied by component i) decreases, one of the fluids destroys the channels, breaking them up until only isolated regions occupied by this fluid remain. This phenomenon is known as residual oil or water saturation, with the corresponding values of s_i designated s_i^0.

However, a mathematical description of a more complex physical process may use the concept of a continuous medium.

Let us regard a two-component fluid as a collection of continua filling the same volume of incompressible pore space. For each of these continua, density ρ_i, flow velocity $\vec{v}_i$ and pressure p_i, let us introduce saturation s_i.

Then, by analogy with (1.1), we can write the continuum equations for each fluid component in the form

$$\frac{\partial}{\partial t}(m\rho_i s_i) + div(\rho_i \vec{v}_i) = 0 \quad i = 1, 2. \tag{1.9}$$

Bearing in mind the qualitative aspects of multiphase flow, Muskat (cf. [31]) proposed the following formal generalization of Darcy's Law for each of the fluids:

$$\vec{v}_i = -K\frac{k_i}{\mu_i}(\nabla \rho_i + \rho_i g) \quad i = 1, 2, \tag{1.10}$$

where K is the flow coefficient of the porous medium's for homogeneous fluid, as before, (or the symmetrical tensor in the case of an anisotropic medium); μ_i is the dynamic viscosity coefficients, and k_i must depend on saturation s_i, as part of the pore space is occupied by another fluid.

The fact that k_i are simply the functions of s_i and are virtually independent of pressure, flow rate and other fluid flow parameters, was repeatedly confirmed by laboratory experiments.

In accordance with its definition, saturation s. varies within the range

$$0 < s_i^0 \leq s_i \leq 1 - s_j^0 < 1, \quad j \neq i \quad (s_1 + s_2 = 1)$$

and when it reaches $s_i = s_i^0$, the movement of the ith component ceases when the condition

$$k_i(s_i^0) = 0, \quad i = 1, 2$$

has been met.

When analysing immiscible multiphase flows, it is important to take into account the effect of forces acting on their interface. When two immiscible fluids (I and II) come into contact with one another and with the solid surface of the pores (Fig. 1.1) the liquid–liquid interface $\Gamma_{1,2}$ approaches the solid wall at a contact angle θ. If θ is an acute angle,

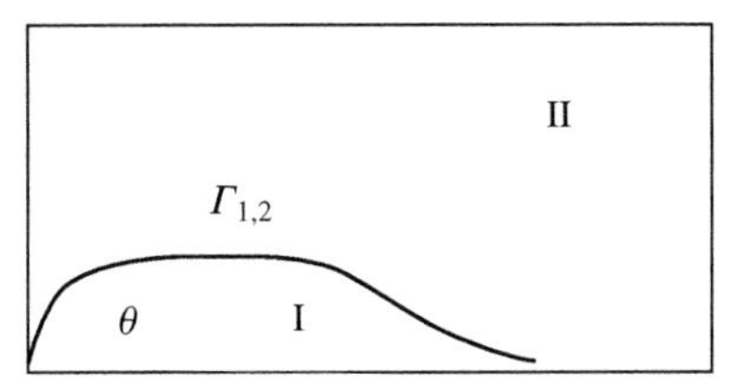

Figure 1.1 A two-component fluid.

fluid I is called the wetting fluid (since it tends to spread over the solid), and fluid II is called the non-wetting fluid. The phase pressure difference which occurs at boundary $\Gamma_{1,2}$ is called capillary pressure:

$$p_2 - p_1 = p_c(x, s) \geq 0, \qquad \frac{s_0 - s_2^0}{1 - s_1^0 - s_2^0} \equiv s \in [0, 1]. \tag{1.11}$$

Capillary pressure p_c depends on the curvature of $\Gamma_{1,2}$, the saturation s_1 of the wetting fluid and the properties of the porous medium, and is expressed by the Laplace formula:

$$p_c(x, s) = \overline{p}_c(x) j(s), \quad \overline{p}_c = \sigma \cos\theta \left(\frac{\overline{m}}{|K_0|}\right)^{1/2} \tag{1.12}$$

where σ is the interfacial tension coefficient; $J(s)$ is the Leverett function, while $|K_0|$ is the determinant of matrix $\{k_{ij}\}$ if K_0 is the symmetrical flow tensor of an anisotropic porous medium.

Experiments have shown that the order in which pore space is filled by the two phases determines the shape of the $J(s)$ curves. This phenomenon is known as capillary pressure hysteresis. As a rule, relative phase permeabilities k_i and Leverett functions $J(s)$ are calculated in saturation experiments, when capillary forces cause the wetting phase to displace the non-wetting phase filling the whole of the porous material, eliminating the hysteresis effect.

In the case of an isothermal (constant temperature) flow, the equation system (1.9)–(1.11) can be closed with respect to $\vec{v}_i, p_i, \rho_i$ and $s = \dfrac{s_1 - s_1^0}{1 - s_1^0 - s_2^0}$ of immiscible fluids flowingin a porous medium, by specifying equations of state for the fluids:

$$\rho_i = \rho_i(p_i), \quad i = 1, 2.$$

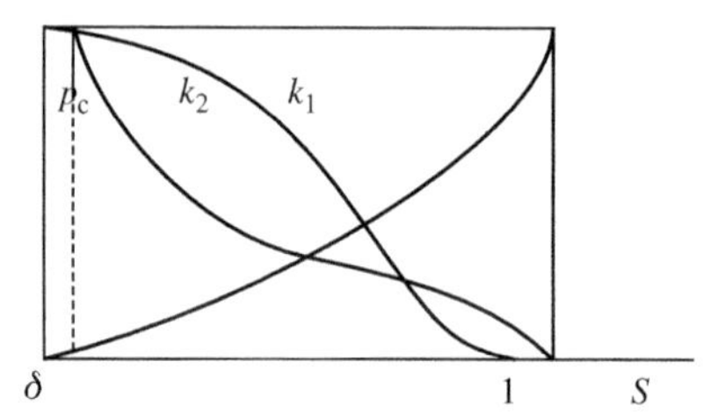

Figure 1.2 A typical curve describing the relationship between relative permeability and the Leverett J function on the one hand and reduced saturation on the other.

In what follows, unless stated otherwise, both fluids are assumed to be incompressible, i.e. $\rho_i = const.$

The resultant mathematical model of multiphase incompressible fluid flow [equations (1.9)−(1.11)] is called the Muskat-Leverett model in honour of Muskat who first proposed the generalization (1.10) of Darcy's and Leverett's Law and was also the first to use Laplace's Law (1.11), (1.12).

The Muskat-Leverett model assumes that parameters $\overline{m}, K_0, \overline{k}_i$ are the prescribed functions of the variables x and s and that all the numerical parameters (μ_i, ρ_i, s_i^0 and others) are fixed. A typical curve describing the relationship between relative permeability $\overline{k}_i$ and the Leverett J function on the one hand and reduced saturation $s \in [0, 1]$ on the other is shown in Fig.2. If we take into account the equality $k_i(s_i^0) = 0$, the phase permeabilities $\overline{k}_i(s) = \overline{k}_i(s)/\mu_I$, will have the following characteristics:

$$k_i(s) > 0, \quad s \in (0, 1); \quad k_1(0) = k_2(1) = 0.$$

To decide how the compressibility of the formation can be taken into account, let us consider the continuity equation (1.9). It should be noted that the assumptions about relative phase permeabilities in regions where

$$s_i = s_i^0 > 0, \text{ produce } \vec{v}_i \equiv 0.$$

Let $m = m(p)$ and $p = p_i + f_i(s_i)$ be average pressure, and let us consider a fixed region Ω with a boundary $\Gamma = \partial\Omega$, in which the saturation of one of fluids reaches its residual value $s_i = s_i^0$. It follows from the properties of phase permeabilities that in this case $v = 0$, hence $\dfrac{\partial m \rho_i s_i}{\partial t} = s_i^0 \dfrac{\partial m \rho_i}{\partial t} = = s_i^0 \dfrac{\partial (p_i \rho_i)}{\partial t} = 0$ and therefore $\dfrac{\partial p_i}{\partial t} = 0$ in Ω. The latter equality shows that the flow in Ω should be stationary irrespective of the values of pi at $\Gamma = \partial\Omega$, and at $\dfrac{\partial p_i}{\partial t} \neq 0$ at Γ.

This contradiction suggests that in this particular case it was wrong to introduce porosity and that instead the residual values of saturation should have been included in effective porosity.

It is obvious that we should consider $\sigma_i = \frac{s_i - s_i^0}{1 - s_1^0 - s_2^0}$ as dynamic saturations and $\sigma_i \in [0, 1]$, $i = 1, 2$ $(\sigma_1 + \sigma_2 = 1)$ at the same time, and assume that the m described by formula (1.1) includes residual saturation.

In that case, the saturation equation will assume the form

$$\frac{\partial(m\rho_i\sigma_i)}{\partial t} + div(\rho_i\,\vec{v}_i) = 0 \quad i = 1, 2. \tag{1.13}$$

For incompressible formations, the continuity equation can also be conveniently presented in the form (1.13).

1.2.3.2 Boundary Conditions and Initial Conditions

Let us begin by considering isothermal flow.

Impermeable boundaries. These are formed by the top and bottom of the formation through which a non-homogeneous (or homogeneous) liquid flows. The impermeability condition of this boundary Γ_0 has the same form for all phases:

$$\vec{v}_i\vec{n} = 0, \quad x = (x_1, x_2, x_3) \in \Gamma_0, \quad i = \overline{1, r+1}, \tag{1.14}$$

where $\vec{n}$ is the vector of the outward normal to the boundary $\Gamma = \partial\Omega$ of the finite multiply-connected flow region Ω and $\Gamma_0 \subset \Gamma$.

Wells. Values specified for injection and production wells can be either the distribution of phase pressures and saturations

$$p_i = p_{i0}(x, t), \quad s_i = s_{i0}(x, t), \quad x \in \Gamma_1, \quad i = \overline{1, r+1}, \tag{1.15}$$

or phase flow rates

$$\vec{v}_i\vec{n} = \frac{k_i}{k}Q(x, t), \quad k = \sum\nolimits_i^{r+1} k_i, \quad x \in \Gamma_i, \quad i = \overline{1, r+1} \tag{1.16}$$

where $Q(x, t)$ is the specified mixed flow rate. Equation (1.16) shows that the mixtures are selected and injected in proportion to phase mobilities. In the case of a two-phase liquid, the conditions described by (1.16) can be obtained if the following conditions are met:

a. The specified mixed flow rate is $(\vec{v}_1 + \vec{v}_2)\vec{n} = Q(x, t)$, $x \in \Gamma_1$;

b. Gravity forces operating in the wells are not taken into account and the capillary pressure gradient p_c is ignored as being negligible in comparison with the phase pressure gradients, i.e. $\nabla p_1\vec{n} = \nabla p_2\vec{n}$.

Boundaries with a homogeneous fluid at rest. It is assumed that the pressure p_1 of a fluid at rest is distributed in accordance with the hydrostatic law, that this boundary is the same as the pressure of the corresponding phase, and that saturation s_1 is continuous at Γ_2:

$$p_1 = p_0 + \rho_1 gh, \quad s_1 = 1 - \sum_1^2 s_i^0, \quad x \in \Gamma_2, \tag{1.17}$$

where p_0 is fluid pressure at some fixed level ($h = 0$); h is height measured from this level; ρ_1 is the fluid density. For the second component, either (1.14) is true at $x \in \Gamma_2$, or, if the fluid at rest is in a permeable medium, then it is natural to assume that

$$p_2 = p_2^0, \quad s_2 = s_2^0, \quad x \in \Gamma_2, \tag{1.18}$$

where p_2^0 is pressure at residual saturation s_2^0 with the phase in question. It should be noted that conditions (1.17), (1.18) are identical to (1.15).

Free boundaries. As a rule, a free (unknown) boundary Γ_3 is a line of contact discontinuity between one or several saturation levels (e.g. when a homogeneous fluid flows into "dry" rock). In addition to the conditions of impermeability of Γ_3 described by (1.14), it is assumed that the flow velocity of all phases is the same as the mixed flow velocity as a whole (a kinematic condition) and that the "average" pressure of the mixed flow is continuous throughout the neighborhood of Γ_3.

Initial conditions. In the case of incompressible fluids ($\rho_i = const$), it is assumed that the specified initial distribution of saturation is

$$s_i|_{t=0} = s_{i0}(x, 0), \quad x \in \Omega, \quad i = 1, 2 \tag{1.19}$$

and that the distribution of the "average" pressure of the mixed flow can also be specified in order to allow for the compressibility of the liquids and the rock:

$$p|_{t=0} = p_0(x, 0), \quad x \in \Omega \tag{1.20}$$

Occasionally, by idealizing the process of flow in various ways, the dimensionality of the space of independent variables may be reduced (in a general case this refers to four-dimensional space: t, x_1, x_2, x_3), e.g. by assuming a stationary flow (independent of time), or a linear flow in an infinite formation of uniform thickness, not subject to the forces of gravity (independent of gravity), or a planar flow in an infinite formation uniform along one of its horizontal coordinates, or a one-dimensional flow, assuming that the formation is infinite and uniform along two of its spatial coordinates.

However, in some flow problems, idealizations of this type may make it impossible for the resultant model to describe important qualitative characteristics of the physical process. For instance, when describing fluid flow to an ideal well using the plane flow model we may lose the connectivity of the flow region, so that the resultant problem becomes mathematically incorrect. In a spatially three-dimensional problem this situation would not present any additional mathematical difficulties.

1.3 THE TRANSFORMATION OF ML MODEL EQUATIONS

1.3.1 The Muskat-Leverett (ML) Model

The ML model of two-phase incompressible fluid flow ($\rho_i = const$) through porous media contains an equation system (1.9)–(1.11) [1.2] describing phase flow velocities v_i, pressures p_i and saturations $s_1, s_2 (s_1 + s_2 = 1)$:

$$\overline{m}\frac{\partial}{\partial t}\rho_i s_i + div\ \rho_i \vec{v}_i = 0, \quad -\vec{v}_i = K_i(\nabla p_i + \rho_i \vec{g}), \quad i = 1, 2,$$
$$p_2 - p_1 = p_c(x, s), \quad \left(\frac{s_{1-s_1^0}}{1 - s_1^0 - s_2^0} \equiv s \in x[0, 1]\right) \tag{1.21}$$

In this system, $K_i = K_0(x) \cdot k_i(s)$ a symmetrical tensor of phase permeability; K_0 is thetensor of homogeneous fluid flow and $k_i = \dfrac{\overline{k}_i(s)}{\mu_i}$ are relative phase permeabilities.

Let us transform system (1.21) into a more convenient form by adding the continuity equations [the first equations of (1.21)], divided by $\rho_i = const$, we obtain

$$div\ \vec{v} = 0, \quad \vec{v} = \vec{v}_1 + \vec{v}_2, \tag{1.22}$$

in which $\vec{v}$ is the mixed flow velocity vector and ($|\vec{v}|$ is the total specific mixed flow rate.

Let us now introduce "reduced" ("average") pressure (first proposed by V.N. Monakhov) as the new target function:

$$p = p_1 - \int_s^1 \frac{\partial p_c}{\partial s}\frac{k_1}{k} d\xi + \rho_i g h, \tag{1.23}$$

where $k = k_1 + k_2$, $\nabla gh = \vec{g}$. To explain this choice of target function, let us begin by using Darcy's Law [the second equations of (1.21)] to express the vector $\vec{v}$ in terms of the gradients of functions p_1 and s:

$$-\vec{v} = \sum_1^2 K_i(\nabla p_i + p_i\vec{g}) = kK_0\left(\nabla p_1 + \frac{\partial p_c}{\partial s}\frac{k_2}{k}\nabla s + \frac{k_2}{k}\nabla p_c\right) +$$

$$+\sum_1^2 K_i\rho_i\vec{g} = kK_0\nabla\left(p_1 - \int_s^1 \frac{\partial p_c}{\partial s}\frac{k_2}{k}d\xi\right) + kK_0\int_s^1 \frac{\nabla\partial p_c}{\partial s}\frac{k_2}{k}d\xi + K_0 k_2\nabla p_c$$

$$+\sum K_1\rho_i\vec{g}.$$

In this way, the substitution (1.23) enables us to express vector $\vec{v}$ in terms of ∇p and s, making it independent of ∇s:

$$\vec{v} = -(K\nabla p + \vec{f}) \equiv \vec{v}(s,p), \quad K = kK_0, \tag{1.24}$$

where $\vec{f} = K\int_s^1 \nabla\frac{\partial p_c}{\partial s}\frac{k_2}{k}d\xi + K_2\nabla p_c + K_2(\rho_2 - \rho_1)\vec{g}$ and ∇ is used only with respect to x, which occurs explicitly, i.e.

$$\nabla p_c(x,s) = \left(\frac{\partial}{\partial x_1}p_c, \frac{\partial}{\partial x_2}p_c, \frac{\partial}{\partial x_3}p_c\right).$$

By analogy with (1.23), we obtain

$$-\vec{v}_1 = K_1(\nabla p_1 + \rho_1\vec{g}) = K_1\left(\nabla p - \frac{\partial p_c}{\partial s}\frac{k_2}{k}\nabla s + \int_s^1 \nabla\frac{\partial p_c}{\partial s}\frac{k_2}{k}d\xi\right).$$

Whence, assuming $a = -\dfrac{\partial p_c}{\partial s}\dfrac{k_1 k_2}{k_1 + k_2}$ and $f_0 = K_1\int_s^1 \nabla\frac{\partial p_c}{\partial s}\frac{k_2}{k}d\xi$, we obtain

$$-\vec{v}_1 = K_0 a\nabla s + K_1\nabla p + \vec{f}_0 \equiv -\vec{v}_1(s,p). \tag{1.25}$$

By using (1.24), we can calculate $K_1\nabla p = -K_1K^{-1}(\vec{v}_1 + \vec{f})$, and noting that this definition yields $K_1 = k_1K_0$ and $K_2 = k_2K_0$, $K_1K^{-1} = k_1/k = b(s)$ we can rewrite equation (1.25) as

$$-\vec{v}_1 = K_0 a\nabla s - b\vec{v} + \vec{F}, \quad \vec{F} = \vec{f}_0 - b\vec{f}. \tag{1.26}$$

By substituting (1.25) in the continuity equation for the first phase, we obtain a system of equations with respect to $\{s, p\}$:

$$m\frac{\partial s}{\partial t} = div(K_0 a\nabla s + K_1 \nabla p + \vec{f}_0) \equiv -div\, \vec{v}_1(s, p), \qquad (1.27)$$

$$div(K\nabla p + \vec{f}) \equiv -div\, \vec{v}(s, p) = 0, \qquad (1.28)$$

and by substituting (1.26) we obtain an equivalent system with respect to $\{s, p, \vec{v}\}$:

$$m\frac{\partial s}{\partial t} = div(K_0 a\nabla s - b\vec{v} + \vec{F}), \quad m = \overline{m}(1 - s_1^0 - s_2^0), \qquad (1.29)$$

$$div(K\nabla p + \vec{f}) = 0, \quad -\vec{v} = K\nabla p + \vec{f}. \qquad (1.30)$$

It should be noted that the flow tensor $K_0(x)$ is assumed tobe symmetrical and positively defined, i.e.

$$k_{ij}^0 = k_{ji}^0, \quad \vartheta|\xi|^2 \leq (K_0\xi, \xi) = \sum\nolimits_{i,j} k_{ij}^0 \xi_i \xi_j \leq \vartheta^{-1}|\xi|^2, \quad \xi > 0, \qquad (1.31)$$

$|\xi|^2 = \sum_i \xi_i^2$, and that capillary pressure (pc) and relative phase permeabilities have the following properties:

$$\frac{\partial p_c}{\partial s} < 0, \quad k = k_1 + k_2 > 0, \qquad (1.32)$$

and therefore, if we include (1.31), we obtain $a(x, s) > 0$ when $s \in (0, 1)$ and $a(x, 0) = a(x, 1) = 0$.

Hence, (1.27), (1.28) form a quasi-linear system consisting of a uniformly elliptical equation for $p(x, t)$ and a parabolic equation for $s(x, t)$, which assumes a degenerate form at $s = 0.1$.

1.3.2 The Initial Boundary Value Problem

Let us consider flow in a specified finite region Ω with a piecewise-smooth boundary $\Gamma = \partial\Omega$. As described in Section 1.2, the different boundary conditions corresponding to the actual physical flow can be used to divide Γ into several coherent components Γ_i.

Let $\Omega_T = \Omega \cdot [0, T], \Gamma_{iT} = \Gamma_i \cdot [0, T]$, and let $\vec{n}$ be the outward normal to Γ. Let us now rewrite the boundary data from Section 1.2 for the

s and p functions. Based on (1.24), (1.25), the no-flow conditions (1.34) $\Gamma_0 \subset \Gamma$ given in 1.2 for both phases are equivalent to

$$\vec{v}(s,p)\vec{n} = \vec{v}_1(s,p)\vec{n} = 0, \quad (x,t) \in \Gamma_{0T}. \tag{1.33}$$

The boundary conditions (1.39) and (1.40) (Section 1.2) now become

$$p = p(p_1, s) = p_0(x,t), s = s_0(x,t), (x,t) \in \Gamma_{2T}, \tag{1.34}$$

$$\vec{v}(s,p)\vec{n} = Q(x,t), \quad (x,t) \in \Gamma_{1T}, \tag{1.35}$$

$$\vec{v}_1(s,p)\vec{n} = bQ(x,t), \quad (x,t) \in \Gamma_{1T}. \tag{1.36}$$

Because (1.35) and (1.36) become equivalent to (1.33) when $Q(x,t) \equiv 0$, then it is natural to include Γ_0 in Γ_1 and assume that Γ_1 consists of several components, in some of which $Q = 0$.

Thus, $\Gamma = \Gamma_1 \cup \Gamma_2$. Clearly, (1.27) and (1.28) cannot satisfy the Cauchy-Kovalevskaya theorem (as the second equation does not contain $\frac{\partial p}{\partial t}$), and therefore the initial condition need only be specified for saturation:

$$s|_{t=0} = s_0(x,t), \quad x \in \Omega \tag{1.37}$$

Note that Γ need not contain $\Gamma_1$or Γ_2, so that it may be the case that $\Gamma \equiv \Gamma_1$ or $\Gamma \equiv \Gamma_2$. When $\Gamma \equiv \Gamma_1$, the law of conservation of mass of the mixed flow in Ω generates the necessary condition

$$\int_\Omega p(x,t)dx = \int_\Gamma Q(x,t)dx = 0, \quad t \in [0,T]. \tag{1.38}$$

We conclude this section with a summary of formulae for the coefficients of equations (1.27)–(1.30) and of the boundary conditions (1.33)–(1.36):

$$a = \left|\frac{\partial p_c}{\partial s}\right| \frac{k_1 k_2}{k}, \quad k = k_1 + k_2, \quad b = \frac{k_1}{k} \equiv K_1 K^{-1},$$

$$\vec{f}_0 = K_1 \int_s^1 \nabla \frac{\partial p_c}{\partial s} \frac{k_2}{k} d\xi, \quad K_i = k_i K_0 \quad (i = 1,2), \tag{1.39}$$

$$\vec{f} = K_1 K^{-1} f_0 + K_2[(\rho_2 - \rho_1)\vec{g} + k_2 \nabla p_c], \quad K = K_1 + K_2,$$

$$\vec{F} = \vec{f}_0 - b\vec{f} = -k_1 k_2 k^{-1} K_0[\nabla p_c + (\rho_2 - \rho_1)\vec{g}].$$

1.3.3 The Independence of Total Flow Velocity from the Distribution of Saturation

If the coefficients $K = K_0(x)k(s)$ and $\vec{f}(x,s)$ in (1.28) do not depend on then the equation system (1.27), (1.28) breaks down and allows us to calculate the velocity field $\vec{v}$ and phase saturations $s_i(x,t)$ one after the other, while formulae (1.39) allow us to state these conditions in terms of the functional parameters used in the Muskat-Leverett model:

1. $k = k_1(s) + k_2(s) = const$: This assumption is true with sufficient accuracy for miscible fluids for which $k_1 = \lambda s$, $k_2 = \lambda(1-s)$, $\lambda = const$. In the case of immiscible fluids, the value shows significant deviation from the constant only in the vicinity of the limiting values of reduced saturation $s = 0, 1$.
2. $\dfrac{1}{\overline{m}(x)} detK_0(x) = const$, in which case $p_c = p_c(s)$, i.e.$\dfrac{\partial p_c}{\partial x_i} = 0$
3. Either gravity forces are not taken into account (for example, in a plane flow), or the fluids have similar densities $\rho_1 = \rho_2$.

It follows from (1.39) that assumptions (1.22) and (1.23) enable the condition $\dfrac{\partial \vec{f}}{\partial s} = 0$ to be satisfied.

1.3.4 Some Shortcomings of the Muskat-Leverett Model

Some of the model's shortcomings were noted above.

1. The model's solutions do not remain robust when its functional parameters change. Experiments to determine relative phase permeabilities $k_i(s)$ and capillary pressures $p_c(s)$, have shown that these functions are poorly specified in the vicinity of $s = s^*, s_*$. At the same time, the values of $\dfrac{dk_i}{ds}$ and $\dfrac{dp}{ds}$ in the vicinity of these saturations s decisively influence the structure of the solution of (1.27), (1.28).

   Indeed, the behaviour of $k_2(s)$ and $p_c(s)$ in the vicinity of $s = s_*$ is well described by

$$k_2(s) = f_1(s)(s - s_*)^{\lambda_1}, \quad p_c(s) = \frac{f_2(s)}{(s - s_*)}\lambda_2, \tag{1.40}$$

where $\lambda_i = const > 0$, and $0 < f_i(s) \in C^1[s_*, s)$. For many formations it may be assumed that $\lambda_1 = 3$ [143, p. 176] and $\lambda_2 = 2$ [110, p. 208].

At these values of $\lambda_1, a(s_*) = \left.\frac{k_1 k_2 |p_c'|}{k_1 + k_2}\right|_{s=s_*} \neq 0$ and, consequently, $a(s)$ and $k_1(s)$, become zero only when $s = s^*$. To provide a full picture, let us consider the case of one-dimensional flow through homogeneous rock at a specified mixed flow rate $v_1 + v_2 = Q(t)$. The system (1.27), (1.28) then becomes equivalent to the Rappoport-Liss equation

$$m\frac{\partial s}{\partial t} = \frac{\partial}{\partial x}\left[a(s)\frac{\partial s}{\partial x} + Q(t)b(s)\right], b(s) = \frac{k_2}{k_1 + k_2}. \tag{1.41}$$

The next task is to find a solution to equation (1.41), which will satisfy the condition

$$s|_{t=0} = s_0(x), \quad x \in [0, X], \quad s|_{x=0} = s^* - \delta, \quad s|_{x=X} = s_*, \tag{1.42}$$

where $s_* \leq s_0(x) \leq s_0 - \delta, \quad \delta > 0$.

Noting that, because equation (1.41) is uniformly parabolic [since $a = (s_*) \neq 0$], and because of the maximum principle $s(x, t) > s$ for all the finite values of $t > 0$ and $0 < x < X$, even when, at $X \geq x > 0$, $s_0(x) \equiv s_*$. On the other hand, if $\lambda_1 - \lambda_2 - 1 > 0$ (and some theories and experiments make the value of λ_1 3,5-4 [110, p. 181]), then $a(s_*) = 0$, and the propagation velocity of perturbations becomes finite (as established by N.V.Khusnutdinova for s^*), and if $s_0(x) \equiv s$ at $x \geq x_0 > 0$ $(x < X)$, then, provided that $X(T)$ is sufficiently large, and in particular, when $X = \infty$, for each finite $T > 0$ there exists a point $x = \xi(T) > 0$ such that $s(X, T) \equiv s$, at $x \geq \xi(T) > 0$.

For the sake of clarity, let us assign water the index 1, and oil the index 2.

It should be noted that if $X < \infty$ (the production well $x = X$ is located at a finite distance) and $s_*(x) \equiv s_*, s \in [x_0, X], x_0 > 0$, as before, then the minimum time $t = T > 0$, during which water saturation $s(X, T) > s$, at $x < X$ and $s(X, T) = s$ is called pure oil withdrawal time. In the first case $[a(s_*) \neq 0]$ there is no such $T > 0$, but in the second case $[a(s_*) = 0]$ it can be uniquely defined. This makes it impossible to trust forecasts of the time of water encroachment of production wells, in particular those based on the Muskat-Leverett model.

2. *The boundary paradox.* This drawback of the Muskat-Leverett model is also conveniently illustrated by a one-dimensional model with a specified mixed (water/oil) flow rate, $v_1 + v_2 = Q(t)$. Alekseev and Khusnutdinova [1] demonstrated that, given specified smoothness conditions, equations (1.41), (1.42) have a unique solution at $t \in [0, T]$ for any finite $T > 0$. However, the existence of a solution within any arbitrary interval $0 \leq t \leq T$ contradicts the physical implications of the initial model, since it must be the case that when water is injected into injection well ($x = 0$), pure oil cannot indefinitely continue to be recovered from production well ($x = X$).

   This paradox makes it difficult to formulate physically acceptable conditions for production wells. Moreover, asymptotics studies suggest that over a period of time these equations produce unrestricted growth of the saturation gradient in production wells. This is a boundary effect which must also be allowed for in numerical calculations.

3. *The initial function becomes indeterminate if the medium is only weakly saturated with the wetting fluid.* Let a wetting fluid (e.g. water) be injected into a formation, and let its initial saturation $s_0(x)$ be lower than its residual saturation ($s_0(x) < s_*, x \in \Omega_0 \subset \Omega$) and, in particular, $s_0(x) \equiv 0, x \in \Omega_0 \subset \Omega$. Because the Muskat-Leverett model is valid only for $0 < s_* \leq s(x, t) \leq s^* < 1$, it can only begin to apply from a time $t_0 < 0$ which satisfies these conditions. The model cannot tell us what the saturation profile would be at a time $t_0 > 0$.

   At the same time, B.I. Pleshchinsky, using an oil formation model, was able to demonstrate experimentally [103] that there is a significant relationship between a formation's initial water saturation and the movement of the water saturation front.

4. *Darcy's laws lose their uniformity in the Muskat form.* The relationship between the phase flow velocities $\vec{u}_i$ at which the fluids flow through pores, and flow velocities $\vec{v}_i$ is described by $\vec{v}_i = m s_i \vec{u}_i$. Consequently, the left hand sides of Darcy's Law equations are proportional to partial velocities ($s_i \vec{u}_i$), whereas the right hand sides contain total rather than partial phase pressures and densities. It will be shown below that this formal non-uniformity of Darcy's Law makes the equivalent Muskat-Leverett model and its saturation s and reduced pressure equations much more complicated than the corrected model. This becomes particularly important when we attempt to take fluid compressibility into account.

1.3.5 The Displacement Model

Because the model we are proposing (V.N. Monakhov [75]) is phenomenological, as is the Muskat-Leverett model, it must be constructed so as to allow experimental determination of its characteristics. The model's characteristics naturally fall into the categories of formation, liquid and flow characteristics.

The Characteristics of (Homogeneous) Formations

a. *Average permeability* $k=\overline{k}(s)$: this characteristic depends on residual water and oil saturation. Its limiting values are determined in standard permeability experiments: $k_* = \overline{k}(s_*)$ is determined on the basis of oil flow in a water saturated sample (Darcy's experiment), while average formation permeability does not change and is the same as k^* i.e.

$$\overline{k}(s) = k^* < k_*, \text{ at } s \geq s^*.$$

Experiments [e.g. 103] have shown that when fluid flows into "dry" rock, permeability can be described by the formula

$$\overline{k}(s) = k^* + k_0\left(\frac{s}{s_*}\right)^{\lambda}, \quad k_0 = const, \quad \lambda \geq 3 \ at\ 0 \leq s \leq s_*,$$

k_0 can then be found from $\overline{k}(s_*) = k_*$, together with $\overline{k}(s)$:

$$\overline{k}(s) = \begin{cases} k^*, & s \geq s^* \\ k^* + (k_* - k^*)\left(\frac{s}{s_*}\right)^{\lambda}, & s \in [0, s_*]. \end{cases}$$

b. *Phase permeabilities:* These are input as required by the specific model.

The Characteristics of (Flowing) Fluids

a. *Average density*: $\rho = \sigma\rho_1 + (1-\sigma)\rho_2 \equiv \overline{\rho}_1 + \overline{\rho}_2$, ρ_i – true densities, and $\overline{\rho}_i$ – partial density, $\sigma = \dfrac{s - s_*}{s^* - s_*}$.

b. *Average pressure*: $p = \sigma p_1 + (1-\sigma)p_2 \equiv \overline{p}_1 + \overline{p}_2$.

c. *Average viscosity*: $\ln \mu = \sigma \ln \mu_1 + (1-\sigma)\ln \mu_2$ (see [113, p. 289]). $\mu(\sigma)$ may also take other forms, determined in independent experiments, or in displacement experiments, in which case it can be determined simultaneously with formation characteristics.

d. *Average phase and mixed flow velocities*: $\vec{v}_1, \vec{v}_2$ and $\vec{v} = \vec{v}_1 + \vec{v}_2$.

Let the fluids be incompressible and let the formation be water-saturated ($s(x,t) \geq s_*$). It is assumed that the mixed flow follows Darcy's Law:

$$\vec{v} = -\frac{k^*}{\mu}(\nabla p + \rho \vec{g}), \quad div\vec{v} = 0, \tag{1.43}$$

where k^* is the effective permeability coefficient, introduced above and based on the assumption that stationary phases (at $s = s_*, s^*$) are classed as pore space.

Here, $p = \sigma p_1 + (1-\sigma)p_2$, $\rho = \sigma\rho_1 + (1-\sigma)\beta$, $\mu = \mu_1 e^{\chi\sigma}$($\chi = \ln \mu_2/\mu_1$) are the pressure, density and viscosity of the mixed flow, $\sigma = \frac{s - s_*}{s^* - s_*}$.

The flow of the displacement fluid (water, assigned the index 1) is the same as in the case of incomplete saturation, and follows the relevant law [40, p. 180]:

$$\vec{v}_1 = -\frac{k^*}{\mu_0}\sigma^\lambda(\nabla p_1 + \rho_1\vec{g}), \quad m_1\frac{\partial\sigma}{\partial t} + div\vec{v}_1 = 0, \tag{1.44}$$

where $m_1 = m(s^* - s_*)$, $\lambda = const \geq 3$, and μ_0 either equals μ_1, or we can assume that $\mu_0 = \mu(\sigma)$. However, we need to add to the system the following Laplace Law:

$$p_2 - p_1 = \overline{p}_c(\sigma) \equiv p_c[s(\sigma)] \tag{1.45}$$

The known properties of p_c [74] give us

$$\overline{p}_c(\sigma) = \frac{1-\sigma}{\sigma^2}\tilde{p}_c(\sigma); \quad 0 < \tilde{p}_c < \infty, \quad \sigma \in [0,1].$$

Let $k = \frac{k^*}{\mu}$, $k_1 = \frac{k^*}{\mu_0}\sigma^\lambda$ and let us express $\vec{v}_1$ in terms of $\vec{v}$ and σ. This gives us

$$p_1 = p + (\sigma - 1)\overline{p}_c \equiv p - p_0(\sigma); \quad p_0 = (1-\sigma)\overline{p}_c$$

(note that $p_0'(\sigma) = -\overline{p}_c + (1-\sigma)\overline{p}_c' < 0$). Hence

$$\vec{v}_1 = -\frac{k_1}{k}\vec{v} + k_1 p_0' \nabla\sigma + k_1(\rho_2 - \rho_1)(1-\sigma)\vec{g}$$

or

$$\vec{v}_1 = -\overline{a}_1\nabla\sigma - \overline{b}\vec{g} + \overline{c}\vec{v},$$

where
$\bar{a}_1 = -k_1 p_0', (\bar{a}_1|_{\sigma=0,} 1 = 0), \bar{b} = k_1(1-\sigma)(\rho_2 - \rho_1)\, \bar{c} = k_1/k = (\mu/\mu_0)\sigma^{\lambda}$.
Substituting $\vec{v}_1$ in the second equation (1.44), we obtain the system

$$m_1 \frac{\partial \sigma}{\partial t} = div(\bar{a}_1 \nabla \sigma - \bar{b}\vec{g} + c\vec{v}), \quad divk(\nabla p + \rho \vec{g}) = 0, \tag{1.46}$$

which has the same form as (1.27), (1.28) in the Muskat-Leverett model.

In the case of an unsaturated formation (containing regions where $s < s_*$), the relevant Darcy's laws can assume the form

$$\vec{v} = -\frac{\bar{k}}{\mu}(\nabla \rho + \rho \vec{g}), \quad v_1 = -\frac{\bar{k}}{\mu_0}(\nabla p_1 + \rho_1 \vec{g}), \tag{1.47}$$

and when $0 \le s \le s_*, p(s) \equiv p_2, \rho(s) \equiv \rho_2, \mu \equiv \mu_2, \mu_0 \equiv \mu_1$, where as $k(s)$ is as determined above.

It follows from the design of displacement models that all their parameters, which should be determined directly from displacement experiments, are constants (1.3.4 point 1). However, equations (1.47) additionally make it possible to describe flow in unsaturated formations (1.3.4 point 4).

It should be noted that the solution of this problem is especially topical, since it relates to coal seam degasification, which usually involves injecting fluid into "dry" rock.

We have also proposed another model describing the flow of immiscible fluids in formations incompletely saturated with the wetting phase. This model will not be discussed here.

1.3.6 A Model Including Partial Pressures and Densities. Provision for the Compressibility of Fluids

Let $s \in [s_*, s^*]$ and let the following analogues of Darcy's Law, flowing phase continuity equation and Laplace Law [76] be satisfied for each of the phases:

$$\vec{v}_i = \bar{k}_i(\bar{p}_i + \rho_i \vec{g}); \; m_1 \frac{\partial(\sigma_i \rho_i)}{\partial t} + div(\rho_i \vec{v}_i) = 0; \; p_2 - p_1 = \bar{p}_c(\sigma), \tag{1.48}$$

where $\bar{p}_i = \sigma_i p_i$ and $\bar{\rho}_i = \sigma_i \rho_i$ are respectively the partial phase pressures and densities, $\sigma_1 \equiv \sigma = \dfrac{s - s_*}{s^* - s_*}, \sigma_2 = 1 - \sigma, p_c(s) \equiv \bar{p}_c[\sigma(s)]$, and $\bar{k}_i = \bar{k}_i(s, x)$ are analogues of phase permeabilities, $m_1 = m(s^* - s_*)$.

It should be noted that p_i are determined with an accuracy of up to constant p_{0i} (average pressures). Therefore, if fluctuations δ_i, of the functions $(p_i - p_{0i}) \in [0, \delta_i]$ are small as compared to their gradients ∇p_i, the first equations in (1.23), (1.27) approach the usual form of Darcy's Laws when

$$\overline{k}_i = \frac{k_i}{\sigma_i} \approx c_i \sigma_i^{\lambda_{i-1}} \quad (\lambda_i \geq 3, c_i = const), \tag{1.49}$$

since, on this assumption, $(p_i - p_{0i}), \overline{k}_i, \sigma_i$ can be regarded as small. To close system (1.32), we need to add equations of state of fluids which we will assume to be compressible:

$$\rho_i = \gamma_i p_i, i = 1, 2 \quad (\gamma_i = const > 0). \tag{1.50}$$

From Laplace Law and equations of state of the fluids we find

$$F(\sigma, \overline{\rho}_1 \overline{\rho}_2) \equiv \frac{\overline{\rho}_2}{\gamma_2(1-\sigma)} - \frac{\rho_1}{\gamma_1 \sigma} - \overline{p}_c(\sigma) = 0,$$

and since $\overline{p}'_c \leq 0$, we have $\frac{\partial F}{\partial \sigma} > 0$, and have therefore established the existence of the implicit function $\sigma = \sigma_1 = f_1(\overline{\rho}_1, \overline{\rho}_2), (\sigma_2 = 1 - \sigma \equiv f_2)$.

It follows therefore that equations (1.48), (1.50) are equivalent to the following parabolic system of equations with respect to the $\overline{\rho}_1$ *functions*:

$$m_1 \frac{\partial \overline{\rho}_i}{\partial t} = divA_1(\nabla \overline{\rho}_i + \gamma_i \overline{\rho}_i \vec{g}), \quad A_i = \frac{\overline{k}_i \overline{\rho}_i}{\gamma_i f_i}, \quad i = 1, 2. \tag{1.51}$$

If, as above, $(p_i - p_{0i})$ are small compared to ∇p_i, it follows from (1.50) that $\rho_i \approx const$.

Consequently, if we take (1.49) into account, this makes the model described by (1.48), (1.50) close to the Muskat-Leverett model. This being the case, we can regard equations (1.51) as a physical regularization of equations (1.21), corresponding to the Muskat-Leverett model.

Continuing to generalize the above modification of Darcy's laws, let us present them in the form

$$v_i = -k_0(x), \quad \nabla(\chi_i \varphi_i), \quad i = 1, 2, \tag{1.52}$$

where $\varphi_i = p_i + \rho_i gh$ is the hydraulic pressure head of the phases $(g\nabla h = \overline{g})$, $k_0 = (x)$ is the average permeability of the medium to homogeneous fluids and $(\mu_i \chi_i)$ are analogues of relative phase permeabilities $k_{0i}(s)$.

If fluctuations δ_i of the functions $(\varphi_i - \varphi_{0i}) \in [0, \delta]$, $\varphi_{0i} = const$ are small by comparison with $\nabla\varphi_i$, it is easy to verify, as before, that equations (1.52) are close to Darcy's laws in the Muskat-Leverett model when $\mu_i \chi_i(s) = k_{0i}(s)$.

Let us now assume that the fluids are incompressible ($\rho_i = const$). The equation $div(\vec{v}_1 + \vec{v}_2) = 0$ then gives us

$$div[k_0(x)(\chi_1\varphi_1 + \chi_2\varphi_2)] \equiv div[k_0(x)\varphi] = 0. \tag{1.53}$$

Conditions (1.34), (1.36) [respectively (1.34) and (1.35)] for functions s and p correspond to the following boundary value problem for the target function $\varphi(x, t)$:

$$\nabla\varphi\vec{n}|_{\Gamma_1} = 0, \quad \varphi|_{\Gamma_2} = \varphi_0(k, t) \quad (k_0\nabla\varphi\vec{n} = q(x, t)), \tag{1.54}$$

where $\varphi_0(x, t)$ is calculated explicitly via $\sigma_0(x, t)$ and $p_0(x, t)$.

Let $\varphi = \Phi(x, t)$ be the solution of the boundary value problem (1.53), (1.54), and let us use Laplace Law to calculate

$$\chi_1\varphi_1 = \frac{\chi_1}{\chi_1 + \chi_2}\left[\Phi - \chi_2 p_c - \chi_2(\rho_2 - \rho_1)gh\right] \equiv F(s, x, t),$$

Substituting this into the first phase continuity equation produces the equation for s:

$$m\frac{\partial s}{\partial t} = div_0[k(x)F(s, x, t)]. \tag{1.55}$$

In this way, by using Darcy's laws in the form shown in (1.52), we have reduced the initial problem to the integration of a decomposable equation system (1.53), (1.54) (in the limiting one-dimensional case, pressure p is explicitly calculable).

Functions $\chi_i(s)$ should be such as to satisfy the condition of parabolicity of (1.55), $\dfrac{\partial F}{\partial s} > 0, s \in (s_*, s^*)$. This is especially true for the similarity of (1.52) to the ordinary Darcy's laws discussed above, with the system (1.53), (1.55) again representing a physical regularization of equations (1.21). The proposed method of regularization of equations (1.21) may also be used locally, in particular to plot their differential approximation.

It is interesting to note that when $k = const$ (in the case of a homogeneous formation), a similar situation occurs even with the ordinary form of Darcy's Law if we consider a linear problem with the condition $\frac{k_1(s)}{\mu_1} + \frac{k_2(s)}{\mu_2} = const$ (see especially [73], formulations of s, ψ). Such splitting of the initial problem seems not to be accidental, since numerical calculations suggest that the link between bulk pressure or flow paths and saturation is weak.

1.3.7 Boundary Layer Approximations

For simplicity, let us consider a two-dimensional problem of the flow of immiscible fluids in the region $\Omega\{0 < x < 1, 0 < y < h\}$, bounded by an impermeable top and bottom of the formation ($y = 0, y = h$), and by an injection ($x = 0$) and a production ($x = 1$) well. Let us write equations (1.27), (1.28) in the following equivalent form:

$$m\frac{\partial S}{\partial t} = div(a\nabla s + b\vec{g} + c\vec{v}), \tag{1.56}$$

$$-\vec{v} = k\nabla p + f\vec{g}, \quad div\vec{v} = 0, \tag{1.57}$$

where $c = \frac{k_2}{k}, \quad b = cf, \quad \vec{v} = (u, v)$.

Let us assume that the formation is thin (h/l is small), and that

$$k_i = \delta^2\tilde{k}_i, \quad x = \tilde{x}, \quad y = \delta\tilde{y}, \quad u = \tilde{u}, \quad v = \delta\tilde{v}. \tag{1.58}$$

We will also assume that the form of Darcy's Law remains the same, whatever the formation thickness [so that equations (1.57) do not change]. Taking into account (1.58), let us transform equation (1.56) to represent saturation, discarding terms proportional to δ and δ^2. Returning to the previous variables and target functions, we obtain

$$\frac{\partial S}{\partial t} - c'u\frac{\partial S}{\partial x} = \frac{\partial}{\partial y}\left(a\frac{\partial S}{\partial y}\right) + c'v\frac{\partial S}{\partial y} \quad (c' < 0). \tag{1.59}$$

When $u \geq 0$, the fact that (1.59) is an evolutionary equation for x in a production well ($x = 1$) makes it unnecessary to specify the boundary conditions for $s(x, y, t)$, while the saturation profile for $x = 1$ takes shape as the non-wetting fluid is displaced. Let us note that the proposed model [76] can also be used locally, but only in the vicinity of production wells.

1.4 THE MLT MODEL FOR CONSTANT RESIDUAL SATURATIONS

In the temperature model of two-phase fluid flow proposed by O.B. Bocharov and V.N. Monakhov [15] (the MLT model), the effect of thermal processes on the nature of fluid flow is taken into account by changing the viscosities and the capillary properties of the various components of the fluid, in accordance with their own temperature and that of the pore space matrix. The mathematical model representing these assumptions consists of a composite equation system which includes a parabolic temperature equation and a system of elliptic and parabolic equations for the saturation of one of the fluid components and the average pressure of the mixed flow. A feature of the MLT model is that all its component equations, except for the equations of Darcy's Law and Laplace Law, follow from the laws of conservation of the mechanics of continuous media. In addition, the MLT model is easy to work with, in the sense that its description uses only functional parameters capable of experimental determination.

1.4.1 Derivation of Equations

Let $s_i (i = 1, 2)$ be phase saturations; m_0 is porosity; $\alpha_i = m_0 s_i \ (i = 1, 2)$ and $\alpha_3 = 1 - m_0$ are the volumes of the fluids and of the solid phase (the pore space matrix) respectively; ρ_i, p_i and $\vec{u}_i$ are respectively the densities of incompressible fluids ($\rho_i = const$), their pressures and velocities; $\vec{v}_i = \alpha_i \vec{u}_i = m_0 s_i \vec{u}_i \ (i = 1, 2)$ is the phase flow velocities, and $\vec{v} = (\alpha_1 \vec{u}_1 + \alpha_2 \vec{u}_2) = \vec{v}_1 + \vec{v}_2$ are mixed flow velocities.

Let us write down the energy balance equation. Let us assume that in each point of the porous medium there exists a thermal equilibrium, i.e. that the phase temperatures θ_i are the same, $\theta_i \equiv \theta \ (i = 1, 2, 3)$. Let $c_{pi} = const \ (i = 1, 2, 3)$ (phase heat capacity at a constant pressure); $e_i = c_{pi}\theta$ is the internal phase energy; $\vec{q}_i = -\alpha_i \lambda_i(\theta) \nabla \theta$ *i*th phase heat inflow vector (from Fourier's Law). If we then divide both parts of the phase energy balance equations by $(\rho_i c_{pi})$ [40]:

$$\rho_i \alpha_i \left(\frac{\partial e_i}{\partial t} + \vec{u}_i \nabla e_i \right) = -div \vec{q}_i, \quad i = 1, 2, 3$$

and add them term-by-term, we will arrive at the equation for θ:

$$\frac{\partial \theta}{\partial t} = div[\lambda(x, s, \theta) \nabla \theta - \vec{v}\theta]. \tag{1.60}$$

In this equation, $\lambda = \sum_1^3 \alpha_i \lambda_i (\rho_i c_{pi})^{-1}, s = (s_1 - s_1^0)(1 - s_1^0 - s_2^0)^{-1}$ is the reduced saturation ($s_i^0 = const$). It is assumed that the rocks are stable and that the phase flow follows the Muskat-Leverett model [40]:

$$\frac{\partial \alpha_i s_i}{\partial t} + div \rho_i \vec{v}_i = 0; \quad v_i = -K_0(x) \frac{\overline{k}_i(s)}{\mu_i(\theta)} (\nabla p_i - \rho_i \vec{g}),$$

$$p_2 - p_1 = \gamma(\theta) \cos \alpha(\theta) \left[\frac{m_0(x)}{K_0(x)} \right]^{1/2}, j(s) \equiv p_c(x, \theta, s),$$

where $K_0(x)$ is the absolute permeability tensor of the medium; $\overline{k}_i(s)$ is the relative phase permeabilities; μ_i is the phase viscosities; $\vec{g}$ is acceleration of gravity; γ is the surface tension coefficient; α is wetting angle; $J(s)$ is Leverett capillary pressure function. The following designations are used below: $k_i(s, \theta) \equiv \overline{k}_i(s) \mu_i^{-1}(\theta)$.

1.4.2 The Transformation of the Muskat-Leverett Equations

Problem statement. Let us introduce a formula for average pressure, based on an analogy with [65]:

$$p = p_2 + \int_s^1 \frac{k_1}{k_1 + k_2} (\xi, \theta) \frac{\partial}{\partial \xi} p_c(x, \xi, \theta) d\xi.$$

After the appropriate transformations, the Muskat-Leverett equations are reduced to the following system:

$$m \frac{\partial s}{\partial t} = div \left[K(a_1 \nabla s - a_2 \nabla \theta + \vec{f}_1) - b_1 \vec{v} \right] \equiv (-\nabla \cdot \vec{v}_1), \tag{1.61}$$

$$(-\nabla \cdot \vec{v}) \equiv div[K(\nabla p + \vec{f}_2 + a_3 \nabla \theta)] = 0. \tag{1.62}$$

In this system, $m = m_0(1 - s_1^0 - s_2^0)$ – effective porosity,

$$b_i = k_i(k_1 + k_2)^{-1} \ (i = 1, 2), \quad K = K_0(k_1 + k_2), \quad a_0 = k_1 k_2 (k_1 + k_2)^{-2},$$

$$a_1 = |p_{cs}| a_0, \quad a_2 = p_{c\theta} a_0; \quad \vec{f}_1 = [-\nabla_x p_c + (\rho_1 - \rho_2) \vec{g}] a_0,$$

$$a_3 = -k_1 p_c \theta - \int_s^1 \frac{\partial}{\partial \theta} (b_1 p_{cs}) ds, \quad \vec{f}_2 = \int_s^1 -\nabla_x p_{cs} b_2 ds - b_1^{-1} \vec{f}_1.$$

Let $\Omega \in R^3$ be a bounded region. Let us assume that $\vec{\theta} = (s, \theta)$ and divide the boundary $\partial\Omega$ into several components in accordance with their boundary conditions:

$$(p, s, v) = (p_0, s_0, v_0), \quad (x, t) \in \Sigma^1 = \Gamma^1 \times [0, T],$$

$$\vec{v}_i \cdot \vec{n} = b_i R, \quad i = 1, 2; \quad v = v_0(x, t), \quad (x, t) \in \Sigma^2 = \Gamma^2 \times [0, T], \quad (1.63)$$

$$v_i \cdot n_i = 0, \quad i = 1, 2; \quad \lambda \frac{\partial v}{\partial n} = \beta(v_0 - v), \quad (x, t) \in \Sigma^3 = \Gamma^3 \times [0, T],$$

where $\vec{n}$ is the unit vector of the outward normal to Γ; $\beta = \sum_1^3 \beta_i (\rho_i \times c_{pi})^{-1}, \beta_i$ is the ith phase heat transfer coefficient. Sections Γ^1, Γ^2 model the injection and withdrawal sections and the contact with homogeneous fluid at rest, while Γ^3 represents contact with the surrounding impermeable rocks.

Boundary conditions (1.63) need to be augmented by the initial condition for $\vec{\theta}$:

$$\vec{\theta}|_{t=0} = \vec{\theta}_0(x, 0), \quad x \in \Omega. \quad (1.64)$$

When $\Gamma^1 = \varnothing$, the law of conservation of mass of the mixed flow in region Ω produces the necessary condition

$$\int_\Omega p(x, t) dx = \int_\Gamma R(x, t) dx = 0, \quad t \in [0, T] \quad (\Gamma^1 = \varnothing). \quad (1.65)$$

1.5 NON-ISOTHERMAL FLOW OF IMMISCIBLE FLUIDS WITH VARIABLE RESIDUAL SATURATIONS (THE MLT MODEL)

The model considered in this section was proposed in Eqn 1.36 and differs from the model considered in Section 1.4 in that it takes into account the fact that residual phase saturations depend on the temperature of the inhomogeneous fluid, which we were able to establish experimentally. The inclusion of this effect enables the model to describe the movement of the boundary between the inhomogeneous fluid and its stationary components (the Stephan problem).

1.5.1 The Model Equations and Problem Statement

Following the assumptions made in Section 1.4, the energy balance equation for an inhomogeneous fluid becomes

$$\frac{\partial\theta}{\partial t} = div[\lambda(x,s,\theta)\nabla\theta - \vec{v}\theta], \tag{1.66}$$

where θ is temperature; λ is the thermal diffusivity; $s = s_1$ is wetting phase saturation; $\vec{v} = \vec{v}_1 + \vec{v}_2$ is average mixed flow velocity, and $\vec{v}_i$ is the phase flow velocities.

The phase flow follows the patterns described by the Muskat-Leverett model [40], where residual saturations are not constant, $s_i^0 = s_i^0(\theta) \geq \bar{s}_i^0 = const \geq 0$. These properties $s_i^0, (i = 1, 2)$ produce the following conditions of wetting phase saturation $s(x, t)$:

$$0 \leq const = \bar{s}_* \leq s_*(\theta) \leq s(x,t) \leq s^*(\theta) \leq \bar{s}^* = const \leq 1, \tag{A}$$

where $s_* = s_1^0(\theta), s^* = 1 - s_2^0(\theta), 0 \leq \bar{s}_* = \inf_\theta s_*(\theta), \bar{s}_* = \sup_\theta s^*(\theta) \leq 1$.

Analysis of experimental and theoretical work [2, 90, 143] shows that it would be natural to consider the functional parameters of the Muskat-Leverett model (relative phase permeabilities $\bar{k}_i$ and the Leverett J function) as dependent on the dynamic saturation of the wetting phase:

$$\sigma = \frac{s - s_*(\theta)}{s^*(\theta) - s_*(\theta)}, \quad s_* \leq s \leq s^*; \quad \sigma = 0, s < s_*; \quad \sigma = 1, \tag{B}$$

$$s > s^*.$$

This condition determines the function $\sigma = \Phi(s, \theta), s \in [0, 1]$.

The Muskat-Leverett equations can be reduced to the following equation system [16]:

$$m\frac{\partial S}{\partial t} = div\left[Ka_0(a_1\nabla\sigma - a_2\nabla\theta + \vec{f}_1) - b_1\vec{v}\right] \equiv div\vec{v}_1, \tag{1.67}$$

$$div\vec{v} \equiv div[K(\nabla p + a_3\nabla\theta + \vec{f}_2)] = 0; \quad \sigma = \Phi(s,\theta), \tag{1.68}$$

where $K(x, \theta, \sigma)$ is a tensor associated with the permeability of the medium:

$$a_0 = a_0(\theta), a_i = a_i(\sigma, \theta), i = 1, 2, 3; \quad b_k = b_k(0, \theta), \bar{f}_k = \bar{f}_k(x, \sigma, \theta), k = 1, 2;$$

$$a_0(0) = a_0(1) = b_1(0, \theta) = 0; \quad \ln f\ a_1 \geq \alpha_0 > 0.$$

Let $\Omega \subset R^3$ be a bounded region whose boundary $\partial\Omega$ is divided into several components on the basis of their boundary conditions:

$$(p, s, \theta) = (p_0, s_0, \theta_0), (x, t) \in \Sigma^1 = \Gamma^1 \times [0, T],$$

$$\vec{v}_i \vec{n} = b_i R, i = 1, 2; \quad \theta = \theta_0(x, t), (x, t) \in \Sigma^2 = \Gamma^2 \times [0, T],$$

$$v_i n = 0, i = 1, 2; \quad \lambda \frac{\partial \theta}{\partial n} = \beta(\theta_0 - \theta), (x, t) \in \Sigma^3 = \Gamma^3 \times [0, T].$$

In this system, $\vec{n}$ is the unit vector of the outward normal to $\partial\Omega$, sections Γ^1 and Γ^2 model the injection and withdrawal sections and the contact with homogeneous fluid at rest, while Γ^3 represents the contact with surrounding impermeable rocks.

Boundary conditions need to be augmented by the initial condition

$$(s, \theta)|_{t=0} = (s_0, \theta_0)(x, 0), \quad x \in \Omega. \tag{1.69}$$

When $\Gamma^1 = 0$, the law of conservation of mass of the mixed flow in region Ω produces the necessary condition

$$\int p(x, t)dx = \int_\Gamma R(x, t)dx = 0, \quad t \in [0, T]. \tag{1.70}$$

Dynamic saturation σ_0 is reestablished with respect to θ_0 and $s_0$at $t = 0$, and at Σ^1, by means of the unique dependence (B): $\sigma_0 = \Phi(s_0, \theta_0)$.

Note: Let σ be a reasonably smooth function, making the boundaries γ_k of sets $G_k = \{\sigma(x, t) = k\}, k = 0$ piecewise smooth curves. For this to be the case, γ_k must satisfy the following conditions:

$$[\vec{v}\vec{n}_k]_k = 0, \quad \left[\lambda \frac{\partial \theta}{\partial n_k}\right]_k = 0; \quad mU_{n_k}[s]_k = -K\left[a_0 a_1 \frac{\partial \sigma}{\partial n_k}\right]_k,$$

where $[\varphi]$ is the difference between the values of $\varphi(x, t)$ to the right and to the left of γ_k; n_k is the outward normal to G_k and U_{n_k} is the velocity of motion of γ_k in the direction n_k. Only one phase flows in regions G_k, so that their saturation s is constant, and depends on the history of their formation. Thus, the velocity of the boundaries $\gamma_k - U_{n_k}$ may be determined both on the basis of Stephan conditions at $[s] \neq 0$, and on the basis of Verigin conditions at $s = s_*(\theta)$ in G_0 and $s = s^*(\theta)$ in G_1.

The validity of the initial boundary value problems formulated here for the models described in Sections 1.4 and 1.5, divided into classes of generalized solutions, is shown in Eqns 1.35 and 1.36.

1.6 THE COMBINATION OF WELL AND RESERVOIR FLOWS OF VISCOUS INCOMPRESSIBLE FLUIDS

Introduction. Mathematical modelling of processes taking place in the near-well zone is complicated primarily by the heterogeneous nature of the multicomponent flow in the well and in the surrounding porous medium.

Let us explain the dynamics of reservoir flow in the near-well zone. For simplicity, we will assume that the fluid enters the well only from its lower section, and that its movement is due to the pressure head generated by the difference between formation and atmospheric pressure (at the wellhead). In such circumstances, the fluid flow is satisfactorily described by the well-known exact solutions of the Navier-Stokes equations (such as the Poiseuille solution), or by their boundary layer modifications. At real fluid pressure head values, the average velocity of fluid flowing in the well is quite high by comparison with the slow flow of fluid through the surrounding porous medium, and this explains the difficulty of describing the combined flow of fluid in a well and in a porous medium. This difficulty is usually resolved by constructing various intermediate models of fluid flow in the transition zone (merging asymptotic expansions, boundary layer smoothing, ignoring the reverse effect of fluid flow inside the well on reservoir flow, etc.).

However, the application of these models to flow dynamics shows that they provide a poor representation of the physical nature of fluid flows in transition zones and that they greatly distort the general characteristics of the flows even far away from the transition zones.

V.N. Monakhov and N.V. Khusnutdinova [94] have proposed a model describing the combined high-velocity flow of a viscous fluid inside a well and the reservoir flow of the same fluid through the surrounding porous medium in terms of boundary layer approximations of both flows.

1.6.1 Problem Statement

Plane stationary flow of an incompressible fluid in a well is described by the Navier-Stokes equation:

$$(\vec{u}\nabla)\vec{u} = \mu\Delta\vec{u} - \nabla p + \vec{F}; \quad \vec{u} = 0, \quad (x, y) \in (D_1),$$

where $\vec{u} = (u, v)$ is the flow velocity vector of a fluid with a density $\rho \equiv 1$; $\mu = const$ is viscosity; $p = p_0 + \rho gh, p$ is pressure, $(g = gh), \vec{F} = 0$.

We will also use the Navier-Stokes equations to describe reservoir flow in region D_2, adjacent to D_1. In these equations, in accordance with the flow theory (1.71), $\vec{F} = -\lambda\vec{u}$, $\lambda(x, y) = \dfrac{m\mu}{k}$ – m is porosity, k is the permeability of the porous medium ($\vec{v} = m\vec{u}$- flow velocity).

We will consider only problems dealing with the combination of reservoir flow in a porous medium (the formation) and in a group of imperfect wells (a "plane well" or simply "well") corresponding to a vertical section of the formation $(g = (-g, 0))$.

Let the direction of axis O x be parallel to the direction of fluid flow in the well. Assuming the well diameter $(2h)$ to be much smaller than its length and assuming that $|\vec{v}| \ll |\vec{u}|$, we can replace the initial equations with boundary layer equations:

$$(\vec{u}\nabla)\vec{u} = \mu\vec{u}_{yy} - p_x \nabla \cdot \vec{u} = 0, \quad (x, y) \in D_1. \tag{1.71}$$

Let the reservoir flow (along Oy) be perpendicular to the fluid flow in the well, and let the thickness of the formation (region D_2) be much smaller than its length. On this basis, assuming that $|\vec{u}| \ll |\vec{v}|$, we can derive boundary layer equations:

$$(\vec{u}\nabla)\vec{v} = \mu\vec{v}_{yy} - p_y - \lambda\vec{v}, \quad \nabla \cdot \vec{u} = 0, \quad (x, y) \in D_2. \tag{1.72}$$

It is assumed that flow velocity vector and pressure p are continuous along the flow transition curve $(\Gamma = \overline{D}_1 \cap \overline{D}_2)$:

$$[\vec{u}] = 0, [p] = 0, (x, y) \in \Gamma, \tag{1.73}$$

where $[f] = f|_{\Gamma_2} - f|_{\Gamma_1}, \Gamma_k = \Gamma \subset D_k, (k = 1, 2)$, and f_{D_k}- boundary values $f(x, y)$ at $(x, y) \in D_k$.

Note that if we take into account the direction of reservoir flow after the substitution of $x=\overline{y}, y=-\overline{x}, u=\vec{v}, v=-\vec{u}$, at $\lambda=0$ equations (1.71), (1.72) become Prandtl's boundary layer equation (1.71) for $\vec{u}(\overline{x},\overline{y}), \vec{v}(\overline{x},\overline{y})$.

Polubarinova-Kochina [105] noted the appearance of a slippage effect in experimental studies of fluid flow in the vicinity of porous surfaces and proposed a simple model to describe it.

Let us assume for clarity that the flow convergence line $\Gamma{:}y=0$ and that therefore $D_1{:}y<0, D_2{:}y>0$.

According to [105], this is a situation which instead of satisfying conditions (1.73), satisfied the conditions of convergence $[v]=[p]=0$:

$$\frac{u}{y}=\frac{\alpha}{\sqrt{k}}(u-Q),$$

Where f $=f(x\pm 0)$; Q_+ is the fluid flow rate through the porous surface; α is a constant, describing the porous medium in the vicinity of Γ; k is permeability.

1.6.2 The Combination of Formation Flow and Free Flow Near the Well Wall

Let $D_1=\{x>0, -h<y<0\}$ be a region corresponding to a symmetrical section of the well (with respect to $y=-h$), and let $D_2=\{x>0, y>0\}$ represent the formation flow region.

Let us assume that near the wall $y=0$, velocity vector $\vec{u}$ and pressure p are continuous, satisfying condition (1.73), where $[f]=f(x+0)-f(x-0)$.

Let us also assume that the condition of no flow is satisfied near the wall, $y=0$, while the condition of symmetry is satisfied along the line $y=-h$, that the fluid flow at the wellhead, $x=0$, is

$$\vec{u}|_{y=0}=(u_y,v)|_{y=-h}=0, \quad u|_{x=0}=u_0(y)\geq 0, \tag{1.74}$$

and that $u_0(y)\equiv 0, y\in[-h,0)$, in the case of a perfect well.

Having solved problems (1.71) and (1.74), and on the basis of condition (1.73) and the existence of no flow conditions at the bottom of the formation, we can calculate the boundary data for equations (1.73):

$$u|_{x=0}=v|_{x=0}=0, \quad v|_{y=0}=v_0(x), \tag{1.75}$$

where $v_0(x)=v(x-0)$.

If we regard the line $y = -h$ as the well wall, (1.74) assumes the form:

$$u|_{y=0} = (u, v)|_{y=-h} = 0, \quad u|_{x=0} = u_0(y) \geq 0. \tag{1.76}$$

1.6.3 The Combination of Formation Flow and Free Flow at the V,well Entry Point

Let us assume the well to be an open hole completion [89, p. 419], so that

$$D_1 = \{0 < y < h, x > 0\}, \quad D_2 = \{x < 0, y > 0\}.$$

The flow in such a well can be calculated by solving the boundary value problem

$$\vec{u}|_{y=h} = (v, u)|_{y=0} = 0; \quad u_x|_{x=0} = 0 \tag{1.77}$$

for equations (1.71).

Having solved problems (1.71) and (1.77), we can calculate the boundary data for the boundary layer equations (1.72) in region D_2, having taken well symmetry (with respect to $y = 0$) into account:

$$\overline{u}|_{x=0} = u_0(y), \quad y > 0; \quad u_y|_{y=0} = 0, \tag{1.78}$$

where $u_0 \equiv 0, y > h; \quad u_0 = q(y), U_0 = 0, 0 \leq y \leq h, q = u(+0, y)$.

In the opposite situation, if the line $y = h$ is the axis of symmetry and the line $y = 0, x > 0$ is the well wall, boundary conditions (1.77), (1.78) are replaced by

$$u|_{y=0} = (v, u)|_{y=0} = 0, u|_{y=h} = u(x), \quad 0 \leq x \leq X, \quad u_x|_{x=0} = 0, \tag{1.79}$$

$$u|_{x=0} = u(+0, y) = u_0(y), 0 < y < h, v|_{x=0} = 0, v|_{y=0} = v_1(x), x < 0, \tag{1.80}$$

$$\lim_{x \to +0}(x, y) = v \equiv const > 0, \quad 0 < y < h.$$

1.6.4 Thermal Boundary Layer in Formation Flow Problems

The flow of fluid through a porous medium in which the two-dimensional fluid flow region is greatly elongated in one direction (a thin formation), can be described by thermal boundary-layer equations

$$\begin{aligned} \rho\vec{u}\nabla u &= (uu_y)_y - p_x + \rho f_1; \quad div(\rho\vec{u}) = 0; \\ \rho\vec{u}\nabla\theta &= (\lambda\theta_y)_y + \mu u_y^2 + up_x, \end{aligned} \tag{1.81}$$

in which the external force $\vec{F}=(f_1,f_2)$ is represented, as suggested by Zhukovsky (see Section 1.2), as $\vec{F}=-\gamma\vec{u}$.

In the above equations, $\vec{u}=(u,v)$ is the velocity vector; $\rho=\rho(\theta,p)$ is density; $\mu=\mu(\theta)$ is viscosity; $p=p(x)$ is pressure; $\theta=\int_{T_0}^{T}c_p(\xi)d\xi$-enthalpy; T is the temperature; $\lambda=\lambda(\theta)$ is thermal conductivity coefficient; $\gamma=m\mu/\rho k, m(x)$ is porosity; $k(x)$ is permeability.

Let us assume that the Prandtl number is 1, that is $\mu\rho=\lambda\rho c_p^1=\sigma(\theta)_0$ and $p_x=0$.

We can then use Mises variables (x,ψ) (ψ is flow function $|\rho u=\psi_y,\rho v-v(x)=-\psi_x, v_0(x)\rho v|_{y=0}$) to write equations (1.81) for the horizontal component $u(x,\psi)$ of the velocity vector $\vec{u}=(u,v)$, and for total energy $h=\theta+1/2u2$, in the following form:

$$L\vec{h}\equiv(\sigma u\vec{h}\psi)_\psi-v_0(x)\vec{h}_\psi-\vec{h}_x=\vec{\gamma}f_0,\quad (x,\psi)\in D, \tag{1.82}$$

where $\vec{h}=(u,h),\vec{f}_0=(I,u),D=\{x,\psi|0<x<X,\psi>0\}$.

We can also consider the following problem of boundary layer continuation (1.82):

$$\vec{h}|_{x=0}=\vec{h}_0(\psi),\psi\geq 0;\quad (u,h-h_1)|_{\psi=0}=0,\quad x\in[0,X]. \tag{1.83}$$

V.N. Monakhov, and N.V. Khusnutdinova [94] have demonstrated the existence of generalized solutions of (1.82), (1.83) which could represent fluid flow accompanied by the development of no flow zones ($u\equiv 0$) and thus demonstrate that $\partial u/\partial\psi$ can tend towards infinity.

1.7 A FORMATION FLOW MODEL OF WAXY, HIGH GAS CONTENT OIL DISPLACEMENT

1.7.1 Process Description

The main distinguishing parameter associated with high gas content oil flow is saturation pressure p_H. If high gas content oil pressure falls below saturation pressure, the gas phase is released from the oil. A characteristic parameter of waxy oil flow is wax crystallization temperature θ_K; when formation temperature falls below this point, crystalline wax is precipitated onto the pore surfaces. These factors need to be taken into consideration in the development of waxy, high gas content oil fields, when it is important to maintain formation pressure p, aiming at $p>p_H$. The essential process requirement is to maintain a high formation temperature, and that is achieved by injecting steam or hot water into the formation. The development of heavy oil deposits where the oil contains large quantities of gas presents an especially difficult problem, because such formations contain areas where both the gas and the wax approach their critical points:

$$|p - p_H| \ll 1, \quad |\theta - \theta_k| \ll 1.$$

Oil production in such doubly (!) critical conditions is associated with physical effects which it is difficult to explain in theoretical terms. For instance, in some wells, production rates can fall virtually to zero, and do not respond to formation pressure increases. When such sections are re-drilled, the pores of core samples are found to contain quite a stable "foam" consisting of gas bubbles plugged with wax crystals. It is clear that in such circumstances oil recovery can be enhanced only by increasing formation temperature, i.e. by steady steam injection into the formation. To calculate the steam flow rates and temperatures required for this process, we need suitable mathematical models, capable of taking the interaction of p_H and θ_k into account. Below, we propose a simple mathematical model describing this process.

1.7.2 The Proposed Mathematical Model

Let p_H be the bubble point pressure of gas-cut heavy oil whose reduced density is $\rho = 1$ at $p > p_H$. The process of gas separation at $p < p_H$ is allowed for by considering a special equation of state for the mixed oil and gas flow:

$$\rho = \rho(p) = \begin{cases} 1, & p \geq p_H \\ \delta(p - p_H) + 1 & p \leq p_H \end{cases} \tag{1.84}$$

where $\delta = const$. In what follows, let $Y = Y(\theta)$ be the internal energy corresponding to wax crystallization, θ is the equilibrium temperature, $m = const$ is porosity and $k = k(\theta, p)$ is permeability. Assuming the fluid flow to be one-dimensional and planar (orthogonal to the gravity vector) we arrive at the following equations describing the formation flow of waxy, high gas content oil:

$$\begin{cases} m\dfrac{\partial \rho(p)}{\partial t} - \dfrac{\partial}{\partial x}\left(\rho k \dfrac{\partial p}{\partial x}\right) = 0, \quad v = -k\dfrac{\partial p}{\partial x}, \\ \dfrac{\partial}{\partial t}Y(\theta) - \dfrac{\partial}{\partial x}\left(\lambda \dfrac{\partial \theta}{\partial x} - v\theta\right) = 0, \quad \lambda = \lambda(\theta) \end{cases} \tag{1.85}$$

Below we describe solutions of (1.85) which take the form of a simple wave: $p = p(\xi), \theta = \theta(\xi), \xi = x - at, a = const.$ They transform (1.85) into the form

$$\begin{cases} \dfrac{d}{d\xi}\left(map + \rho k \dfrac{dp}{d\xi}\right) = 0, \\ \dfrac{d}{d\xi}\left(aY + \lambda \dfrac{d\theta}{d\xi} - \nu\theta\right) = 0 \end{cases} \tag{1.86}$$

$dE, vdE,$ Integrating equations (1.85) and introducing a new function

$$\varphi = \int_{p_0}^{p} K(\theta, \eta)\eta d\eta \equiv \Phi(\theta, p),$$

we obtain

$$\varphi_\xi - \Phi_\theta \theta_\xi = c_1 \rho^{-1} - am \equiv \psi_1(\theta, p),$$

$$(\lambda - \Phi_\theta)\theta_\xi + \varphi_\xi = c_2 - aY \equiv \psi_2(\theta), \tag{1.87}$$

and finally

$$\varphi_\xi = f_1(\theta, \varphi, \xi), \quad \theta_\xi = f_2(\theta, \varphi, \xi), \tag{1.88}$$

where $\lambda f_1 = (\lambda - \Phi_\theta)\psi_1 + \Phi_\theta \psi_2, \lambda f_2 = \psi_2 - \psi_1, c_1$ and c_1 are arbitrary real constants.

1.7.3 The Development of High Gas Content Heavy Oil Displacement Algorithms and the Validation of Numerical Models

Equation (1.88), which describes the formation flow of waxy, high gas content oil using simple wave analytical variables, has a standard structure, which allows us to use the classical iteration system

$$\varphi_\xi^{n+1} = f_1(\theta^n, \varphi^n, \xi), \quad \theta_\xi^{n+1} = f_2(\theta^n, \varphi^n, \xi). \tag{1.89}$$

The only unusual feature of (1.88) is that internal energy $Y(\theta)$ which forms part of coefficient (1.88), is not a smooth function. However, this does not prevent us from using the iterative function (1.89).

We have also developed a full description of the solution of boundary-value problems for (1.88) and have carried out a computer implementation.

1.8 FORMATION FLOW OF TWO IMMISCIBLE INHOMOGENEOUS FLUIDS

The practical implementation of the Muskat-Leverett model describing two-phase (two-component) fluid flow (1.4) requires the determination of three functional parameters (relative phase permeabilities and the Leverett J function), a process which presents considerable experimental difficulties. The flow models of an n-component fluid when $n \geq 3$ require even more functional parameters which it is virtually impossible to determine experimentally.

We have proposed a mathematical model of the flow of two immiscible inhomogeneous fluids (such as water/steam or oil/gas), using only the Muskat-Leverett functional parameters, and replacing the usual constant density condition with the conditions of incompressibility of fluids. Similar models have proved useful in oceanography and hydrology.

A flow model for one inhomogeneous fluid was first proposed in [91] and the validity of the application of initial boundary value problems to this model was examined in [91, 124]. In this book, we present a study of these problems in relation a flow model of two inhomogeneous fluids.

1.8.1 Problem Statement

It is assumed that the flow of inhomogeneous fluids in a porous medium is described by the laws of two-phase fluid flow which form the Muskat-Leverett model (1.2): the law of conservation of mass, Darcy's laws and Laplace laws for capillary pressure discontinuity:

$$\begin{cases} m(s_i\rho_i)_t + div(\rho_i\vec{v}_i) = 0 \quad (s_1 + s_2 = 1); \\ \vec{v}_i = -K_0 k_i(\nabla p_i - \rho_i\vec{g}); \quad p_2 - p_1 = p_c(x, s). \end{cases} \tag{1.90}$$

In these equations, $s_i (i = 1, 2)$ are phase saturations; $\rho_i, p_i, \vec{v}_i$ is the corresponding densities, pressures and flow velocity vectors; $p_c(x, s)$ is the capillary pressure; $k_0 = k_0(x)$ is the absolute permeability tensor of the medium; $k_i = \vec{k}_i(s_i) \cdot \mu_i^{-1}$ is the relative phase permeabilities; $\mu_i = const$ is fluid viscosities; $\vec{g}$ is the gravity acceleration vector.

Instead of using equations of state of the fluids to close equation (1.90), we have used the conditions of their incompressibility:

$$\rho_{it} + \vec{u}_i\nabla\rho_i = 0 \quad \vec{v}_i = ms_i\vec{u}_i, \tag{1.91}$$

which indicate that the fluids retain their densities p, along the trajectories of their flow ($\vec{u}_i$ are their flow velocity vectors).

Using equations (1.90) and (1.91), we obtain

$$\rho_i[ms_{it} + div\vec{v}_i] + ms_i[\rho_{it} + \vec{u}_i\nabla\rho_i] = 0, \tag{1.92}$$

and this gives us the equations for s_i, which coincide with the relevant equations when $\rho_i = const$:

$$ms_{it} + div\vec{v}_i = 0, \quad i = 1, 2.$$

Therefore, if we introduce the average pressure from 1.2 ($s \equiv s_1$):

$$p = p_2 + \int_s^{s^*} b\frac{\partial}{\partial\xi}p_c(x, \xi)d\xi, \quad b = k_1/k \tag{1.93}$$

equations (1.90) are transformed into

$$\begin{cases} ms_t = div[k_0(a_1\nabla s + \vec{f}_1) - b\vec{v}] \equiv -div\vec{v}_1, \\ 0 = div[k(\nabla p + \vec{f}_2)] \equiv -div\vec{v}, \end{cases} \tag{1.94}$$

where $\vec{v} = \vec{v}_1 + \vec{v}_2$ – mixed flow velocity; $s_* > 0, 1 - s^* > 0$ – residual phase saturations.

$$a_1 = |p_{cs}|a_0, a_0 = k_1k_2/k, \; k = k_1 + k_2, k = k_0k,$$

$$\vec{f}_1 = a_0[-\nabla_x p_c + (\rho_1 - \rho_2)\vec{g}], \quad \vec{f}_2 = \int_s^{s^*} \nabla_x p_{cs} k_2 k^{-1} ds - \vec{f}_1 b^{-1},$$

where $k_1(s_*) = k_2(s^*) = p_c(s^*) = 0$.

System (1.94) is closed by density equations ρ_I, which may be equations (1.91) or, in view of (1.92), their equivalent equations (1.93).

To calculate the pressure of the target functions (s, ρ_1, ρ_2, p), we examine the initial boundary-value problem

$$s|_{t=0} = s_0(x), \quad v_i \cdot \vec{n}|_{\partial\Omega} = 0,$$

$$(\rho_1, \rho_2)|_{t=o} = \rho_{i0}(x), \quad x \in \Omega,$$

where $\Omega \subset R^3, \Omega_T = \Omega[0, T]$.

1.8.2 The Thermal Model Equation

Let s_i $(i = 1, 2)$ be the phase saturations of pore space $(s_1 + s_2 = 1)$, m is porosity, $\alpha_i = ms_i$ $(i = 1, 2)$, $\alpha_3 = 1 - m$ is the exchange concentrations of the fluids and the solid phase (the pore space matrix), $\rho_i, p_i, \vec{u}_i$ is density, pressure and fluid flow velocity, $\vec{v}_i = \alpha_i \vec{u}_i = m s_i \vec{u}_i$ $(i = 1, 2)$ is phase flow velocities, $\vec{v} = \vec{v}_1 + \vec{v}_2$ is the mixed flow velocity. It is assumed that the rocks are not subject to deformation, and that each point of the porous medium is in a state of thermal equilibrium, i.e. that their phase temperatures θ_i are the same: $\theta_i = \theta$ $(i = 1, 2, 3)$.

It is further assumed that the phase flow is as described by the Muskat-Leverett model (1.2):

$$\frac{\partial m s_i \rho_i}{\partial t} + div \rho_i \vec{v}_i = 0, \tag{1.95}$$

$$\vec{v}_i = -K_0(x) \frac{\sigma_i(s)}{\mu_i(\theta)} (\nabla p_i - \rho_i \vec{g}), \tag{1.96}$$

$$p_2 - p_1 = \gamma(\theta) \cos \alpha(\theta) \left(\frac{m(x)}{|K_0(x)|} \right)^{1/2} J(s) \equiv p_c(x, \theta, s), \; s \equiv s_1 \tag{1.97}$$

In the above equation system, $K_0(x)$ is the absolute permeability tensor of the medium; $\sigma_i(s)$ is relative phase permeabilities; $\sigma_1(s_*) = \sigma_2(s^*) = 0, s_*, 1 - s^*$ is residual phase saturations; μ_i is phase viscosities; $\vec{g}$ is the acceleration of gravity; γ is the surface tension coefficient; α is the wetting angle; $J(s)$ is the Leverett capillary pressure function. In what follows, $k_i(s, \theta) = \sigma_i(s) \mu_i^{-1}(\theta), k = k_1 + k_2$.

We now replace the fluid equations of state, (1.95)–(1.97), with the conditions for an incompressible fluid:

$$\frac{\partial \rho_i}{\partial t} + \vec{u}_i \nabla \rho_i = 0, \quad (i = 1, 2), \tag{1.98}$$

which indicate that the fluids retain their densities ρ_i along the trajectories of their flow. If we take into account only convective heat transfer and thermal conductivity, we can write down the mixed flow energy balance equation in the form

$$\frac{\partial \theta}{\partial t} + div(\vec{v}\theta - \lambda(x, \theta, s) \nabla \theta) = 0. \tag{1.99}$$

In this equation, λ is the thermal diffusivity of the mixed flow (consisting of the two fluids and the porous matrix). In [15], we describe equation (1.99), derived from general energy balance equations for components with constant phase densities and heat capacities, for a mixed flow which is in thermal equilibrium, where

$$\lambda = \sum_{i=1}^{3} \alpha_i \lambda_i. \tag{1.100}$$

In a more generalized case of an inhomogeneous incompressible fluid, equations (1.99) and (1.100) derived in [15] remain valid, provided that the product of multiplication of phase densities by their heat capacity remains constant.

1.8.3 Transformation of Equations. Problem Statement

By expanding the differentiation in (1.95) and using (1.96), we obtain

$$\rho_i(ms_{it} + div\vec{v}_i) + ms_i(\rho_{it} + \vec{u}_i \nabla \rho_i) = 0,$$

and this provides us with equations for s_i, whose form is the same as that of the corresponding equations for $\rho_i = const$ 1.2:

$$ms_{it} + div\vec{v}_i = 0 \quad i = 1, 2. \tag{1.101}$$

Because $s_2 = 1 - s_1$ this system is equivalent to the system for $s, v_1$and v:

$$\left.\begin{aligned} ms_t + div\vec{v}_1 &= 0, \\ div\vec{v} &= 0. \end{aligned}\right\} \tag{1.102}$$

Following the principle used in Section 1.2, let us now introduce average pressure

$$p = p_2 + \int_s^{s^*} b_1 \frac{\partial}{\partial \xi} p_c(x, \theta, \xi) d\xi, \quad b_i = k_i/k, \quad i = 1, 2.$$

By transforming (1.96), (1.97) and (1.102), we arrive at the following system of equations for$s, p, \theta, \rho_1, \rho_2$:

$$\begin{cases} ms_t = div[K_0(a_1 \nabla s - a_2 \nabla \theta + f_1) - b_1 \vec{v}] \equiv -div\vec{v}_1(s, p, \rho_i, \theta), \\ 0 = divK_0 k(\nabla p + f_2 + a_3 \nabla \theta) \equiv -div\vec{v}(s, p, \rho_i, \theta), \\ \dfrac{\partial \theta}{\partial t} + div(\vec{v}\theta - \lambda(x, \theta, s) \nabla \theta) = 0, \\ \dfrac{\partial m s_i \rho_i}{\partial t} + div \rho_i \vec{v}_i = 0, \quad i = 1, 2 \end{cases} \tag{1.103}$$

where

$$a_1 = |p_{cs}|a_0,\ a_2 = p_{c\theta}a_0,\ a_0 = b_1 k_2,\ a_3 = -k_1 p_{c\theta} - \int_s^{s^*} \frac{\partial}{\partial\theta}(b_1 p_{cs})ds,$$

$$f_1 = a_0[(\rho_1 - \rho_2)g - \nabla_x p_c],\ f_2 = \int_s^{s^*} b_2 \nabla_x p_{cs} ds - g(b_1\rho_1 + b_2\rho_2).$$

Let $\Omega \subset R^3$ be a bounded region, $\Omega_T = \Omega \times [0, T], \partial\Omega = s, \Gamma = s \times [0, T]$, $s = s_1 \cup s_2$, and $\Gamma_i = s_i \times [0, T]$. To calculate the target functions, let us consider the following initial boundary value problem:

$$\begin{cases} (s, \theta, \rho_1, \rho_2)|_{=0} = (s, \theta, \rho_1, \rho_2)_0(x), \quad x \in \Omega, \\ \vec{v}_i \vec{n}|_\Gamma = 0, \quad \theta|_{\Gamma_1} = \theta_0(x, t), \quad \lambda \left.\frac{\partial\theta}{\partial n}\right|_{\Gamma_2} = \beta(\theta_0 - \theta) \end{cases} \tag{1.104}$$

In this equation, $\vec{n}$ is the unit vector of the outward normal to Γ; $\beta(s)$ is the heat transfer coefficient for a three-component mixed flow. In the above exact derivation of (1.99), this component had the form $\beta = \sum_{i=1}^{3} \alpha_i \beta_i (\rho_i c_i)^{-1}$, β_i being the heat transfer coefficient of the ith phase.

CHAPTER 2

Analytical and One-Dimensional Models of Thermal Two-Phase Flow

2.1 INTRODUCTION

Virtually all methods of calculation of oil recovery parameters assume that the field can be divided into so-called basic elements of symmetry, corresponding to specific regular well patterns. As a result, the first stage of all methods of evaluation of field development systems consists of analyzing the process of oil recovery from this basic element. It is usually assumed that the basic element can be regarded as a closed system with symmetry conditions at its boundaries, boundary conditions specified for the wells (production rate, pressure or saturation) and uniform formation parameters within the element itself. The figures for all the basic elements are then added together in order to calculate overall field performance.

The flow of inhomogeneous fluid through the porous medium of each basic element can be described by one-dimensional models of subsurface fluid flow, which form the basis of all computer-aided oil production management systems. Accordingly, we have devoted considerable attention to the development and validation of one-dimensional numerical models of subsurface fluid flow (Sections 1.7—1.8).

In reservoir engineering, there is a special role for simplified models of fluid flow, based on providing a range of exact (analytical) solutions of one-dimensional models: stationary solutions, analytical parabolic or traveling wave solutions and others.

Muskat's displacement laws, as well as Polubarinova-Kochina and Charnov's assessments of near-wellbore zones, based on the parabolic self-similarity of the simplest flow models, still provide us with the tools of assessing some development parameters.

Analytical solutions of flow equations are widely used for the following purposes:

1. They are of independent interest as special solutions of the initial equations;

2. They are used as reference standards (tests) of the various approximate methods of solving more general equations;
3. They are used for preliminary numerical or analytical studies of the singularities of the initial equations;
4. In many cases, they represent asymptotic forms of the solutions of a broad range of problems, in particular those in which the detailed structure of boundary conditions as well as initial conditions becomes less important, although they are frequently of the greatest interest;
5. Combined with comparison theorems, they provide an efficient theoretical tool for the study of the properties of solutions for the initial variables;
6. In some applied fields (e.g. in oil production), they can serve as a forecasting tool.

In this chapter, we focus on the theory and numerical construction of analytical solutions of subsurface flow dynamics equations (1.2–1.6).

2.2 BOUNDARY VALUE PROBLEMS FOR ORDINARY DIFFERENTIAL EQUATIONS

2.2.1 Functional Spaces

Let us consider the Banach spaces described below, using an arbitrary measurable set $E \subset [0, l], l > 0$:

$$L_p(E) = \left\{ f(x) \left(\int_E |f(x)|^p dp \right)^{\frac{1}{p}} \equiv \|f\|_{p,E} < \infty \right\}, \quad p \geq 1;$$

$$\|f\|_{2,E}^2 = (f, f)_E = \int_E |f(x)|^2 dx \quad [L_2(E) - \textit{Hilbert space}];$$

$$W_p^1(E) = \left\{ f(x) \|f\|_{p.E} + \|f_x\|_{p.E} \equiv \|f\|_{p.E}^{(1)} < \infty \right\}, \quad p \geq 1;$$

$$W_p^k(E) = \left\{ f(x) \sum_{s=0}^{k} \|f_x^{(s)}\|_{p.E} \equiv \|f\|_{p.E}^{(k)} < \infty \right\}, \quad p \geq 1; \quad k - \textit{integer}, k \geq 1;$$

$$C(E) = \left\{ f(x) \max_E |f(x)| \equiv |f|_E < \infty \right\},$$

$$C^a(E) = \left\{ f(x) \max_{x,y \in E} \left(f(x) + \frac{|f(x) - f(y)|}{|x - y|^a} \right) \equiv |f|_{a,E} < \infty \right\};$$

$$C^{k+a}(E) = \left\{ f(x) \sum_{s=0}^{k} |f_x^{(s)}|_{a.E} \equiv |f|_{a.E}^{(k)} < \infty \right\}, \qquad k - \textit{integer}, a \in [0, 1).$$

In space $L_p(E)$, the functions on $C^\infty(E)$ are dense, and we can replace functions approximating $f(x)\in L_p(E)$ functions with their *averaged forms*:

$$f_p(x)=\int_E \omega_\rho(|x-y|)f(y)dy, \quad \rho\to 0, \tag{2.1}$$

where $\omega_\rho(r)=\rho^{-1}\omega(r/\rho)$, $\omega(\tau)$ is a non-negative infinitely differentiable function, $\tau\geq 0$, which equals zero at $\tau\geq 1$ and is such that $\int_{\tau\leq 1}\omega(\tau)d\tau=1$.

Space $W_p^1(E)$ (as is $L_p(E)$), $p\geq 1$ consists of *a class* of equivalent functions $\tilde{f}(x)$, $\|\tilde{f}(x)-f(x)\|_{p,E}^{(1)}=0$, which can always be represented by

$$\bar{f}(x)=\lim_{p\to 0}mes^{-1}E\rho\int_{Ep(x)} f(y)dy, \quad \chi\in E,$$

$$E_p(x)=I_p(x)\cap E, \quad I_p(x)=\{y\ |x-y|<\rho\}, \|\bar{f}(x)-f(x)\|_{\rho,E}=0$$

$\bar{f}(x)$ is absolutely continuous for almost all $x\in E$, and its derivative is calculated as an ordinary derivative of an absolutely continuous function and coincides almost everywhere in E with the generalized derivative f_x, i.e. $\bar{f}_x=f_x\in L_p(E)$.

Therefore, in what follows we will always understand $f(x)\in W_p^1(E)$ to mean $\bar{f}(x)$, being a representative of a class of equivalent $f(x)$ functions. This is also how the further properties of functions $f(x)\in W_p^1(E)$ should be understood.

Let $f(x)\in W_p^1(E), p\geq 1$. Let us extend $f_x=\frac{df}{dx}, x\in E$ by a section $[0,l]\supset E$, assuming $\frac{df}{dx}=0, x\in([0,l]\backslash E)$ and representing

$$f(x)=\int_{x_0}^{x}\frac{df}{dt}dt+f(x_0), \quad x_0\in E. \tag{2.2}$$

From expression (2.2), we obtain

$$\max|f-f_0|\leq\int_E\left|\frac{df}{dt}\right|dt, \text{ that is } |f-f_0|_E\leq\|f_x\|_{1,E},$$

in this case $W_1^1(E)\subset\subset C(E)$ ($W_1^1(E)$ fits compactly into $C(E)$).

Let now $f(x)\in W_p^1(E), \frac{1}{p}+\frac{1}{q}=1$. Then

$$|f(x_2)-f(x_1)|\leq\int_{x_1}^{x_2}1\cdot\left|\frac{df}{dx}\right|dx\leq\|1\|_{q,[x_1,x_2]}\cdot\|f_x\|_{p,E}=(x_2-x_1)^{1/q}\|f_x\|_{p,E},$$

$$(f_x=0, \quad x\in[0,l]\backslash E),$$

Therefore,

$$|f|_{\alpha,E} \leq \|f\|^{(1)}_{p,E}, \quad \alpha = \frac{p-1}{p} \geq 0, \tag{2.3}$$

and $W^1_p(E) \subset\subset C^{\alpha}(E)$.

Let $f(x_0)=0, x_0 \in E$. Then, representation (2.2) gives us

$$|f(x)| \leq \int_0^x \left|\frac{df}{dt}\right| dt \leq \int_0^l \left|\frac{df}{dt}\right| dt = \int_E 1 \cdot \left|\frac{df}{dt}\right| dt \leq (mesE)^{1/q} \|f_x\|_{p,E},$$

$$\frac{1}{p} + \frac{1}{q} = 1 \quad \left(\frac{df}{dx} = 0, x \in E\right).$$

Consequently,

$$\left\{\int_E |f(x)|^s dx\right\}^{1/s} \leq (mesE)^{1/q} \|f_x\|_{p,E} \left\{\int_E 1 dt\right\}^{1/s} = (mesE)^{1/s+1/q} \|f_x\|_{p,E}.$$

In this way, we arrive at the Poincare inequality

$$\|f\|_{s,E} \leq (mesE)^{\beta} \cdot \|f_x\|_{p,E}; \quad f(x_0) = 0, \quad x_0 \in E, \tag{2.4}$$

where $p>1, s>1, \beta = \dfrac{p-1}{p} + \dfrac{1}{s}$.

2.2.2 Convergence in $L_p(E), p \geq 1$

In this section, we consider the following types of convergence for the sequence $\{f_k(x)\}$ of functions $f_k(x) \in L_p(E), p \geq 1$:

1. $f_k \xrightarrow[L_p]{} f$ (f_k *converge* to f *strongly* in L_p), if $\|f_k - f\|_{p,E} \to 0, k \to \infty$;
2. $f_k \xrightarrow[a.e.]{} f$ (f_k *converge* to f *almost* everywhere on E), if in the set $\tilde{E} \subset E, mes\tilde{E} = mesE$ all $f_k(x)$ and $f(x)$ assumes finite values and $|f_k(x) - f(x)| \to 0, k \to \infty, x \in \tilde{E}$;
3. $f_k \xrightarrow[C\pi]{} f$ (f_k *converge* to f weakly), if

$$\int_E [f_k(x) - f(x)]\varphi(x)dx \to 0, \forall \varphi \in L_p(E), \frac{1}{p} + \frac{1}{q} = 1, \quad p>1.$$

Theorem 1

a. if $f_k \xrightarrow[L_p]{} f(x)$, then there exists a subsequence $\{f_{k_n}\}, f_{k_n} \xrightarrow[a.e.]{} f$, $n \to \infty$;

b. if $f_k \xrightarrow[a.e.]{} f$, then $\forall \varepsilon > 0$ there exists $E_\varepsilon \subset E, mesE_\varepsilon \geq mesE - \varepsilon$ and $f_k \xrightarrow[eq.]{} f$ on E_ε (Egorov theorem);

c. if $f_k \xrightarrow[L_1]{} f$ and $\|f_k\|_{p,E} \leq M, p > 1$ (M doesn't depend on k), then $f_k \xrightarrow[L_{\bar{p}}]{} f, \forall \bar{p} < p$ and $f_k \xrightarrow[L_p]{} f$;

d. if $f_k \xrightarrow[C\pi.]{} f$ and $\|f_k\|_{2,E} \to \|f\|_{2,E}$, then $f_k \xrightarrow[L_2]{} f$.

2.2.3 The Properties of Truncations in $W_p^1(E), p \geq 1$

Let us consider the measurable sets

$$E_m = \{x | f(x) > m\}, E_m^0 = \{x | f(x) = m\} \text{ and } \overline{E}_m = E_m \cup E_m^0, f(x) \in W_p^1(E).$$

And let us note the following properties of these sets:

$$E_m = \cup_{\forall \varepsilon > 0} E_{m+\varepsilon}, \quad \overline{E}_m = \cap_{\forall \varepsilon > 0} E_{m-\varepsilon};$$

$$mes(E_m / E_{m+\varepsilon}) \to 0, \quad mes(E_{m-\varepsilon} / \overline{E}_m) \to 0 \text{ at } \varepsilon \to 0.$$

Let us now introduce a truncation $f^{[m]}(x)$ of function $f(x)$, assuming $f^{[m]}(x) = \max\{f(x) - m; 0\}$.

Theorem 2

a. If $f(x) \in W_p^1(E), p \geq 1$, then $f^{[m]}(x) \in W_p^1(E)$, in this case $\frac{df^{[m]}}{dx} = \frac{df}{dx}, x \in E_m$ and $\frac{df^{[m]}}{dx} = 0, x \in (E / E_m)$;

b. If $f_k \xrightarrow[L_p]{} f$, then also $f_k^{[m]} \xrightarrow[L_p]{} f^{[m]}$, in this case

$$mes\{(E_m / E_m^k) \cap E_m\} \to 0, mes\{E_m^k / (E_m^k \cap \overline{E}_m)\} \to 0 \text{ at } k \to \infty,$$

where $E_m^k = \{x | f_k(x) > m\}$;

c. If $\|f_k - f\|_{p,E}^{(1)} \to 0$ at $k \to \infty$, then $\|f_k^{[m]} - f_k^{[m]}\|_{p,E}^{(1)} \to 0$ at $k \to \infty$.

2.2.4 Maximum Principles

Theorem 3 (the extremum principle).

Let $u = u(x), x \in E = \{x | 0 < x < l\}$ be regular (classical) solution of the differential equation

$$Lu \equiv au_{xx} + bu_x - cu = 0 \quad (c \geq 0, a > 0), \tag{2.5}$$

with continuous coefficients$(a, b, c) \in C(E)$.

In this case,$u(x)$cannot reach either a negative relative minimum or a positive relative minimum in any point $x \in E$.

Proof Let us introduce the function $u(x) = (\gamma - e^{-\beta x})v(x) \equiv \alpha(x)v(x)$, which, by virtue of (2.5), satisfies the equation

$$\alpha a v_{xx} + (b\alpha + 2\beta a e^{-\beta x})v_x - [\beta e^{-\beta x}(a\beta - b) + c\alpha]v = 0.$$

Let us select a β sufficiently large to ensure that $a\beta - b > 0, x \in E$, which is possible, since $a > 0$, and let us then select a γ such that $\alpha = \gamma - e^{-\beta x} > 0, x \in E$.

This will give us the equation

$$\overline{L}u \equiv \overline{a}v_{xx} + \overline{b}v_x - \overline{c}v = 0, \ (\overline{c} > 0, \overline{a} > 0). \tag{2.5*}$$

Let there be a point $x_2 \in E$ at which a negative minimum of $v(x)$ can be reached. In that case

$v_x(x_2) = 0, v_{xx}(x_2) \geq 0$ and $\overline{L}v(x_2) \geq \overline{c}(x_2)|v(x_2)| > 0$, which contradicts the assumption that equation (2.5*) can be satisfied. The statement relating to the maximum can be proved in the same way.

Theorem 4 *(generalized extremum principle). Let* $u = u(x)$*be the regular solution of equation* (2.5)*:*

$$L_0 u + f \equiv a u_{xx} + b u_x + f(x, u) = 0 \quad (a > 0, c = 0). \tag{2.6}$$

Here $(a, b) \in C(E); f(x, u) \in C[E \times \overline{U}], \overline{U} = \{u | 0 < u < 1\}$,

$f(x, 0) = f(x, 1) = 0, \forall x \in E$, in this case $0 \leq \min_{\partial E} u(y)$, $1 \geq \max_{\partial E} u(y)$.

This satisfies the estimates

$$(\min_{\partial E} u \geq)0 \leq u(x) \leq 1(\geq \max_{\partial E} u), \quad x \in E. \tag{2.7}$$

Proof Let $\gamma > 0, \beta > 0$ be arbitrary constants which obey the condition $\gamma > 1 (x \in E = (0, 1))$. Let us assume that $u = v \cdot (\gamma - e^{-\beta x})$, and extend the function $f(x, u)$ along the continuity

$$\overline{f}(x, v) = \begin{cases} f(x, u), & u > 1; \\ 0, & u \leq 0 \end{cases}$$

and introduce it into (2.6) in place of $f, \overline{f}(x, v)$. This transforms (2.7) into a form analogous to (2.5*):

$$\overline{L}_0 v + \overline{f}(x, v) = 0,$$

where $\beta > 0$ is selected from condition $\overline{c} > 0$.

Let the negative minimum of function $v, v(x_2) < 0$ be achieved at point $x_2 \in E$. Then $v_x(x_2) = 0, v_{xx}(x_2) \geq 0, \overline{f}(x_2, v(x_2)) = 0$, whence

$$\overline{L}_0 v(x_2) + \overline{f}(x_2, v(x_2)) \geq \overline{c}(x_2)|v(x_2)| > 0,$$

which contradicts the assumption that equation (2.6) can be satisfied at $f = \overline{f}$ at point $x_2 \in E$. The lower estimate in (2.7) makes it possible to remove the truncation from function $\overline{f}$, i.e. to set $\overline{f} \equiv f$.

The upper estimate in (2.7) can be proved in the same way by introducing the function $v = (1 - u)(\gamma - e^{-\beta x})^{-1}$.

Theorem 5 *Let the coefficients of equation (2.5) have the properties* $a > 0$, $(a, b) \in C(E), E = [-1, 1]$ *and* $u(x) \in C^2(E) \cap C(\overline{E})$, *at* $c \equiv 0$, *with* $L_0 u = au_{xx} + bu_x \geq 0$; $u(x) < u(x_0)$, $x \in E$, $x_0 \in \partial E$. *In that case,* $(\mathrm{d}u/dx)(x_0) = \sigma > 0$.

Proof For the sake of clarity, let us assume that $x_0 = 1$. Let us then consider the sets

$$E_0 = \left\{\left|x - \frac{1}{2}\right| < \frac{1}{2}\right\}, \quad E_1 = \left\{|x - 1| < \frac{1}{4}\right\}, \quad E^* = E_0 \cap E_1.$$

and introduce the auxiliary function

$$\omega = e^{-\alpha\rho^2} - e^{-\alpha/4}, \rho = \left|x - \frac{1}{2}\right|, x \in E_0.$$

If we select a sufficiently large $\alpha \gg 1$, it is obvious that

$$L_0\omega = 2\alpha e^{-\alpha\rho^2}\left[a2\alpha\left(x - \frac{1}{2}\right)^2 - a - b\left(x - \frac{1}{2}\right)\right] > 0, \quad x \in E^*,$$

$$\text{as} \quad x - \frac{1}{2} \geq \frac{1}{4}, x \in E^*.$$

Since $u(x) < u(1)$ at $x \in E \supset E^*$, then, provided that $0 < \varepsilon \ll 1$ is sufficiently small, we have

$$v(x) = u(x) + \varepsilon\omega(x) < u(1), x = \frac{3}{4} \in \partial E^*,$$

$$L_0 v = L_0 u + \varepsilon L_0 \omega > 0, x \in E^*.$$

Therefore, $v(x)$ cannot reach its absolute maximum in E^*, and consequently

$$v\left(\frac{3}{4}\right) < u(1), v(1) = u(1)\ (\omega(1) = 0).$$

As a result, at point $x = 1$

$$0 \leq \frac{dv}{dx} = \frac{du}{dx} + \varepsilon \frac{d\omega}{dx},$$

whence $\frac{du}{dx}(1) \geq -\varepsilon \frac{d\omega}{dx}(1) = \varepsilon \alpha e^{-\alpha/4} \equiv \sigma > 0$

Note If in the assumptions of Theorem 5

$$L_0 u \leq 0, u(x) > u(x_0), x \in E, x_0 \in \partial E, \text{ then } \frac{du}{dx}(x_0) = \sigma < 0.$$

To prove this assertion, it is sufficient to consider function $\bar{u} = u(x_0) - u(x)$, for which $L_0 \bar{u} \geq 0, \bar{u}(x) < \bar{u}(x_0) = 0$.

We can use Theorem 5 to state the next proposition, which is analogous to Theorems 3 and 4.

Theorem 6 *Each inconstant solution* $u(x) \in C^2(E) \cap C(\overline{E})$ *of equation (2.5) at* $c \equiv 0, a > 0, (a, b) \in C(\overline{E})$ *satisfies the inequalities*

$$\min u(s) < u(x) < \max u(s), x \in E, s \in \partial E. \tag{2.7*}$$

Proof Let $u(x) const$ have an internal maximum point $x_0 \in E$ and $E_0 = \{|x - (1 - \varepsilon)x_0| < \varepsilon x_0\} \subset E$ at a sufficiently small $0 < \varepsilon \ll 1$, and let $u(x) < u(x_0), x \in E_0$.

By virtue of Theorem 5, in that point $\frac{du}{dx}(x_0) > 0$, which contradicts the equality $\frac{du}{dx}(x_0) = 0$, which is true for an internal maximum point.

The next theorem follows as a direct consequence of the assertions made by Theorems 5 and 6:

Theorem 7 *In the conditions postulated by Theorem 5, the following inequalities are satisfied in boundary points* $x_k \subset \partial E, k = 0, 1$

$$\left|\frac{du}{dx}(x_k)\right| = \sigma_k > 0.$$

Theorem 8 *Let the conditions of Theorem 4 be satisfied for the coefficients of equation (2.6).*

In that case, inequalities (2.7) will be satisfied for a regular solution of equation (2.6), which satisfies a second boundary-value problem of the type

$$[u_x - \varphi(u)]_{x=0} = 0, \quad u|_{x=1} = u_1 \in [0, 1], \tag{2.8}$$

where $\varphi(u) \in C[0, 1]$, $\varphi(0) = \varphi(1) = 0$ *satisfy the inequalities (2.7).*

Proof Let us extend the functions $f(x, u)$ and $\varphi(u)$ along a continuity, as we did in the proof of Theorem 4, assuming that

$$\overline{\varphi} = \varphi, \overline{f} = f, u \in [0, 1]; \overline{\varphi} = \overline{f} = 0, \overline{u} \in [0, 1],$$

and substitute them into (2.6) and (2.8). Let us assume that $u(x) > 1$. We can then find a neighbourhood $E_0 = (0, x_*), x_* \leq 1$, for which $u(x) \geq 1$, $x \in \overline{E}_0$, $u(x_*) = 1$, in which case $\overline{f} = \overline{\varphi} = 0, x \in \overline{E}_0$.

If $u(x) = const, x \in \overline{E}_0$, then in accordance with Theorem 7 extremum cannot be reached in the point $x_0(u_x(x_0) = 0!)$, whereas Theorem 6 makes it impossible for it to be reached in the internal points E_0. However, in that case $u(x) = u(x_0) = \text{const}, x \in \overline{E}_0$, and since $u(x_*) = 1$, then $u(x) = 1$, $x \in \overline{E}_0$.

Theorem 4 applies in region $\{E/E_0\}$, and thereby $u(x) \leq 1, x \in \overline{E}$. The lower inequality of (2.7) $u(x) \geq 0, x \in \overline{E}$ is determined in a similar way.

2.2.5 Generalized Solutions of Differential Equations

Let us now consider the quasi-linear ordinary differential equations

$$(a_1 u_x + a_2)_x + a_3 + a_4 u_x = 0, x \in E(\partial E = x_0, x_1) \tag{2.9}$$

and examine the boundary-value problems

$$u|_{x=x_k} = u_k, k = 0, 1; \tag{2.10}$$

$$(a_1 u_k + a_2)|_{x=x_k} = 0, u|_{x=x_j} = u_j, j \neq k. \tag{2.11}$$

Coefficients $a_k = a_k(x, u), (x, u) \in E \times [0, 1]$ obey the conditions

$$\begin{cases} a_1 > 0, (x, u) \in E \times (0, 1); \quad a_k(x, u)|_{u=0,1} = 0, k = 1, 2, 3; \\ |a_k| \leq M_0, k = 1, 2, 3, 4. \end{cases} \tag{2.12}$$

Note that by virtue of (2.12), $u \equiv 0$ and $u \equiv 1$ are the solutions of (2.9), (2.10) and (2.9), (2.11).

Let us agree to consider that $a_k(x, u)$ have been determined for all $u \in R^1$ extending them with the boundary values of $a_k(x, 0)$ and $a_k(x, 1)$ at $\overline{u} \in [0, 1]$. Let us also introduce the notation $\int_E a(x)b(x)dx = (a, b)_E$.

Calculation *Let us define the generalized solution $u(x)\in V$* of problems (2.9), (2.10) and (2.9), (2.11) as represented by a function $u(x)$ having the following properties:

i. $|u|\leq M, |a_1 u_x|\leq M_1, x\in \overline{E}$;

ii. It satisfies conditions (2.10) in (2.9), (2.10) and the condition $u(x_j)=u_j, j\neq k$ in (2.9), (2.11);

iii. It satisfies the integral identity

$$(a_1 u_x + a_3, \eta_x)_E - (a_3 + a_4 u_x, \eta)_E = 0, \forall \eta \in C^1(E) \tag{2.13}$$

when, in addition, in (2.9), (2.10), $\eta(x_k)=0, k=0,1$, and in (2.9), (2.11) $-\eta(x_j)=0, j\neq k$.

Theorem 9 *(the maximum principle). The generalized solutions $u(x)\in V$ of boundary value problems (2.9), (2.10) and (2.9), (2.11) satisfy the inequalities*

$$0\leq u(x)\leq 1, x\in \overline{E}, \tag{2.14}$$

if in (2.10) $u_k\in[0,1]$, and in (2.11) $u|_{x=x_j}=u_j\in[0,1], j\neq k$.

Proof The proof follows the determination of $u\leq M$ and demonstrates that $M\leq 1$. Let us begin by assuming the opposite, i.e. that $M>1$, and let us select a v for which $1<v<M$. Let us consider the problem for a subsidiary equation of the form (2.9), in which coefficient a_1 is replaced by $\overline{a}_1 = a_1 + \varepsilon, \varepsilon>0$.

Let us consider the truncation $u^{[v]}=\max\{u-v,0\}$. Obviously, $u^{[v]}(x_k)=0, k=0,1$ in (2.9), (2.10) and $u^{[v]}(x_j)=0, j\neq k$ in (2.9), (2.11)

Function $u^{[v]}(x)$ can therefore be used as a testing function ($\eta = u^{[v]}$) in equation (2.13) at $\overline{a}_1 = a_1+\varepsilon$ instead of a_1. This will give us

$$(\overline{a}_1 u_x + a_2, u_x)_* = (a_3 + a_4 u_x, u^{[v]})_*, \tag{2.15}$$

where $(u,v)_* = \int_{E_*} u(t)v(t)dt, E_v = \{x|u-v>0\}, E_* = E_v/E_M^0$.

Let us note that since on $E_v, u>v>1$, therefore $a_k=0, k=1,2,3$ and $u_x^{[v]}=u_x$ when $x\in E_v$ [Theorem 2, Property (a)], and when $x\in E_M^0 - u_x = u_x^{[v]} = 0$ ($u|_{E_M^0}=\max_E u$). Thus, we can use (2.15) to find

$$\varepsilon\|u_x\|^2_{2,E_*} = (a_4 u_x, u-v)_* \leq M_0\|u_x\|_{2,E_*} - \|u-v\|_{2,E_*}.$$

Inequality (2.4) states that when $s=p=2$,

$$\|u-v\|_{2,E_*} \leq \text{mes}E_*\|u_x\|_{2,E_*}$$

and that provides us with the inequality

$$\varepsilon \leq M_0 mes E_* = M_0 mes(E_\nu / E_M^0),$$

This inequality is true for each fixed $\varepsilon > 0$ and arbitrary $\nu \in (0, M)$. However, when $\nu \to M$ $mes E_* = mes(E_\nu / E_M^0) \to 0$ [Theorem 2, Property (b)], and the inequality is no longer true. Consequently, the assumption that $M > 1$ is wrong, and therefore $u(x) \leq 1$ is right. The inequality $u(x) \geq 0$ can be proved in the same way.

2.3 NUMERICAL AND ANALYTICAL METHODS OF INVESTIGATION OF THERMAL TWO-PHASE FLOW PROBLEMS

In this section, we suggest numerical and analytical methods of investigating the thermal flow of two-phase fluids in oil formations.

2.3.1 Thermal Recovery Methods

Thermal methods are the most commonly used and best understood methods of developing high-viscosity and waxy oil deposits, and the depleted sections of light (low-viscosity) oil deposits. Thermal recovery is based on the fact that heating rapidly reduces the viscosity of the oil, and thus increases its mobility, and in the case of waxy oils also prevents wax crystallization in the pores.

Two principal thermal recovery methods are currently in use: steam drive, which consists of injecting a heating medium (steam or hot water) through injection wells, and steam treatment of production wells. A variation of this method is to inject a heating and a cooling medium (water) alternately into either injection or production wells, shutting in some of the wells for some of the time, or to combine the methods, and in particular to convert production wells into injection wells and vice versa.

Thermal recovery methods produce a range of different displacement flow dynamics: unidirectional displacement in an injection well-production well system, with alternating heating and cooling of sections of the formation, thermocapillary saturation of shut-in wells, flow around no-flow zones (bypassed oil) and others. This is why it is virtually impossible to arrive at a realistic forecast of the effectiveness of complex thermal recovery methods, using solely engineering approaches (e.g. material balance calculations or statistics), and why up-to-date methods of mathematical modelling need to be applied.

2.3.2 The Muskat-Leverett Thermal Model (MLT Model)

Bocharov and Monakhov [15] have proposed and investigated a mathematical model of thermal two-phase flow (the MLT model). It differs from earlier models [140] in that it uses experimentally-determined relationships between viscosity, capillary properties and temperature, while its energy equation follows from the laws of conservation of energy of the fluids and the porous medium.

The equations of the one-dimensional MLT model for a homogeneous isotropic porous medium, transformed with respect to dynamic water saturation $\theta = \theta_1$, equilibrium temperature $\theta(x, t)$ and average pressure x, tp assume the form (Chapter 1, 1.5)

$$s_t = [a(\lambda_1 s_x + a_1 \theta_x) + a_2 v]_x, \theta_t = (\lambda_2 \theta_x - v\theta)_x, \ - v_x \equiv (\lambda_3 p_x + a_3 \theta_x)_x = 0, \tag{2.16}$$

where $v = v(t)$ is the two-phase fluid flow rate. The coefficients $a(s)$, $\vec{a}(s, \theta) \equiv (a_1, a_2, a_3)$ and $\vec{\lambda}(s, \theta) = (\lambda_1, \lambda_2, \lambda_3)$ in (2.16) are expressed explicitly by thefunctional parameters of the initial MLT model, and have the following properties, which allow for the physical implications of these parameters [15]:

a. $a(s) > 0, s \in (0, 1), a(0) = a(1) = a_2(0) = 0; \quad \lambda_i(s, \theta) \geq m_0 > 0;$
b. $(a(s), \vec{a}(s, \theta), \vec{\lambda}(s, \theta)) \in C^1(\overline{Q}), Q = \{s, \theta | 0 \leq s < 1, \theta_* < \theta < \theta^*\}.$

Assumptions (a) ensure respectively the parabolicity and uniform parabolicity of the equations for $s(x, t)$ and $\theta(x, t)$ and the non-degeneracy (non-simplification) of the ordinary equation for $p(x, t)$ (in which t acts as a parameter).

2.3.3 Self-similar (Analytical) Solutions Theory

The use of mathematical models, such as the MLT model or the Muskat-Leverett model of isothermal two-phase flow [2], requires complex mathematical techniques, and therefore, approaches which make use of simpler and more practical methods are especially valuable. One such approach, still successfully used in practice, is to describe the process of oil displacement by means of approximate formulae derived from exact solutions of the initial model equations. These include stationary solutions, dependent only on the variable x, analytical (self-similar) parabolic solutions dependent on $y = x(t+1)^{-1/2}$, analytical traveling wave solutions, dependent on $z = x + ct$ ($c = const$) and some others. Simple Muskat formulae (displacement laws), Charny's formulae (near-wellbore zones) and others [139], based on parabolic self-similarity remain reliable tools of engineering analysis of oil field development by isothermal methods.

This being the case, let us begin our examination of thermal recovery methods by constructing analytical (self-similar) MLT model equations, described by ordinary differential equations. Let us consider the case of equation (2.16) with a known total mixed flow rate $v = v(t)$, which corresponds to a unidirectional mixed flow in a steam drive or steam treatment. It is clear that in this case the initial functions $s(x,t), \theta(x,t)$ can be found independently of function $p(x,t)$, which can then be reconstructed from them. Provided that $v = \tilde{v}(t+1)^{-1/2}$ (in what follows, the tilde over the v has been omitted), is specified, we can find solutions to (2.16) which depend on only one independent variable $y = x(t+1)^{-1/2}$ - they are parabolic self-similar solutions $s(y), \theta(y), p(y)$ which satisfy the transformed equations (2.16):

$$[a(\lambda_1 s_y + a_1\theta_y) + a_2 v]_y + \frac{1}{2} y s_y = 0, (\lambda_2\theta_y - v\theta)_y + \frac{1}{2} y\theta_y = 0; \quad (2.17)$$

$$\lambda_3 p_y + a_3\theta_y = -v = const. \quad (2.18)$$

Let the two-phase fluid flow take place between two wells (groups of wells), located in points $y = y_k, k = 0, 1$, of which let $y = y_0 = 0$ be the location of the injection well, and $y = y_1$ the location of the production well ($y_1 = \infty$ is also a possible case). The values specified for wells $y = y_k$ can be either s, θ, p:

$$(s, \theta)|_{y=y_k} = (s_k, \theta_k), \quad k = 0, 1, \quad (2.19)$$

$$p|_{y=y_k} = p_k, \quad k = 0, 1, \quad (2.20)$$

or streams $\partial s = a(\lambda_1 s_y + a_1\theta_y) + a_2 v, \partial\theta = \lambda_2\theta_y - v\theta, \partial p = \lambda_3 p_y + a_3\theta_y$ of these values:

$$\partial s|_{y=y_k} = S_k; \quad \partial\theta|_{y=y_k} = T_k, \quad k = 0, 1, \quad (2.21)$$

$$\partial p|_{y=y_k} = P_k, \quad k = 0, 1, \quad (2.22)$$

where S_k, T_k and P_k - constants.

Let us examine the first boundary value problem (2.17), (2.18) for $s(y)$ and $\theta(y)$. Because equation (2.17) for θ is homogeneous with respect to the derivatives of ($v = const$), we can use the linear substitution $\theta = \gamma_1\tilde{\theta} + \gamma_2, \gamma_k = const$ to reduce the boundary conditions (2.19) for θ to $\theta_0 = 1$, $\theta_1 = 0$, and we will assume that this has been done.

Assuming the coefficient $\lambda_2(y) \equiv \lambda_2[s(y), \theta(y)]$ to be a known function, we can arrive at the following expression for $\theta(y)$:

$$\theta = 1 - NF(y); F = \int_0^y \lambda_2^{-1}(\xi)e^{-\Lambda(\varepsilon)}d\xi, \Lambda = \int_0^y \lambda_0(\xi)d\xi. \qquad (2.23)$$

Here $\lambda_0 = (0,5y - \nu)\lambda_2^{-1}(y), N = [F(y_1)]^{-1}$.

Expression (2.23) leads directly to the estimates

$$0 \le \theta(y) \le M_1 \exp(-\alpha_2 y^2) \le 1, \quad |\theta_y| \le M_2 \exp(-Q_2 y^2), \quad y \in (0, y_1) \quad (2.24)$$

where the constants $M_i > 0$ and $\alpha_i > 0$ do not depend on the value of $y_1 > 0$.

Let us assume that $y_1 = \infty$, and let us calculate the approximate value of the constants α_i and M_i in the inequality (2.24), so as to arrive at a more exact definition of the asymptotics of $|\theta(y)|$ at $y \to \infty$. Let $y \ge \eta_0 = 8|\nu|$.

In that case,

$$\lambda_0 = \left(\frac{1}{2}y - \nu\right)\lambda_2^{-1} \ge \frac{1}{4}\nu_0^{-1}y, \quad \nu_0 = \max\lambda_2,$$

$$\Lambda = \int_0^y \lambda_0(t)dt \ge \int_0^{\eta_0} \lambda_0(t)dt + \alpha_0(y^2 - \eta_0^2), \quad \alpha_0 = \frac{1}{8}\nu_0^{-1},$$

and it can easily be seen that inequalities (2.24) occur when

$$M_1 = \left(\lambda^* \sqrt{8\lambda^*}\right)/\lambda_*, M_2 = (\lambda^*)^2\left(\lambda_* \pi \sqrt{\lambda_*}\right)^{-1},$$

$$\alpha_1 = \alpha_2 = (8\lambda^*)^{-1},$$

where $\lambda_* = \min_{s,\theta} \lambda_2(s, \theta), \lambda^* = \max_{s,\theta} \lambda_2(s, \theta)$.

Let us now consider equation (2.17) for $s(y)$ and note that irrespective of the value of $s_i \in [0, 1], i = 0, 1$ in (4), 1.0 suggests that $s(x, t)$ can be estimated as

$$0 \le s(y) \le 1, y \in [0, y_1] \ \forall y_1 > 0. \qquad (2.25)$$

Let us introduce a new function $u = \int_0^s \alpha(\xi)d\xi$ and, assuming that the coefficients of equation (2.17) for $s(y)$ are known, let us express this equation in the form

$$(\lambda_1 u_y + \varphi)_y = 0, u(0) = u_0, u(y_1) = u_1. \qquad (2.26)$$

In this expression,

$$u_i = \int_0^{s_i} \alpha(\xi)d\xi, i = 0, 1, \varphi = 0,5\left(\gamma s + \int_y^{y_1} s(\xi)d\xi\right) + aa_1\theta_y + a_2 v.$$

Integrating (2.26), we obtain

$$-\lambda_1 u_y = \varphi + C, u(0) = u_0, \tag{2.27}$$

where $C = C_0^{-1}\left[u_0 - u_1 - \int_0^{y_1} \lambda_1^{-1}(\xi)\varphi(\xi)d\xi\right], C_0 = \int_0^{y_1} \lambda_1^{-1}(\xi)d\xi$. Since, on the basis of the assumption made in (a), $\lambda_1 \geq m_0 > 0$, we can use (2.27) to find

$$|u_y| = a(s)|s_y| \leq M_0(y_1). \tag{2.28}$$

Let us note that the properties of the MLT model's functional parameters satisfy the inequalities [15] (Chapter 1, 1.5):
c. $(a, |a_2|) \leq Ks^\gamma, (a, |a_2 - a_2(1,\theta)|) \leq K(1-s)^\gamma, \gamma \geq 1.$

Theorem 1 (The finite velocity of propagation of perturbations).

Let us assume that assumptions (a), (b) and (c) have been satisfied. In that case, when $s(y_1) = 0$[*or* $s(y_1) = 1$], $y_1 \gg 1$, *there exists a finite value of* y_* , *such that*

$$s(y) \equiv 0[\text{or } s(y) \equiv 1] \text{ at } y \geq y_*, \tag{2.29}$$

i.e. the front $s = 0$ $(s = 1)$ *propagates at a finite velocity.*

Proof Let us assume that

$$y_*^2 = Y^2 + K_0\delta^{-1}, Y = 2K\max(M_2|a_1|, |v|), \tag{2.30}$$

where $K_0 = \int_0^1 a(t)t^{-1}dt, \delta = 0,125 \min \lambda_1$. Since the constants K, K_0, M_2 and δ do not depend on y_1 then at $y_1 \gg 1$ the constant y_* also does not depend on y_1.

Let us consider equation (2.26) for $u = \int_0^s a(t)dt$. Since $u(y_1) = 0$ and in the vicinity $y = y_1$ we have $u(y) \geq 0$, clearly $u_y(y_1) \leq 0$ and therefore $C = -\lambda_1(0,0)u_y(u_1) \geq 0$, since $\varphi(0,0) = 0$. In that case,

$$-\lambda_1 u_y = \varphi + C \geq \frac{1}{2}ys - a|a_1||\theta_y| - |v||a_2| \geq \frac{1}{4}ys \text{ at } y \geq Y,$$

where Y is determined in (2.30). In this way, we arrive at the inequality

$$[\Phi(s)]_y + 2\delta y \leq 0, y \in [Y, y_1], s(Y) = s_2 \geq 0, \tag{2.31}$$

where $\Phi = \int_0^s a(t)t^{-1}dt.$

Integrating (2.31), we find

$$-\int_{s}^{s_2} a(t)t^{-1}dt + \delta(y^2 - Y^2) \leq 0, y \geq Y.$$

And since

$$\Phi(s) - \Phi(s_2) = -\int_{s}^{s_2} a(t)t^{-1}dt \geq \int_{0}^{1} a(t)t^{-1}dt \equiv -K_0,$$

then it follows from the preceding inequality that $\delta(y^2 - Y^2) - K_0 \leq 0, y \geq Y$, which is possible only if $Y \leq y \leq y_*, y_*$ having been taken from (2.30). Consequently, for (2.31) to be true at $y \geq y_*$, (2.29) must be satisfied. The case of $s(y_1) = 1$ can be reduced to the case we have considered, by substituting $\sigma = 1 - s$.

Let us write down the expressions (2.23), (2.27) in the form of the following Cauchy problem for the vector function $\vec{u} = (u, \theta)$:

$$\vec{u}_y = \vec{\psi}(s, \theta, y), \vec{u}(0) = \vec{u}_0 \tag{2.32}$$

Here,

$$\psi_1 = -\lambda_1^{-1}(\varphi + C), \varphi = 0,5\left[ys + \int_{y}^{y_1} s(\xi)d\xi\right] + aa_1\psi_2 + a_2v,$$

$$\psi_2 = -\lambda_2^{-1}N\, exp(-\Lambda(y));$$

$s = s(u)$ is the inverse function of $u = u(s) = \int_0^s a(t)dt$; the constants C, N and function $\Lambda(y)$ are determined in (2.23), (2.27). If the functions $\psi_k(s, \theta, y)$ are constructed taking into account assumptions (a) and (b), they will be continuous for a set of arguments in any $y_1 \in (0, \infty)$.

The properties of the MLT model parameters [15] imply the following analogue of conditions (c):

d. $a(s) \geq m_1[s(1-s)]^{\gamma_0}, \gamma_0 > 0.$

Theorem 2 (Holder continuity). *Let us assume that conditions (a)–(d) have been satisfied. If that is the case, then, the following a priori estimates are satisfied for the solutions of $\vec{u}(y)$ in (2.32):*

$$\| \vec{u}(y)\|_{C^{1+\alpha}(l)} \leq N_0(y_1), \alpha = (1+\gamma_0)^{-1}, l = [0, y_1]. \tag{2.33}$$

If $s(y_1) = 0$ [or $s(y_1) = 1$], then the constant N_0 will depend only on y_* for all $y_1 \geq y_*$ and, in particular, we can assume that $y_1 = \infty$ [y_* from (2.30)].

Proof Let us first establish Holder continuity of the transformation $s = s(u)$, which is the inverse of $u = \int_0^s a(t)dt$:

$$|s(u_2) - s(u_1)| \leq K|u_2 - u_1|^{\alpha}, (u_1, u_2) \in [0, p], p = \int_0^1 a(t)dt.$$

Obviously, for this to be true, it is sufficient that $|u(s_2) - u(s_1)| \geq K_0|s_2 - s_1|^{1+\gamma_0}$ at $(s_1, s_2) \in \left[0, \frac{3}{4}\right]\left[\frac{1}{4}, 1\right]$.

For the sake of clarity, let us assume that $0 \leq s_1 \leq s_2 \leq \frac{3}{4}$. Then

$$u_2 - u_1 = \int_{s_1}^{s_2} a(s)ds \geq K_0 \int_{s_1}^{s_2} s^{\gamma_0}(1-s)^{\gamma_0} ds \geq K(s_2^{\gamma_0+1} - s_1^{\gamma_0+1})$$
$$\geq K(s_2 - s_1)^{\gamma_0+1}, K = K_0 4^{-\alpha}(1+\gamma_0)^{-1},$$

The last of the series of inequalities follows from the consideration of $f(\sigma) = (1 - \sigma^{\gamma})(1-\sigma)^{-\gamma}, \gamma = 1 + \gamma_0, \sigma = \frac{s_1}{s_2}$, for which $\min f(\sigma) = f(0) = 1 (f_\sigma > 0, 0 < \sigma < 1)$.

Thus, we have proved that $s(u) \in C^{\alpha}[0, p]$. Since $|u_y| \leq M_0$, therefore $s[u(y)] \in C^{\alpha}[0, y_1]$. Now, the expressions (2.23) and (2.27), in which the coefficients $\lambda(y) \equiv \lambda[s(y), \theta(y)], a_2(y)$ and others display Holder continuity, produce the inequalities (2.33).

In the same way, $s(y_1) = 0$ [or $s(y_1) = 1$] and $y_1 \geq y_*$ produce the estimate $[u_y] \leq M_1(y_*)$ and because of the finite velocity of propagation of perturbations $u(y) \equiv 0$ $(u \equiv u_1)$ at $y \geq y_*$, which clearly confirms the truth of the theorem.

Theorem 3 (Existence). *If assumptions (a)—(d) are satisfied, the Cauchy problem (2.32) has at least one solution,* $\vec{u}(y) \in C^{1+\alpha}[0, y_1]$ *, for all* $y_1 > 0$ *.*

If $u(y_1) = 0$ $\left(u(y_1) = \int_0^1 a(t)dt \equiv u_1\right)$*, then the solution is continuable at* $y \to \infty$*, and* $u(y) \equiv 0$ $(u(y) \equiv u_1)$*, when* $y \geq y_*$ *[*y_* *from (2.30)].*

Proof of the first part of the theorem is based on estimates (2.33) and properties $\vec{\psi}(s, \theta, y)$, and follows from the classical results of the theory of ordinary differential equations [138, p. 498].

The last proposition of the theorem follows as a simple consequence of Theorems 1 and 2.

Let us regard the functions $\theta(y)$ and $s = f[u(y)]$, where $\vec{u}(y) \equiv (u(y), \theta(y)) \in C^{1+\alpha}[0, y_1]$ is the solution of (2.32), and $s = f(u)$ is an inverse mapping of $u = \int_0^s a(t)dt$, as representing a generalised solution of (2.17), (2.19).

The existence of such a generalized solution follows from the theorem proved above. Clearly, in points y for which $s(y) \neq 0, 1$, the generalized solution $(s(y), \theta(y))$ is also the classical solution of (2.17), (2.19).

Note The second boundary value problem (2.17), (2.21) and a mixed problem in which condition (2.19) is specified for one of the points $y_k, k = 0, 1$, while condition (2.21) is specified for the other, are both examined exactly as for (2.17) and (2.19). It is also not difficult to prove that (2.17)–(2.20) are capable of solution when the flow rate is unknown (v in (2.17), (2.18) is the target value).

2.3.4 Computational Analysis

Computational analysis was carried out for the following three problems:

I. Unidirectional displacement [problem (2.17), (2.19)];

II. A known pressure drop – A near-wellbore zone problem [problem (2.17)–(2.20)];

III. Countercurrentthermocapillary saturation ($v = 0$).

While each case has its own special characteristics, the common difficulties of construction of the algorithms required for their numerical solution are as follows:

– The range of variation of the independent variable $y \in [0, \infty]$ is not limited;
– The equations forming the system are not linear, and their matrix is not diagonal for the higher derivatives;
– The physical properties of oil formations mean that the coefficients of s_{yy} and θ_{yy} are quite small, while those of s_y and θ_y change their sign, i.e. that there exist transition points which depend on the solution, and this results in the appearance of internal boundary layers [which are regions of high saturation and temperature gradients $s(y)$ and $\theta(y)$].

Estimate (2.24) enables us to calculate a y_2 such that $|\theta| \leq \varepsilon$ at $y \geq y_2$, where ε is a reasonably small number which depends on the required accuracy of solution. Thereafter, since in the case of $s(y)$ perturbations propagate at a finite velocity (Theorem 1), the semi-infinite interval boundary value problem is reduced to a finite interval problem $[0, y_1]$, where $y_1 = \max(y_2, y_*)$. The exact value of y_1, can be determined during the process of computation by introduction iteration with respect to the unknown parameter y_1 (free boundary) and using a priori estimates as the initial approximations of y_1. By normalizing the interval ($\bar{y} = y/y_1$) of the solution, we can solve the problem for the intercept $[0, 1]$, but the target parameter y_1 then becomes included in the coefficients of the equation system (in what follows, the vinculum over the y has been omitted).

We then arrive at a finite-difference equation, by means of integration and interpolation [114], using equations for s and θ which give a conservative view of flow:

$$[a(\lambda_1 s_y + a_1\theta_y) + a_2 v + 0,5ys]_y - 0,5s = 0, \tag{2.34}$$

$$[\lambda_2\theta_y + (0,5y - v)\theta]_y - 0,5\theta = 0. \tag{2.35}$$

The non-linear finite-difference temperature equation is made linear by simple iteration. In the finite-difference saturation equation, which is the equivalent of (2.34), $a_2 v$ is linearized with respect to s using the Newtonian method, since in the $s = 0, 1$ degeneration zones this is in fact the highest term. In all remaining non-linear equations, simple iteration is used. Our computation involves two iteration processes: y_1 iteration (outer iteration cycle) and coefficient nonlinearity iteration (inner iteration cycle). The latter process was combined with iterations leading to the fragmentation of the equations.

In oil field practice, the key parameter is water saturation $s(y)$, whose equation appears to be the most complex: it becomes degenerate at $s = 0, 1$, but even in the absence of degeneration, the coefficient of s_{yy} is very small in real field conditions.This corresponds to singular degeneration of the equation, requiring special approximation methods [42]. In our main experiments, we used a monotonic conservative directed difference approximation ("System 1") and a conservative, variant of Samarsky's monotonic equation [115] constructed as proposed in [14] ("System 2"). The first system is simpler, but includes only first order approximation. The second system provides better approximation, and converges uniformly to the degeneration of [42].

Let us consider for example, the abstract quasi-linear degenerating operator $[a(u) = 0$ at $u = 0, 1]$ which corresponds to the principal part of the saturation problem

$$Lu \equiv [a(u)u_x + b(u)]_x,$$

which has been linearized as described above:

$$L^n u^{n+1} \equiv [a(u^n)u_x^{n+1} + b(u^n) + b'(u^n)(u^{n+1} - u^n)]_x$$

where ($u = u(x), x \in [0, 1]$, and n-iteration number). In [117], the approximation of Lu in an arbitrary internal node x_i of a uniform grid $\overline{\omega} = \{x_i = ih, i = 0, \ldots, N, h = 1/N\}$ (with a spacing h) along the intercept [0, 1] is defined as

$$\Lambda_h u^{n+1} \equiv \mu_{0,5} u_x^{n+1} - \mu_{-0,5} u_{\bar{x}}^{n+1} + f(\alpha, u^{n+1}) - f(\alpha, u^n) + f(\beta, b^n).$$

and introduces the following designations:

$$\varphi_{\pm 0,5} \equiv \varphi_{i \pm 0,5} = \varphi(0, 5(u_i^n + u_{i \pm 1}^n)),$$

$$u_i \equiv u(x_i), \mu = (a+\varepsilon)^2 / h(a + \varepsilon + 0, 5\xi hr), r = |b'(u^n)|,$$

$$\alpha_1 = \overset{+}{c}_{0,5}, \alpha_2 = \overset{-}{c}_{0,5}, \alpha_3 = \overset{+}{c}_{-0,5}, \alpha_4 = \overset{-}{c}_{-0,5},$$

$$\varphi^{\pm} = 0, 5(\varphi \pm |\varphi|), c = b'(u^n), \beta_i = \text{sign } \alpha_i,$$

$$f(\alpha, u) = h^{-1}(\alpha_1 u_{i+1} + (\alpha_2 - \alpha_3)u_i - \alpha_4 u_{i-1}),$$

where ε is the regularization parameter of degeneration. $\xi = 0$ produces System 1, while $\xi = 1$ produces System 2.

Within the specified accuracy range (grid spacing approximately 0.01 and accuracy of iteration 10^{-4}), there was virtually no difference between the solutions produced by Systems 1 and 2. If the available computer power makes it impossible to calculate for a grid spacing of 0.01, System 1 should be preferred, as it needs fewer arithmetical operations. In addition, computational analysis [137] shows that this system can describe integrated oil recovery and water saturation indicators withgood accuracy even if the grid spacing is large. In a general case, Newtonian iteration requires reasonably accurate initial approximations, and they are difficult to calculate. Much attention is usually devoted to this problem in practical calculations, and in particular in solving formation flow problems in the complex conditions of real oil fields. We were able to calculate initial iterations for problems I-III by using estimated locations of transition points [from equations (2.34), (2.35)] and combining analytical solutions of the equation for θ at a constant λ_2, corresponding to $\theta = \theta_0, \theta = \theta_1, s = 0; 1; 0, 5,$ with exact solutions of the Buckley-Leverett equation (no capillary interaction between phases), and thus to reduce the number of iterations by a whole order of magnitude.

It should be noted that in the case of problems I-III, nonlinear iteration is fairly stable, and there is convergence at virtually all initial approximations satisfying the boundary conditions, although the number of iterations needs to be increased as appropriate.

We have considered two types of solution of the linear finite difference equation system: scalar runs with iterations between equations and matrix runs. Computational analysis has shown them to be equivalent to one another in terms of the required number of iterations: this was due to the rapid convergence of the temperature iterations with the solution.

The fact that this section is no more than an overview of the problem means that we cannot provide detailed descriptions of the results of computational analysis for each of the problems. We will therefore briefly highlight some of the singularities identified by the calculations.

Unidirectional displacement (Problem 1). Fig. 2.1 shows the estimated distribution of s and θ. The fine line shows the distribution of saturation in the isothermal case, at $\theta = \theta_1$, (in situ conditions).

Computational analysis has shown that if the temperature front lags behind the saturation front, heating ($\theta_0 > \theta_1$) results in increasing only the final oil recovery. In calculations, this takes the form of an additional displacement front (representing regions of high saturation gradients) which corresponds to a temperature front (representing regions of high temperature gradients). Thus, the fact that in these circumstances the formation flow is non-isothermal (the water phase is either heated or

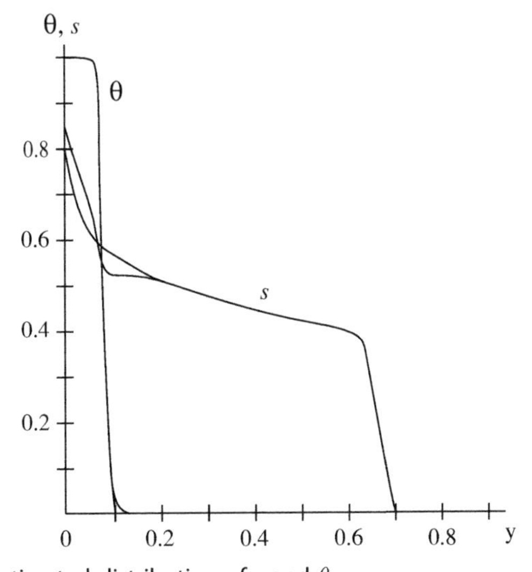

Figure 2.1 The estimated distribution of s and θ.

cooled) allows the distribution of saturation to be controlled in the region $[0, y_2 + \delta_2]$ where $\delta_2 > 0$ is specified as part of the problem and, as will be seen from Fig. 2.1, can be easily calculated numerically.

Near-wellbore zone (Problem II). The singularity of Problem II is that flow rate v is a functional of s, θ, p.

In our case, it can be expressed explicitly as

$$v = -\left(p_1 - p_0 + \int_{y_0}^{y_1} a_3(s, \theta)\lambda_3^{-1}(s, \theta)\theta_y d\xi\right)\left(\int_{y_0}^{y_1} \lambda_3^{-1}(s, \theta)d\xi\right)^{-1} \tag{2.36}$$

This is the expression we have used in our calculations. The distribution of s, θ, p is shown in Fig. 2.2.

Fig. 2.3 shows the values of v calculated from (2.17) (fine line) and from (2.36). It will be seen that in the regions of high s and θ gradients, flow rate v calculated from (2.18) contains a large error. In a non-one-dimensional case, in the absence of the integral function (2.36), special care must be taken in the saturation and temperature front areas. In addition, Problem II calculations take much longer, due to the need to calculate the integrals in (2.36).

In Problem III, which deals with the thermocapillary saturation of low-permeability streaks, the issue of boundary conditions [114] remains unsolved. We have therefore focused our computational analysis of Problem III on the behaviour of the solution at $y = 0$. Fig. 2.4 shows families of solutions for a sequence of left hand side boundary conditions

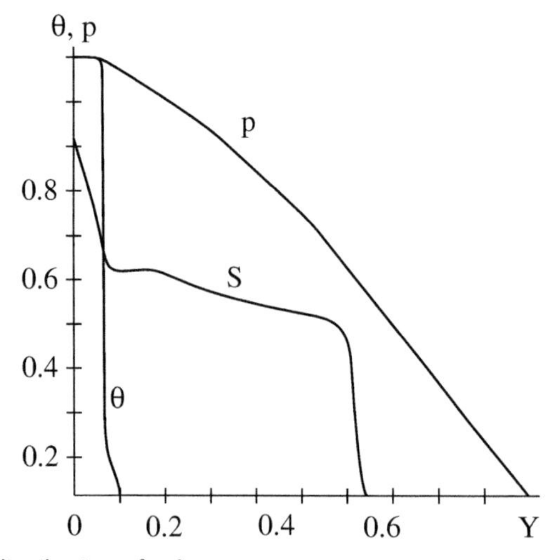

Figure 2.2 The distribution of s, θ, p.

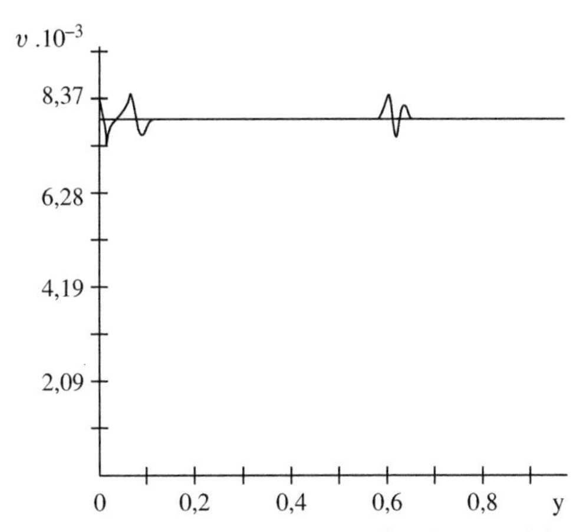

Figure 2.3 The values of v calculated from (2.17) (fine line) and from (2.36).

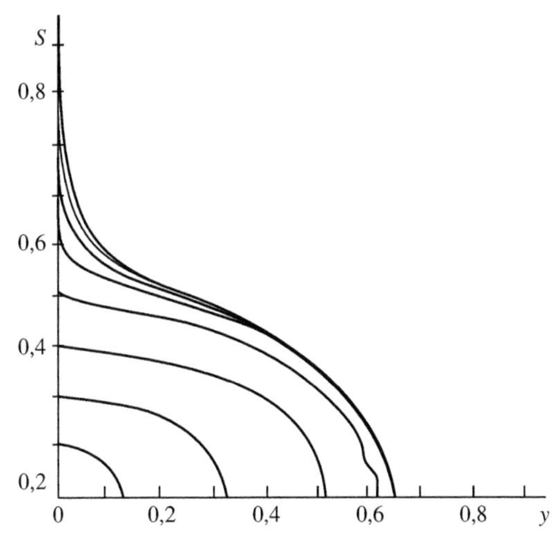

Figure 2.4 Families of solutions for a sequence of left hand side boundary conditions $s_0 = 1 - 0,1j, j = 0, 1, \ldots, 9$.

$s_0 = 1 - 0,1j, j = 0, 1, \ldots, 9$. It will be seen that when $s_0 \in [0,6;1]$, the position of the saturation front y_* remains virtually unchanged.This suggests that in this interval the exact value of s0 does not have a significant effect on the saturation rate.

It should be noted that the value of y_* is important from the process point of view, since it determines the displacement rate, the saturation rate of low-permeability sections, and the near-wellbore zone.

2.4 ANALYTICAL PROBLEMS OF THERMAL TWO-PHASE FLOW IN THE CASE OF VARIABLE RESIDUAL SATURATIONS

The need for enhanced oil recovery means that thermal field development methods occupy an important place in the industry. This need and the increasing power of computer and information systems have generated an interest in the mathematical model of nonisothermal two-phase fluid flow in porous media. The model adds the energy equation [15, 16] to the classical equations of the Muskat-Leverett model.

2.4.1 Problem Statement

This section is devoted to analytical solutions of the model proposed in [16], which describes one-dimensional flow of a two-phase fluid with variable residual saturations:

$$s_t = (aa_1\sigma_x + aa_2\theta_x - va_3)_x, \theta_t = (\lambda\theta_x - v\theta)_x, \sigma = \Phi(\theta, s). \tag{2.37}$$

Assuming that $v(t) = v(t+1)^{-1/2}, v = const$, we can move forward to the analytical variable $y = x(t+1)^{-1/2}$ and derive a system of ordinary differential equations for temperature θ, saturation s and dynamic (relative) saturation σ of one of the phases:

$$(aa_1\sigma_y + aa_2\theta_y - va_3)_y + \frac{1}{2}ys_y = 0, (\lambda\theta_y - v\theta)_y + \frac{y}{2}\theta_y = 0, \sigma = \Phi(\theta, s). \tag{2.38}$$

Here $\lambda = \lambda(s, \theta), a = a(\sigma), a_i = a_i(\sigma, \theta), i = 1, 2, 3$ are the given functions; function $\sigma = \Phi(s, \theta)$ is calculated from the formula:

$$\Phi(s, \theta) = \begin{cases} 0, & if\ s < s_*, \\ \dfrac{s - s_*}{s^* - s_*}, & if\ s_* \leq s \leq s^*, \\ 1, & if\ s > s^*, \end{cases} \tag{2.39}$$

where $s_*(\theta), 1 - s^*(\theta)$ - variable residual water and oil saturations.

Let $\Omega = \{y|0 = y_1 < y < y_2\}$, and let us consider the first boundary value problem

$$(\sigma, \theta)|_{y=0} = (\sigma_0, \theta_0), (\sigma, \theta)|_{y=y_2} = (\sigma_1, \theta_1), \tag{2.40}$$

in this case $y_2 \gg 1$ and, possibly, $y_2 = \infty$.

2.4.2 The Solvability of the Regularized Problem

Let the functional parameters of system (2.38) satisfy the following assumptions of smoothness, boundedness and the existence of a fixed sign:

i. $\|a, a_i; \lambda\|_{C^1(\overline{\Omega})} \leq M_0, i = 1, 2, 3, |\ln m, \ln a_1, \ln\lambda, a_{3\theta}a^{-1}| \leq M_0,$

$$(s^*(\theta), s_*(\theta)) \in C(\overline{\Omega}); \Omega = \{\sigma, \theta | (0, 1) \times (\theta_*, \theta^*)\};$$

$$a_0(\sigma) > 0$$

at

$$\sigma \in (0, 1), a \leq M\sigma(1 - \sigma).$$

Let us assume that

$$u = \int_0^\sigma a(\xi) d\xi. \tag{2.41}$$

Determination 1 Let us designate the set of functions $(u, \theta) \in C^1(\overline{\Omega})$ and $s \in L_\infty(\overline{\Omega})$ *a generalized solution of problem* (2.38)–(2.40), if it meets the following conditions:

1. Almost everywhere in Ω $\sigma = \Phi(s, \theta)$;
   $\forall \eta, \psi \in C^1(\overline{\Omega})$ satisfies the integral identities

$$\begin{gathered} (aa_1\sigma_y, \psi_y) + (aa_2\theta_y - aa_3\psi_y) + \frac{1}{2}\left(ys + \int_y^{y_1} s(\xi)d\xi, \psi_y\right) = 0, \\ (\lambda\theta_y, \eta_y) + \left(\left(v - \frac{y}{2}\right)\theta_y, \eta\right) = 0; \quad (.,.) = (.,.)_\Omega; \end{gathered} \tag{2.42}$$

2. The functions (σ, θ) satisfy boundary conditions (2.40).
3. Let us designate $\delta_0 = \min_\theta s^*(\theta) - \max_\theta s_*(\theta)$, select $\varepsilon \in [0, (\delta_0/4)]$ and construct a piecewise-linear function on s

$$\sigma^\varepsilon \equiv \Phi^\varepsilon(s, \theta) = \begin{cases} \dfrac{s(\bar{s}_* + \varepsilon - s_*)}{\delta(\theta)(\bar{s}_* + \varepsilon)}, & \textit{if } s \in [0, \bar{s}_* + \varepsilon), \\ \dfrac{s - s_*}{\delta(\theta)}, & \textit{if } s \in [\bar{s}_* + \varepsilon, \bar{s}^* - \varepsilon], \\ \dfrac{(s - 1)(\bar{s}^* - \varepsilon - s^*)}{\delta(\theta)(s^* - \varepsilon - 1)}, & \textit{if } s \in (\bar{s}^* - \varepsilon, 1], \end{cases} \tag{2.43}$$

$$0 < \delta_0 \leq \delta(\theta) = s^*(\theta) - s_*(\theta) \leq 1.$$

Let us extend the values of σ, s, and θ beyond the intercepts $[0,1], [\bar{s}_*, \bar{s}^*]$ and $[\theta_*, \theta^*]$; without restriction of generality, let us take as their boundary values (having normalized the functions s, s^*, s_*), $\bar{s}_* = \min_\theta s_* = 0; \bar{s}^* = \max_\theta s^* = 1$. Let us now substitute the resultant functions $\tilde{\sigma}(\sigma), \tilde{s}(s), \tilde{\theta}(\theta)$ in the coefficients of (2.38) and replace $a_0 a_1$ with $\bar{a}_0 a_1, \bar{a}_0 = a_0 + \varepsilon$. Let us extend the piecewise linear function $\Phi^\varepsilon(s, \theta)$ to the exterior of the set $[0,1]$, so that $\Phi^\varepsilon = 0, s \leq 0$ and $\Phi^\varepsilon = 1, s \geq 1$, and having replaced θ with $\tilde{\theta}$, derive a biunique relationship between σ and s. We have thus obtained a regularized problem which retains boundary conditions (2.40), and which we will designate I^*.

Determination 2 Let us designate the aggregate of functions $(u, \theta) \in C^1(\overline{\Omega}), s \in L_\infty(\overline{\Omega})$, which satisfy conditions (2.38), (2.39) (Determination 1) and include identities (2.42) transformed as described above, *the generalized solution of problem I^*.*

Theorem 1 If conditions (i) are satisfied, then the regularized problem I^* has at least one generalized solution.

The proof of validity of this statement can be reduced to the application of the Birkhoff-Kellogg theorem [100, p. 498].

2.4.3 Lemma 2 (The Maximum Principle)

In the generalized solutions of the auxiliary problem I^* the estimates shown below apply to virtually the whole of Ω

$$\theta_* \leq \theta \leq \theta^*; \bar{s}_* \leq s \leq \bar{s}^*; 0 \leq \sigma \leq 1. \tag{2.44}$$

Proof Since equation (2.38) for θ does not degenerate, then, by analogy with 2.2, we can arrive at an estimate on θ. Estimates for s, and therefore also for σ, follow from the results obtained in 2.2.

Note In follows from the estimates of the lemma that identity (2.42) is satisfied for I^* when it includes real coefficients, but not with truncated $\tilde{\sigma}, \tilde{s}, \tilde{\theta}$.

Lemma 2 *For generalized solutions of I^*, the estimates shown below are true if they are uniform with respect to ε*

$$(|\theta_y|, |(\lambda\theta_y)_y|) \leq M\exp(-\alpha y^2 + \beta y); \tag{2.45}$$

$$(|a_0\sigma_y|, |(a_0\sigma_y)_y|) \leq M. \tag{2.46}$$

Indeed, both (2.45) and the weighted estimate on s and its derivative, identical to (2.46), have been derived in 2.3. The estimate on σ follows from the equation

$$(a_1 u_y - F(s, \theta, y))_y = 0, u = \int_0^\sigma a(t)dt, \tag{2.47}$$

where $F(s, \theta, y) = -\left[0,5m\left(ys + \int_y^{y_1} s(t)dt\right) + aa_2\theta_y - a_3v\right]$.

Finally, in view of the monotonicity of function Φ^ε in s, it can be easily demonstrated that s, σ, θ are linked by the relationship $\sigma = \Phi(s, \theta)$ [16]. Indeed, for any θ we have $(\Phi^\varepsilon(v_1, \theta) - (\Phi^\varepsilon(v_2, \theta), v_1 - v_2) \geq 0$. Let us assume that $v_1 = s^\varepsilon$; in that case, $\varepsilon \to 0$, (a passage to the limit), and this leads to $(\sigma - \Phi(v_2, \theta), s - v_2) \geq 0$. Let us assume $v_2 = s - r\omega, r = const > 0, \omega \in C^1(\overline{\Omega})$, reduce it by r and let $r \to 0$ (a passage to the limit). Since Φ is continuous, this leads to $(\sigma - \Phi(s, \theta), \omega) \geq 0$. Since ω was selected arbitrarily, $\sigma = \Phi(s, \theta)$ virtually throughout Ω. Allowing $\varepsilon \to 0$ (a passage to the limit) confirms the assertion.

Theorem 2 *If conditions (i) are satisfied, then the problem (2.38)–(2.40) has at least one generalized solution* $(u, \theta) \in C^1[0, y_2], \forall y_2 < \infty$ *, which satisfies the inequalities (2.44)–(2.46).*

2.4.4 The Numerical Model

The construction of numerical solution algorithms for problem (2.38)-(2.40) is more difficult than in the case of the problems in 2.3, for the following reasons:

1. Residual saturations depend on temperature,
2. The function $s = \Phi(\theta, \sigma)$ with respect to σ and s is not biunique along the intercept $[\tilde{\sigma}_1^0, 1 - \tilde{\sigma}_2^0]$, where $\tilde{\sigma}_1^0 = \min_\theta \sigma_1^0(\theta), \tilde{\sigma}_2^0 = \min_\theta \sigma_2^0(\theta)$.

Let us formulate our principal ideas on how these problems might be solved. We discussed (2.38)–(2.40) in the case of constant residual saturations in (2.3). It was solved by non-linear iteration (representing a combination of Newtonian and simple iterations) which converged, given a wide range of initial approximations. With this in mind, let us extend the iteration process to $\psi(\theta, s)$ (Newtonian iteration with respect to s), and therefore also to variable residual saturations as a function of temperature (thus solving Problem 1). This will result in each iteration representing a problem which has already been solved for constant residual saturations.

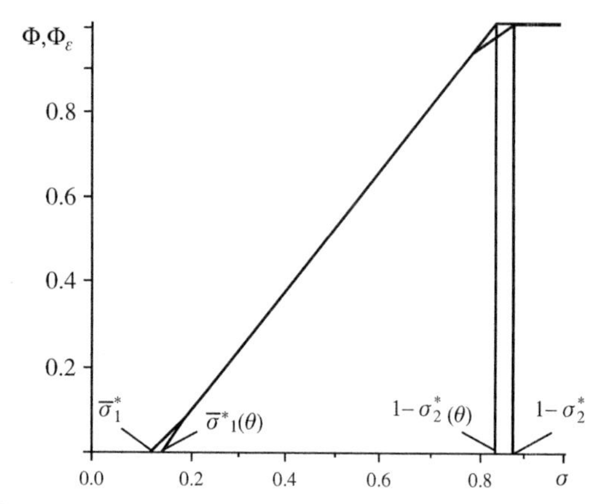

Figure 2.5 The function $s = \Phi(\theta, \sigma)$ and its regularization $s = \Phi_\varepsilon(\theta, \sigma)$.

Problem 2 is solved by the regularization of $s = \Phi(\theta, \sigma)$ described below.

Let us designate $\Delta = 1 - \max_\theta \sigma_2^0(\theta) - \max_\theta \sigma_1^0(\theta)$ and let us then select a $\varepsilon > 0$ sufficiently small to ensure that $\varepsilon < \Delta/4$, and construct a piecewise-linear function on σ

$$s = \Phi_\varepsilon(\theta, \sigma) \equiv \begin{cases} \dfrac{\varepsilon(\sigma - \tilde{\sigma}_1^0)}{\delta(\varepsilon + \sigma_1^0 - \tilde{\sigma}_1^0)}, & \text{if } \sigma \in [\tilde{\sigma}_1^0, \sigma_1^0 + \varepsilon), \\ \Phi(\theta, \sigma), & \text{if } \sigma \in [\sigma_1^0 + \varepsilon, 1 - \sigma_2^0 - \varepsilon], \\ 1 - \dfrac{\varepsilon(1 - \tilde{\sigma}_1^0 - \sigma)}{\delta(\varepsilon - \tilde{\sigma}_2^0 + \sigma_2^0)}, & \text{if } \sigma \in (1 - \sigma_2^0 - \varepsilon, 1 - \tilde{\sigma}_2^0], \end{cases} \tag{2.48}$$

where $0 \leq \Delta < \delta(\theta) \equiv 1 - \sigma_2^0(\theta) - \sigma_1^0(\theta) \leq 1$. The resultant function will be biunique, so that it can be used to recreate uniquely the inverse function $\sigma = \psi_\varepsilon(\theta, s)$.

Fig. 2.5 shows typical plots of $s = \Phi(\theta, \sigma)$ (fine line) and $s = \Phi_\varepsilon(\theta, \sigma)$ (bold line) for a fixed value of $\theta = \theta_1$.

2.4.5 Numerical Calculation Results

Fig. 2.6 shows the calculated distribution of dimensionless temperature $\overline{\theta} = (\theta - \theta_2)/(\theta_1 - \theta_2)$ (fine line) and saturation σ (the vinculum over dimensionless values has been omitted in the diagram).

For comparison purposes, Fig. 2.7 shows the values of σ for nonisothermal flow with variable (fine line) and constant residual saturations.

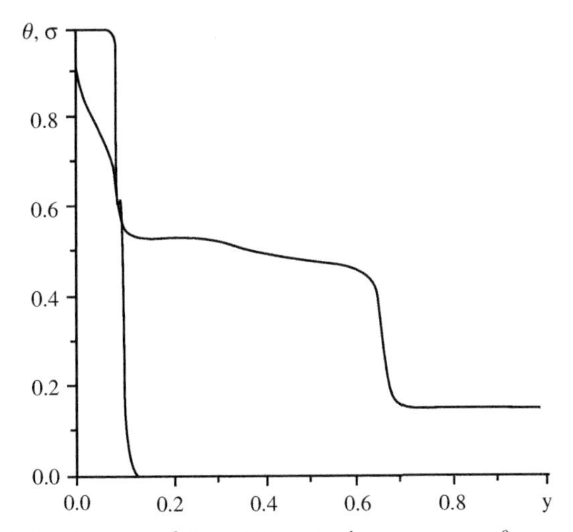

Figure 2.6 Final distributions of saturation σ and temperature θ.

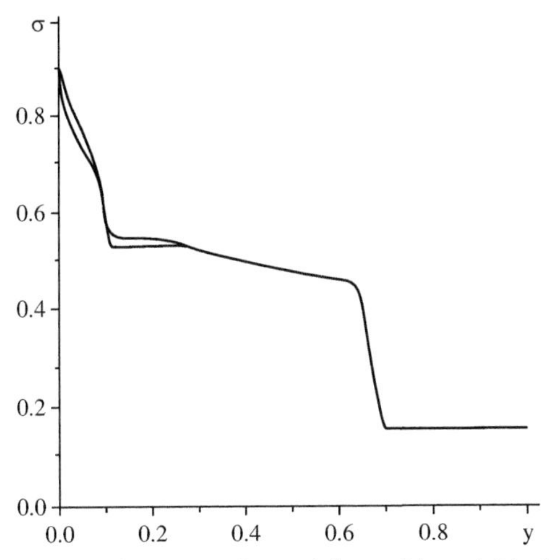

Figure 2.7 The values of σ for nonisothermal flow with variable (fine line) and constant residual saturations.

Let us now consider the boundary-value problem for equation systems which describe two-phase fluid flow in thermocapillary saturation $q=0$, and take into account the variability of residual saturation:

$$(k_0 a a_1 s_y + k_0 a a_2 \theta_y)_y + \frac{1}{2} y \psi_y = 0, (\lambda \theta_y)_y + \frac{1}{2} y \theta_y = 0, \tag{2.49}$$

$$\sigma(y_1) = 1 - \sigma_2^0(\theta(y_1)), \sigma(y_2) = \sigma_1^0(\theta(y_2)), \theta(y_1) = \theta_1, \theta(y_2) = \theta_2, \tag{2.50}$$

where $y_1 = 0, y_2 = \infty$.

The solution algorithm for this problem in the case of constant residual saturations σ_1^0, σ_2^0 was described in Section 2.4.2 and for $q \neq 0$ in this section.

The distribution of saturation σ in the case of variable and constant saturation

The problem of the non-biunique relationship $s = \Phi(\theta, \sigma)$ was solved for $q \neq 0$ by regularization. However, the inverse function $\sigma = \psi_\varepsilon(\theta, s)$ is not continuously differentiable, and this created difficulties with the numerical implementation of the algorithm for thermocapillary saturation, and required the use of the derivative $\frac{d\psi_\varepsilon(\theta,s)}{ds}$ for the segment $[\tilde{\sigma}_1^0, 1 - \tilde{\sigma}_2^0]$. At the same time, the discontinuity of the derivative resulted in quite strong oscillations of the numerical solution ahead of the front.

For this reason, we resorted to the following regularization of function $s = \Phi(\theta, \sigma)$, which was smooth (continuously differentiable) along the segment $[\tilde{\sigma}_1^0, 1 - \tilde{\sigma}_2^0]$:

$$s = \Phi_\varepsilon(\theta, \sigma) \equiv \begin{cases} P_3(\theta, \sigma), & if\, \sigma \in [\tilde{\sigma}_1^0, \sigma_1^0 + \varepsilon), \\ \Phi(\theta, \sigma), & if\, \sigma \in [\sigma_1^0 + \varepsilon, 1 - \sigma_2^0 - \varepsilon], \\ Q_3(\theta, \sigma), & if\, \sigma \in (1 - \sigma_2^0 - \varepsilon, 1 - \tilde{\sigma}_2^0]. \end{cases} \tag{2.51}$$

Here, $P_3(\theta, \sigma)$ and $Q_3(\theta, \sigma)$ are third degree polynomials, which are defined by the conditions of continuous differentiability of $\Phi_\varepsilon(\theta, \sigma)$ at σ along the segment $[\tilde{\sigma}_1^0, 1 - \tilde{\sigma}_2^0]$.

Fig. 2.8 shows a typical plot of $s = \Phi_\varepsilon(\theta, \sigma)$ (bold line) for a fixed value of θ. For comparison purposes, the fine line shows $s = \Phi(\theta, \sigma)$.

We used the above reasoning to perform numerical calculations, using the following forms of variable residual saturations and boundary values

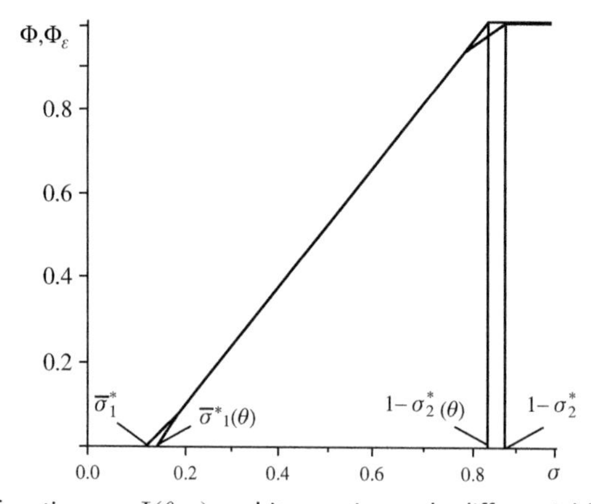

Figure 2.8 The function $s = \Phi(\theta, \sigma)$ and its continuously differentiable regularization.

of temperature: $\sigma_1^0(\theta) = 0,15 - 0,001 \cdot (\theta - 60), \sigma_2^0(\theta) = 0,15 - 0,002 \cdot (\theta - 60)$.

Fig. 2.9 shows the calculated distribution of dimensionless temperature $\overline{\theta} = (\theta - \theta_2)/(\theta_1 - \theta_2)$ (fine line) and saturation σ (the vinculum over dimensionless values has been omitted in the diagram).

For comparison purposes, Fig. 2.10 shows σ in the case of nonisothermal flow, for variable (fine line) and constant residual saturations. It will

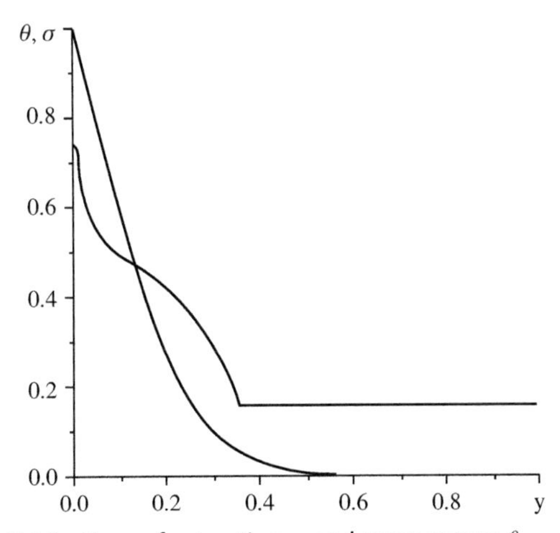

Figure 2.9 Final distributions of saturation σ and temperature θ.

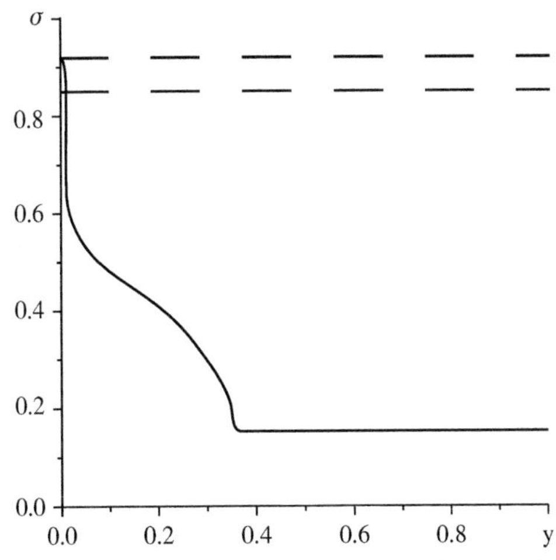

Figure 2.10 Distributions of saturation σ at variable and constant residual saturations.

be seen that the only difference between the plots occurs in a part of the left hand side of the graph, but the high rate of saturation change in this area makes it impossible to represent the difference graphically with sufficient accuracy. The crucial aspect of the diagram is that the plot for the solution with variable residual saturations is higher than the plot for constant saturations.

It was noted in 2.3 that a change of the left boundary value of s_1 in the region $[0,6;1]$ does not significantly affect the saturation rate. In our case, this means that if the variability of residual saturations is taken into account, the value of σ increases at the left hand side end of the segment (in the small neighborhood of the injection well) if hot water is injected, but does not lead to an additional advance of the right saturation front.

The dashed lines show the majorants of the distribution of saturation σ for variable (the upper line) and constant (the lower line) residual saturations. The rectangular area between the dashed lines is an area of potential additional oil displacement obtained when flow modelling takes into account the variability of residual saturations.

2.5 THE QUALITATIVE PROPERTIES OF ANALYTICAL MLT MODEL SOLUTIONS

In this section, we examine the qualitative properties of the exact solutions of the thermal two-phase flow equations (the MLT model) constructed in 2.3, paying particular attention to the physical interpretation of the various properties of these solutions which are of practical importance to oil production: finite velocity of propagation of perturbations, the monotonicity of two-phase mixed flow characteristics (uniform water encroachment and formation heating), and the finite stabilization time of the process.

2.5.1 The Muskat-Leverett Thermal (MLT) Model

The transformed equations of the one-dimensional MLT model of a homogeneous isotropic porous medium for dynamic water saturation $s(x,t)$, equilibrium temperature $\theta(x,t)$ and average pressure $p(x,t)$ have the form (Chapter 1, 1.2)

$$s_t = (a_0\lambda_1 s_x + a_1\theta_x + a_2 v)_x;\ \theta_t = (\lambda_2\theta_x - v\theta)_x;\ -v_x \equiv (\lambda_3 p_x + a_3\theta_x)_x = 0, \tag{2.52}$$

where $v = v(t)$ is the total flow rate of the two-phase fluid. The system (2.52) coefficients $a(s), \vec{a}(s,\theta) \equiv (a_0, a_1, a_2, a_3)$ and $\vec{\lambda}(s,\theta) = (\lambda_1, \lambda_2, \lambda_3)$

are expressed explicitly by the functional parameters of the initial MLT model, and if we take into account their physical reality, have the properties (Chapter 1, 1.2):

a. $a_0(s) > 0, s \in (0,1), a_k(0) = a_k(1) = a_2(0) = 0; \lambda_k(s,\theta) \geq m_0 > 0, k = 0, 1;$

b. $(\vec{a}(s,\theta), \vec{\lambda}(s,\theta)) \in C^1(\overline{Q}), Q = \{s, \theta | 0 \leq s < 1, \theta_* < \theta < \theta^*\}.$

Assumptions (a) ensure respectively the parabolicity and uniform parabolicity of the equations for $s(x,t)$ $s(x,t)$ and $Q(x,t)\theta(x,t)$ and the nondegeneracy of the ordinary equation for $p(x,t)$ (in which t acts as a parameter).

The use of complex mathematical models, such as the MLT model or the Muskat-Leverett model of isothermal two-phase flow [2] ($\theta \equiv const$), requires the creation of mathematical techniques accessible at least to experienced oilfield engineers. One such approach, successfully used in practice, is to describe the process of oil displacement by means of approximate formulae derived from exact solutions of the initial model equations. These include stationary solutions, dependent only on the variable x, analytical (self-similar) parabolic solutions dependent on $y = x(t+1)^{-1/2}$, analytical traveling wave solutions, dependent on $z = x + ct (c = const)$ and some others. Simple Muskat formulae (displacement laws), Charny's formulae (near-wellbore zones) and others [139], based on parabolic self-similarity remain reliable tools of engineering analysis of oil field development by isothermal methods.

This being the case, we propose to begin our examination of thermal recovery methods by constructing analytical (self-similar) MLT model solutions, described by ordinary differential equations.

2.5.2 Parabolic Analytical Solutions

Provided that $v = \tilde{v}(t+1)^{-1/2}, \tilde{v} = const$ is specified (in what follows, the v is used without the tilde), it is possible to find solutions of (2.52) dependent only on one independent variable $y = x(t+1)^{-1/2}$ - parabolic analytical solutions $s(y), \theta(y), p(y)$ which satisfy the following transformed equations (2.52):

$$(a_0\lambda_1 s_y + a_1\theta_y + a_2 v)_y + \frac{1}{2} y s_y = 0, v = const; (\lambda_2\theta_y - v\theta)_y + \frac{1}{2} y\theta_y = 0. \tag{2.53}$$

In this equation, pressure $p(y)$ is found from equation $\lambda_3 p_y + a_3\theta_y = -v = const$ after solving the system (2.53) for $s(y), \theta(y)$.

Let the two-phase fluid flow take place between two wells (groups of wells), located in points $y = y_k, k = 0, 1$, of which let $y = y_0 = 0$ be the location of the injection well, and $y = y_1$ the location of the production well ($y_1 = \infty$ is also a possible case). Let the values of (s, θ) be specified for wells $y = y_k$:

$$(s, \theta)|_{y=y_k} = (s_k, \theta_k), k = 0, 1. \tag{2.54}$$

Assuming that the coefficient $\lambda_2(y) \equiv \lambda_2[s(y), \theta(y)]$ is a known function, we can derive the following expression for $\theta(y)$:

$$\theta = 1 - NF(y); F = \int_0^y \lambda_2^{-1}(\xi)e^{-\Lambda(\xi)}d\xi, \Lambda = \int_0^y \lambda_0(\xi)d\xi. \tag{2.55}$$

Here $\lambda_0 = (0,5y - v)\lambda_2^{-1}(y), N = [F(y_1)]^{-1}$.

Expression 2 (2.55) leads directly to the estimates

$$0 \leq \theta(y) \leq M_1 e^{-\alpha_1 y^2} \leq 1, |\theta_y| \leq M_2 e^{-\alpha_2 y^2}, y \in [0, y_1], \tag{2.56}$$

where the constants $M_i > 0$ and $\alpha_i > 0$ do not depend on value $y_1 > 0$.

Now let us consider equation (2.53) for $s(y)$ and note that irrespective of the value of $s_i \in [0, 1], i = 0, 1$ (2.54) provides the following estimates (2.3)

$$0 \leq s(y) \leq 1, y \in [0, y_1], \forall y_1 > 0. \tag{2.57}$$

Let us introduce a new function $u = \int_0^s a_0(\xi)d\xi$ and assuming that the coefficients of equation (2.53) for $s(y)$ are known, let us express this equation in the form

$$(\lambda_1 u_y + \varphi)_y = 0, u(0) = u_0, u(y_1) = u_1. \tag{2.58}$$

Here $u_i = \int_0^{s_i} a_0(\xi)d\xi, i = 0, 1; \varphi = 0,5\left(ys + \int_y^{y_1} s(\xi)d\xi\right) + a_1\theta_y + a_2 v$.

Integrating (2.58), we obtain

$$-\lambda_1 u_y = \varphi + C, u(0) = u_0, \tag{2.59}$$

where $C = C_0^{-1}\left[u_0 - u_1 - \int_0^{y_1} \lambda_1^{-1}(\xi)\varphi(\xi)d\xi\right], C_0 = \int_0^{y_1} \lambda_1^{-1}(\xi)d\xi$. Since, on the basis of the assumption made in (a), $\lambda_1 \geq m_0 > 0$, we can use (2.59) to find

$$|u_y| = a_0(s)|s_y| \leq M_0(y_1). \tag{2.60}$$

Estimate (2.60) leads directly to the Holder continuity of $s(y) \in C^\alpha(\overline{\Omega})$, $\alpha > 0, \Omega = (0, y_1)$ so that equations (2.53) give us $(u, \theta) \equiv \vec{v} \in C^{1+\alpha}(\overline{\Omega})$, $\alpha > 0, u = \int_0^s a_0(t)dt(2.3)$.

Therefore, evidently, the construction of generalized solutions of (2.53), (2.54) is the equivalent of solving the Cauchy problem

$$-\lambda_1 u_y = \varphi + C; \ -\lambda_2 \theta_y = \psi + K; \vec{v}(0) = \vec{v}_0. \tag{2.61}$$

where the function φ and the constant C are as determined in (2.59):

$$\psi = \frac{1}{2}\left(y\theta + \int_0^{y_1} \theta(t)dt\right) - v\theta, K = K_0^{-1}\left(1 - \int_0^{y_1} \psi(t)\lambda_2^{-1}(t)dt\right),$$

$$K_0 = \int_0^{y_1} \lambda_2^{-1}(t)dt.$$

The classical solvability of (2.61) was demonstrated in 2.3.

2.5.3 The Theorem (of Finite Velocity)

Let the following inequalities be satisfied in addition to conditions (a) and (b):

$$(a_0(s), |a_2(s,\theta)|) \leq Ms^\gamma, \text{or } (a_0, |a_2 - a_2(1,0)| \leq M(1-s)^\gamma), \gamma \geq 1. \tag{2.62}$$

In that case, when $s(y_1) = 0(or\ s(y_1) = 1), y_1 \gg 1$ *there exists a value* $y^* < \infty$ *such that*

$$s(y) \equiv 0(or\ s(y) \equiv 1) at\ y \geq y^* = (y^2 + M_0\delta^{-1})^{1/2}, \tag{2.63}$$

Where $M_0 = \int_0^1 a_0(t)t^{-1}dt, \delta = 1/4 \min \lambda_1, y = 2\ M \max(M_2|a_1|, |v|)$.

The proof of this theorem will be found in 2.3.

Note 1 Let the boundary-value problem be solved for system (2.53) in the interval $[-y_0, y_1], (y_0, y_1) \gg 1$ and let $s(-y_0) = 1, s(y_1) = 0$. Then, according to the theorem, we have a wave-type solution:

$$s(y) \equiv 1\ at\ y \leq -y^*; s(y) \equiv 0\ at\ y \geq y^*. \tag{2.64}$$

2.5.4 Finite Velocity Interpretation

The family of parabolae shown in Fig. 2.11 corresponds to the analytical variable $y = (t+1)^{-1/2}$ in phase plane (x, t).

According to the constructions arrived at in the preceding subsection, the component $u(s) = \int_0^s a_0(\xi)d\xi$ which forms part of the solution of (2.61), as well as $s = s(y)$ has a finite velocity of propagation of perturbations $- s(y) = u(y) \equiv 0, y \geq y^*$ (Fig. 2.12).

In Fig. 2.11, this property of $s(y)$ corresponds to region D^+ (in the wave-type solution these are regions $D^\pm$, where $s \equiv const$).

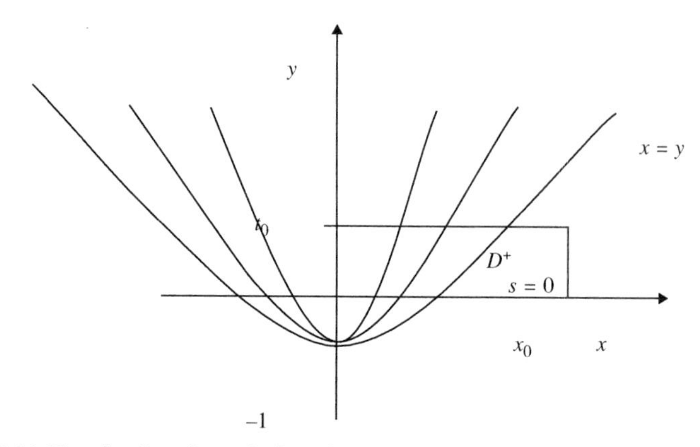

Figure 2.11 The family of parabolae shown corresponds to the analytical variable $y=(t+1)^{-1/2}$in phase plane (x,t).

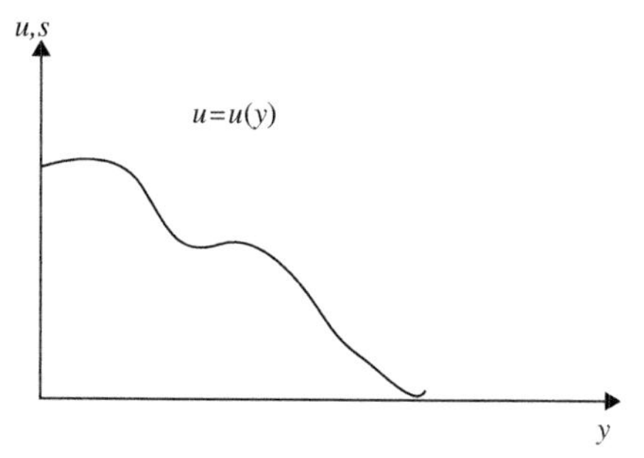

Figure 2.12 A finite velocity of propagation of perturbations $-s(y)=u(y)\equiv 0, y\geq y*$.

The structure of D^+ allows us to interpret the property of finite velocity of the function $s(x,t)\equiv s\left(\frac{x}{\sqrt{t+1}}\right)$ in the following way:

1. For each point of time t_0 there exists a point $x_0=y_*\sqrt{t_0+1}$, such that $s(x,t_0)\equiv 0$ at $x\geq x_0$;
2. For each point $x_0>y^*$ there exists a time interval $[0,t_0],(x_0=y_*\sqrt{t_0+1})$, such that $s(x_0,t)\equiv 0, t\in[0,t_0]$, i.e. for each $x_0>y^*$ the movement of the front $s=0$ is delayed by a time $t_0=t_0(x_0)$.

2.5.5 The Near-wellbore Zone

Note that in the case of one-dimensional flow, the term "well" should be understood to mean a group of wells orthogonal to the OX axis.

Let a fluid-filled formation be completed in a well and let the fluid begin to flow into the well. In this situation, the near-wellbore zone is

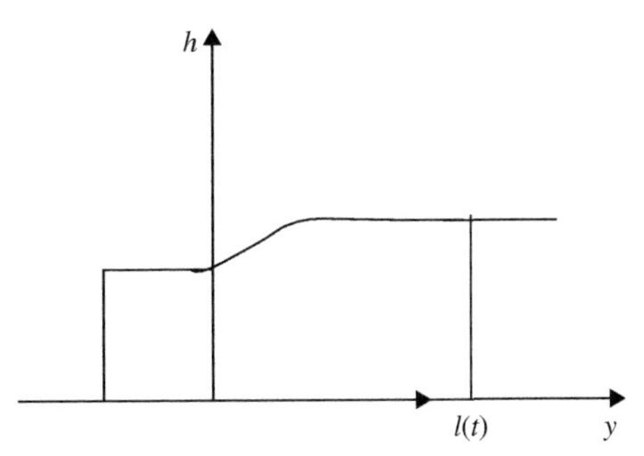

Figure 2.13 The Boussinesq equation $h_t = k_0(h_2)_{xx}$ for the ordinate $h = h(x, t)$ of the free surface defines this zone.

determined by the distance $l(t)$ from the well beyond which the fluid level is regarded as undisturbed. We have used the Boussinesq equation $h_t = k_0(h_2)_{xx}$ for the ordinate $h = h(x, t)$ of the free surface (Fig. 2.13) to define this zone.

Kochina, Charny, Muskat and others have found, using a variety of methods, that $l(t) = K\sqrt{t}, K$ being expressed in terms of the parameters of the porous medium and the fluid flowing through it [110, p. 193-195].

The near-wellbore zone is therefore the parabola $x = K\sqrt{t}$ on the phase plane (x, t).

If we apply this reasoning to oil production problems and consider that the near-wellbore zone also obeys the analytical law $\frac{l}{\sqrt{t+1}} = const$, the solution of (2.61) maybe sought in a finite region. If this is so, then the injection well can be located at a point $y = 0$ (and therefore $x = 0$), and there will be a point $y = y_1 < \infty$ corresponding to the production well's near-wellbore zone $\frac{x}{\sqrt{t+1}} = y_1$ and the boundary conditions $s(y_1) = s_1 \geq 0$ may be specified within this parabola in the plane (x, t).

2.5.6 The Thermodynamic Properties of Coefficients

In order to examine further properties of analytical solutions, we must examine the properties of the coefficients of (2.53) in greater detail, proceeding from the physical meaning of the MLT model's functional parameters.

2.5.6.1 Coefficient Expressions

$$a_0(s) = \bar{k}_1(s)\bar{k}_2(s), \bar{k}_1(0) = \bar{k}_2(1) = 0; \lambda_1 = \left|\frac{\partial p_c}{\partial s}\right| v(s, \theta);$$

$$v^{-1} = k_0\mu_1\mu_2(k_1 + k_2), k_i = \overline{k}_i(s)\mu_i(\theta), i = 1, 2;$$

$$a_1 = -a_0 v\frac{\partial p_c}{\partial \theta}; \; -a_2 = k_1(k_1 + k_2)^{-1}; \lambda_2 = \sum_1^3 m_0 s_i \lambda^i;$$

$$\lambda^i = \overline{\lambda}^i(\theta)(\rho_i c_{p_i})^{-1}.$$

In the above equations, $\overline{k}_i = \overline{k}_i(s), \mu_i = \mu_i(\theta), s_i(s_1 \equiv s, s_2 = 1 - s), \rho_i = const, i = 1, 2$ are respectively phase permeabilities, viscosities, dynamic saturations and densities; m_0 is porosity; $c_{p_i} = const; \lambda_i(\theta), i = 1, 2, 3$ is the phase coefficients of heat capacity and thermal conductivity of the fluids ($i = 1, 2$) and the porous medium($i = 3$); $s_3 = \frac{1 - m_0}{m_0}$. Capillary pressure is expressed by $p_c = \gamma \cos \alpha_j(s)$, where $\gamma = \gamma(\theta)$ is the interfacial tension coefficient; $\alpha = \alpha(\theta)$ is the wetting angle; $J(s)$ is the Leverett J function (which also includes the strain constant associated with average formation permeability).

2.5.6.2 The Properties of a_2

Let $\mu = \mu_2/\mu_1, -a_2 = \overline{k}_1(\overline{k}_1 + \mu\overline{k}_2)^{-1}$. Then $-a_{2\theta} = a_0(\overline{k}_1 + \mu\overline{k}_2)^{-2}\mu_\theta$, $\mu_\theta = \mu_1^{-1}(\mu_{2\theta} - \mu\mu_{1\theta})$. Since $(\mu_{1\theta}) \ll 1$ (the viscosity of water is only slightly dependent on temperature) and $\mu = \mu_2\mu_1^{-1} \ll 1$, therefore

$$a_2 \leq 0; \mu_\theta \leq 0; a_{2\theta} \geq 0; a_{2s} \leq 0 \tag{2.65}$$

The latter inequality results from the properties of phase permeability: $k_{1s} \geq 0, k_{2s} \leq 0$.

2.5.6.3 The Properties of a_1

Interfacial tension decreases as temperature increases, i.e. $\gamma_\theta \leq 0$. The wetting angle increases as temperature increases (in a state of equilibrium, when $|\theta| \gg 1, \alpha = \pi/2$), i.e. $\alpha\theta \geq 0$. Therefore $\frac{d}{d\theta}(\gamma \cos \alpha) \leq 0$, and consequently $\frac{\partial p_c}{\partial \theta} = (\gamma \cos \alpha)_\theta \cdot j(s) \leq 0$ and hence $a_1 = -\frac{\partial p_c}{\partial \theta} v a_0 \geq 0$; $a_1 \equiv 0$ at $p_{c\theta} = 0$.

Note 2 Normalization of the boundary conditions for θ may produce $\theta = (T_0 - T)\cdot\delta, \delta = const > 0, T_0 = const > 0$ and T is temperature, in which case

$$a_1\theta_x = -va_0\frac{\partial p_c}{\partial \theta}\theta_x = -va_0\frac{\partial p_c}{\partial T}T_x,$$

i.e. the properties of $a_{1\theta x}$ do not depend on the normalization of temperature.

2.5.6.4 The Properties of λ_i

We know that $\frac{\partial v}{\partial \theta} = -v^2(\overline{k}_1\mu_{2\theta} + \overline{k}_2\mu_{1\theta})$ and since $\mu_{i\theta} \leq 0$, therefore $v_\theta \geq 0$. In that case

$$\lambda_{1\theta} = |p_{cs}|v_\theta \geq 0 \text{ at } p_{c\theta} = 0. \tag{2.66}$$

It is often assumed that the thermal characteristics of the fluid and the porous medium do not depend on temperature, i.e.

$$\lambda^i_\theta = \lambda_{2\theta} = 0, i = 1, 2, 3. \tag{2.68}$$

2.5.7 Monotonicity Conditions of Water Saturation *s*(*y*)

Let $\overline{a}_1 = a_1\lambda_2^{-1}, q = v - 1/2y$ and let us transform the regularized equation (2.53) for $s(y)$, and substitute $a_\varepsilon = a_0 + \varepsilon$ for a_0. Using equation (2.53) for $\theta(y)$, let us solve

$$z_y + d_z = -g, z \equiv a_\varepsilon\lambda_1 s_y, a_\varepsilon = a_0 + \varepsilon, \varepsilon > 0, \tag{2.69}$$

$$d = \left(\frac{1}{2}y + va_{2s} + \overline{a}_{1s}\lambda_2\theta_y\right)(a_\varepsilon\lambda_1)^{-1}, g = \lambda_2\overline{a}_{1\theta}\theta_y^2 + \overline{a}_1 q\theta_y + va_{2\theta}\theta_y.$$

Having integrated (2.69), we obtain

$$z = e^{-D(y)}\left(z(0) - \int_0^y e^{D(t)}g(t)dt\right), D = \int_0^y d(t)dt. \tag{2.70}$$

Since $s(0) = 1$ and $s(y) \leq 1$, then $s_y(0) \leq 0$ and, consequently, $z(0) = (a_\varepsilon\lambda_1 s_y)(0) \leq 0$. Therefore, if $g(y) \geq 0, y\in[0, y_0]$ for some $y_0 > 0$, then it follows from equation (2.70) that $s_y \leq 0$ at $g(y) \geq 0, y\in[0, y_0]$.

If it proves that $y_0 \geq y^*$, where y^* is a point of the saturation front ($s(y) \equiv 0, y \geq y^*$), $s(y)$ will be monotonic at $y\in[0, \infty)$. It follows from the property (2.65), (2.66) of coefficients a_1, a_2 that a sufficient condition of monotonicity of $s(y)$ is:

$$s_y \leq 0 \text{ at } p_{c\theta} = 0 \text{ and } \theta_y \geq 0, y\in[0, y^*]. \tag{2.71}$$

The requirement $\theta_y \geq 0$ means that the temperature of the injection water is $\theta(0) = \theta_0 \leq \theta_l$- average formation temperature ($\theta_l = \theta(\infty)$), in which case in equation (2.55) we have $\theta_y = NF_y \geq 0, y\in[0, \infty)$, where $N_0 = (\theta_l - \theta_0)F - 1(y_1)$, and $F(y)$ as determined in (2.55).

Note 3 Conditions (2.71) are satisfied in the case of isothermal flow ($\theta \equiv const$), i.e. when $s_y \leq 0, y\in[0, \infty)$.

The introduction of new stimulation methods and processes is intended to provide a solution to one of the main oilfield problems, that of creating a uniform flood front (uniform water encroachment) and thus achieving more complete oil recovery. Bearing this in mind, we can interpret the monotonicity conditions of $s(y)$ derived above as follows.

In each fixed section of the formation $x = x_0 > 0$, its water encroachment $s(x_0, t) = s(\frac{x_0}{\sqrt{t+1}})$ increases monotonically with time: $s(\frac{x_0}{\sqrt{t+1}}) \to s(0) = 1$ at $t \to \infty$.

2.5.8 Local Monotonicity of *s*(*y*)

Let the thermal conductivity coefficients λ_1, λ_2 (of the fluids) and λ_3 (of the porous medium) change slightly, i.e let conditions (2.68) be satisfied. Let us also make one of the following assumptions:

A. The function $\sigma(\theta) = \gamma \cos \alpha$ is linear: $\sigma = \sigma(\theta^*) + \sigma\theta(\theta^*)(\theta - \theta^*)$, $\sigma\theta(\theta^*) \leq 0, \theta^* = const.$

In this case

$-p_{c\theta} = |\sigma_\theta(\theta_*)|j(s), \frac{\partial^2 p_c}{\partial \theta^2} = 0$, from which $a_{1\theta} \geq 0$.

B. The function $\gamma(\theta)$ is linear and $\alpha = const$:

$$\gamma = \gamma(\theta^*) + \gamma\theta(\theta^*)(\theta - \theta^*), \gamma\theta(\theta^*) \leq 0.$$

In that case again $\frac{\partial^2 p_c}{\partial \theta^2} = 0$ and therefore $a_{1\theta} \geq 0$.

Now, let condition (2.68), and either assumption A or assumption B, be satisfied. Consequently, $\bar{a}_{1\theta} = \lambda_2^{-1} a_{1\theta} \geq 0$, i.e. $g(y) \geq 0$ at $y \leq 2v$, whence

$$s_y(y) \leq 0 \text{ at } y \in [0, 2v]. \tag{2.72}$$

Note 4 Let us introduce the function $T = \int_0^y \lambda_2 \theta_t dt$ and express equation (2.53) for θ in the form $T_{yy} = \lambda_2^{-1}\left(v - \frac{1}{2}y\right)T_y \equiv R(y)$.

Where, because of the assumption, $\theta_y \geq 0$ and therefore $T_y \geq 0$. We obtain $R(y) \geq 0$ when $y \leq 2v$ and $R(y) < 0$ when $y > 2v$, i.e. the point $y_0 = 2v$, for which the inequality (2.72) is satisfied, is the inflexion point of function $T(y)$.

Similarly, having introduced the variable $\bar{y}(y) = \int_0^y \lambda_2^{-1}(t)dt$, let us write down (2.53) in the form $\theta_{\bar{y}\bar{y}} = \left[v - \frac{1}{2}y(\bar{y})\right]\theta_{\bar{y}} \equiv \overline{R}(\bar{y})$.

Now, the point $\bar{y}_0 = \bar{y}(y)_0$ is the inflexion point of function $\theta(\bar{y})$.

In the isothermal case ($\theta \equiv const$), the assertion is stronger than in (2.72): $s_y < 0, y \in (y_0, y_1)$, where y_i are the boundary points where $u_y(y_i) = a(y_i)s_y(y_i) = 0, s(y_0) = 1, s(y_1) = 0$ (at $y > y_1, s(y) \equiv 0$ at $y < y_0$, $s(y) \equiv 1$).

Indeed, let us assume that there is an interval $[\alpha, \beta] \subset (y_0, y_1)$ containing a point $y = y_2 \in (\alpha, \beta)$, where $s_y(y_2) = 0$, and $0 < s(y) < 1, y \in [\alpha, \beta]$ and $s_y(\alpha) \neq 0$. To examine the function $\varphi(y) = u_y(y) = a_0 s_y$ let us consider the Cauchy problem $\varphi_y = q\varphi, \; -q = a_0^{-1}(\frac{1}{2}y + va_{2s})$; $\varphi(\alpha) = \varphi_0 \neq 0$, whose solution takes the form $as_y \equiv \varphi = \varphi_0 e^{Q(y)}, \; Q = \int_0^y q(t)dt$.

Because $s \neq 0, 1$ when $y \in [\alpha, \beta]$, then $|q| \leq M < \infty$. Consequently, based on the expression derived for $\varphi = a_0 s_y, s_y \neq 0$ at $y \in [\alpha, \beta]$, and this contradicts the initial assumption $s_y(y_2) = 0$. We have therefore established that $s_y < 0$ at $y \in (y_0, y_1)$.

2.5.9 Local Extrema of *s*(*y*)

Let us assume that

$$\bar{a}_{1\theta} = (a_1 \lambda_2^{-1})_\theta = 0, a_{2\theta} = 0 \text{ (that is } \mu_\theta = 0). \tag{2.73}$$

In that case, the regularized equation (2.53) for $s(y)$ will take the form

$$Ls = a_\varepsilon \lambda_1 s_{yy} + Cs_y + \bar{a}_1(v - 1/2y)\theta_y = 0, a_\varepsilon = a_0 + \varepsilon, \varepsilon > 0, \tag{2.74}$$

where $C = (\lambda_1 a_\varepsilon)_y + \bar{a}_{1s}\theta_y + va_{2s} + 1/2y$.

Let us assume that $R = (v - 1/2y)\bar{a}_1\theta_y$ and let $\theta_y < 0$. In that case $R < 0$ when $y < y_0 = 2v$ and $R > 0$ when $y > y_0$.

Let $\underline{y}$ be the minimum point of $s(y)$, $\min s(y) = s(\underline{y}) \equiv \underline{s} > 0$, and $\bar{y} - \max s(y) = s(\bar{y}) \equiv \bar{s} < 1$. If $\bar{y} < y_0$, then $R(\bar{y}) < 0, s_{yy}(\bar{y}) < 0, s_{yy}(\bar{y}) \leq 0$, $s_y(\bar{y}) = 0$ and the equality $Ls(\bar{y}) = 0$ is impossible. At $\bar{y} > y_0$, there is no contradiction with the equality $Ls(\bar{y}) = 0$. If $\underline{y} > y_0$ then $R(\underline{y}) > 0$, $s_{yy}(\underline{y}) \geq 0, s_y(\underline{y}) = 0$, which contradicts the equality $Ls(\underline{y}) = 0$.

Thus, the relationship between the minimum and maximum points $\underline{y}$ and $\bar{y}$ of $s(y)$ is expressed by the inequalities $\underline{y} \leq y_0 \leq \bar{y}$. In addition, as there can be no maxima of $s(y)$ to the left of $y_0 = 2v$, only one minimum of $s(y)$ is possible. Similarly, only one maximum of $s(y)$ may occur to the right of y_0 (Fig. 2.14). Let us note that this property of analytical MLT

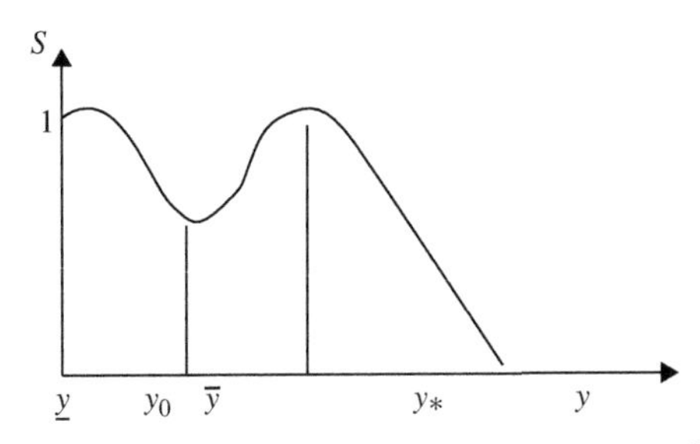

Figure 2.14 Only one maximum of $s(y)$ may occur to the right of y_0.

model solutions was also identified in the course of numerical calculations (2.3, Figs. 2.1 and 2.2).

Let us now assume that $s(\underline{y}) = 0$ and consider the Cauchy problem

$$Ls \equiv a_\varepsilon \lambda_1 s_{yy} + Cs_y + Rls = 0; s(\underline{y}) = 0, s_y(\underline{y}) = 0,$$

where $l = \overline{a}_1 s^{-1} \geq 0$. At $\underline{y} < y_0 R(y) < 0, y \in [0, \underline{y}]$, and the resultant Cauchy problem has only the trivial solution $s(y) \equiv 0, y \leq \underline{y}$, which is impossible.

Let there be a point $\overline{y}$ such that $s(\overline{y}) = 1$. By replacing $1 - s = z$, we arrive at the above problem for z. Therefore, $\min s(y) > 0, \max s(y) < 1$, $y \in (0, y^*)$.

The case of $\theta_y > 0, y \in (0, y_*)$, can be analysed in exactly the same way as that of $(\theta_y < 0)$.

2.5.10 No Flow Zones

Let us formulate the regularity conditions of (2.52) for $s(x, t)$ derived in [44] as they apply to the one-dimensional case of two-phase flow in homogeneous rock:

$$(\mu_{1\theta}, \mu_{2\theta}, p_{c\theta}) = 0 \text{ at } s \notin [\delta_0, 1 - \delta_1], \tag{2.75}$$

where $0 < \delta_0 < 1 - \delta_1 < 1$. The physical condition expressed by (2.75) means that if a flow is predominantly the flow of one fluid $(s_1 = s \leq \delta_0, s_2 = 1 - s \leq \delta_1)$ the equilibrium temperature θ of the porous medium and the fluids should be close to the temperature of the flowing fluid ($\theta \equiv const$). In these circumstances, the thermal characteristics should be constant: $\gamma \cos \alpha = const, \mu_i = const, i = 1, 2$.

The maximum principle for generalized solutions of equations (2.53) for $s(y)$, allows us to arrive at the inequalities shown below, when conditions (2.75) are satisfied:

$$0 < \delta_0 \leq s(y_1) \leq s(y) \leq s(0) \leq 1 - \delta_1 < 1. \tag{2.76}$$

Thus, if $(s(0), s(y_1)) \neq 0, 1$, then, based on (2.76), $s(y) \neq 0, 1$ at $y \in [0, y_1]$, indicating the absence of no flow zones ($s \equiv 0$ or $s \equiv 1$).

2.5.11 Traveling Waves

Up to this point, we have considered parabolic analytical solutions of system (2.52), in which there is a parabolic relationship between the coordinates x and $t, x = y\sqrt{t + 1}$, where y is the analytical variable. Another class of analytical solutions which is of importance to applications consists of (simple) traveling wave solutions, in which the analytical variable y is a linear function of x and t, $y = x - qt$, where $q = const$ is the wave velocity. Formation flow of this kind occurs, for instance, during early field development, in bottom-hole zones treated with chemicals, in thermal development methods and in many other cases. If we introduce the variable $y = x - q$, system (2.52) will be transformed into

$$(a_0\lambda_1 s_y + a_1\theta_y + a_2 v + qs)_y = 0; (\lambda_2\theta_y - v\theta + q\theta)_y = 0. \quad (2.77)$$

The physical reality of the water drive process means that the solutions of equation (2.53) should have the following properties:

1. $\frac{\partial s}{\partial x} \leq 0$ at a fixed t, i.e. water encroachment decreases with distance fromthe injection well;

2. $\frac{\partial s}{\partial x} \geq 0$ at fixed x, i.e. the water encroachment of the section increases withtime.

Thus, transition to the analytical variable $y = x - qt$ should produce $\frac{\partial s}{\partial x} \leq 0$, which corresponds to the assumption $q = const > 0$, where q is wave propagation velocity: $s_x = s_y \leq 0, s_t = -qs_y \geq 0$.

Let us note that, formally, when $q = 0$ we arrive at a stationary problem in which $s = s(x), \theta = \theta(x)$.

2.5.12 Isothermal Traveling Waves ($\theta = const$)

In the isothermal case, the $s(y)$ problem assumes the following simple form:

$$(a(s)s_y - vb(s) + qs)_y = 0; s(0) = 1, s(l) = 0. \quad (2.77)$$

In this equation, $a = a_0\lambda_1; b = -a_2 = k_1(k_1 + k_2) - 1 \geq 0; bs \geq 0, b(0) = 0, b(1) = 1, q = const > 0$ - wave propagation velocity; $v = const > 0$ - total mixed flow rate.

2.5.12.1 Wave Velocity Greater than Flow Rate ($q > v$)

Let us additionally require $vbs \leq q, s \in [0, 1]$, which is equivalent to the inequality

$$C(s) = q(s) - vb(s) > 0, s \in (0, 1). \quad (2.78)$$

Clearly, for (2.78) to be satisfied, it is sufficient that $q \gg v$ (Fig. 2.15).

Let us integrate equation (2.78)

$$a(s)s_y \equiv u_y = -[C(s) + K], K = const \geq 0$$

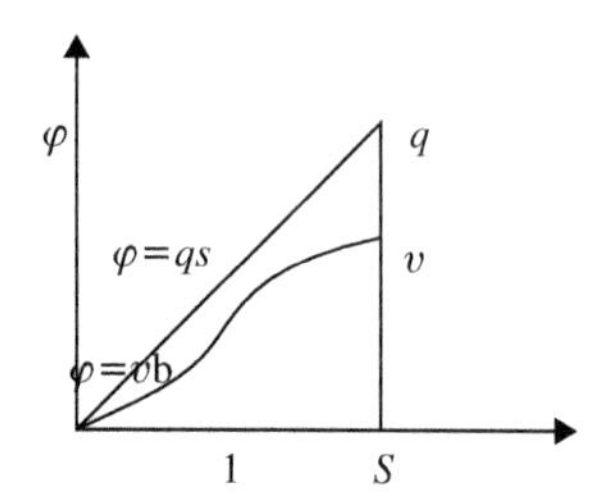

Figure 2.15 $q \gg v$.

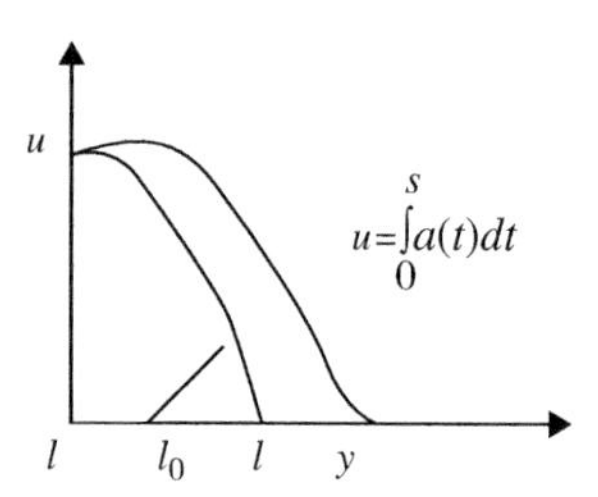

Figure 2.16 The shaded area in this figure is expressed by a formula inverse to 2.80: $s = \psi(y) \equiv [1 - \Phi(s, K)]^{[-1]}$.

and solve the function for $y = y(s)$:

$$y = l - \int_0^s \frac{a(t)dt}{C(t) + K} \equiv l - \Phi(s, K); \Phi(1, K) = l. \tag{2.80}$$

Let us assume that $\Phi(1, 0) = \int_0^1 \frac{a(t)dt}{C(t)} \equiv l_0$.

The properties of the coefficients of (2.78) provide $a = \gamma_0(s)s^\alpha(1-s)^\beta$, $(\alpha, \beta) > 1, b_s = a[s(1-s)^{-1}]\gamma_1(s), |\ln \gamma_i| \leq M, i = 0, 1; C_s = q - vb_s$. Therefore $\Phi(s, K) < \infty$ at $s \in [0, 1], K < \infty$ (including the stationary case $-q = 0$), $\Phi(s, \infty) = 0, \frac{\partial \Phi(s,K)}{\partial K} < 0, s \in (0, 1]$.

A. Let $l \leq l_0$. In this case, based on the properties of $\Phi(s, K)$, there exists a K such that $\Phi(1, K) = l$. This means that in these circumstances there exists a classical solution of (2.78), expressed by a formula inverse to (2.80): $s = \psi(y) \equiv [1 - \Phi(s, K)]^{[-1]}$ (the shaded area of Fig. 2.16).

B. Let $l > l_0$. In this case, the generalized solution of (2.78) takes the form

$$\begin{cases} s = \psi(y) = [l - \Phi(s, 0)]^{[-1]}, y \in [0, l_0], \\ s = 0, y \in [l_0, l]. \end{cases} \tag{2.81}$$

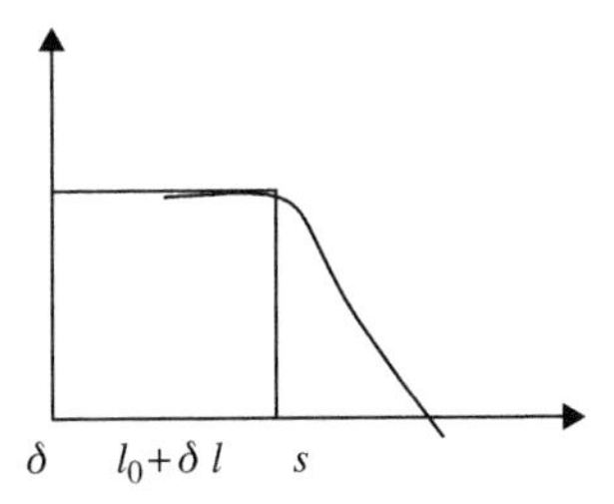

Figure 2.17 The shaded area in this figure can be moved to the right $\delta \in [0, l - l_0]$.

As $\frac{du}{dy}|_{y=l_0} = C[s(l_0)] = 0$, the function $u = \int_0^{s(y)} a(t)dt$, where s = s(y) is expressed by the formula (2.81), is in fact a generalized solution of (2.78).

2.5.12.2 Wave Velocity Equal to Flow Rate($q = v$), The Soliton Equation

Let the condition (2.79) be satisfied as before. This condition is equivalent to the assumption $b_s < 1, s \in (0, 1)$ when $q = v$. In this case $C(1) = q - b(1)v = 0$, and the solution of (2.78) can be written down as

$$\begin{cases} s \equiv \psi(y - \delta); y \in [\delta, l_0 + \delta]; \\ s \equiv 1, y < \delta \leq l - l_0; s = 0, y > l_0 + \delta, \end{cases} \tag{2.82}$$

where $\psi(\xi)$ is as determined in (2.81). This means that the shaded area in Fig. 2.17 can be moved to the right $\delta \in [0, l - l_0]$ and that the corresponding function $u(y) = \int_0^{s(y)} a(t)dt$, with $s(y)$ determined in (2.81), is a generalized solution of (2.78), since $as_y|_{s=0,1} = u_y|_{s=0,1} = C(s)|_{s=0,1} = 0$.

The solution of $u(s)$ is a soliton equation (a bank) which is represented in (2.82) by a moving profile $s = \psi(y - \delta), y \in [\delta, l_0 + \delta], \forall \delta > 0$.

2.5.12.3 Flow Rate Greater than Wave Velocity ($v > q$)

In this case, $C(1) = q - v < 0$. Taking $b_s \geq 0, b_s(0) = 0$ and bearing in mind that usually $b_s(1) = 0$, in the neighborhood of points $C_s = q - vb_s \geq 0$. Consequently, function $\varphi = C(s)$ has the appearance shown in Fig. 2.18.

Let $\min_{0 \leq s \leq 1} C(s) = C(s_0) < 0, 0 < s_0 < 1$. Let us assume that $\overline{C}(s) = C(s) - C(s_0) \geq 0, s \in [0, 1]$. In that case, the solution of (2.78) for the inverse function $y = y(s)$ can be written down as

$$y = l - \int_0^s \frac{a(t)dt}{C(t) + K} \equiv l - \overline{\Phi}(s, K); \overline{\Phi}(1, K) = l. \tag{2.83}$$

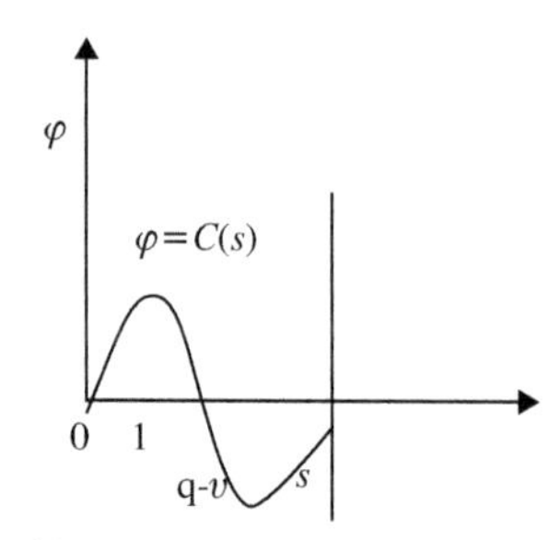

Figure 2.18 Function $\varphi = C(s)$.

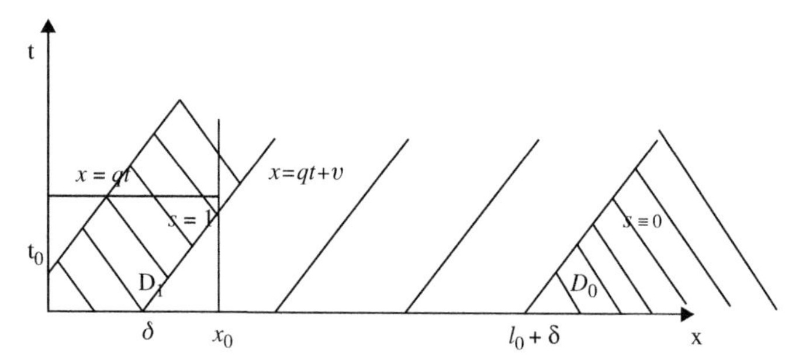

Figure 2.19 A bank.

As $\overline{\Phi}(1,0) = \infty\,(\overline{C}(s_0) = 0)$ and $\overline{\Phi}(1,\infty) = 0$, there exists a $K > 0$ such that $\overline{\Phi}(1,K) = l$ in which case $s(y) = [1 - \overline{\Phi}(s,K)]^{[-1]}$ represents a classical solution of (2.78).

2.5.13 Traveling Wave Solutions

Unlike parabolic analytical solutions, in traveling wave solutions the plane of the variables (x, t) can be covered by a family of straight lines $x = qt + y$, where y is the analytical variable. Fig. 2.19 shows a bank, which is one of the most interesting traveling wave families plotted in Section 2.5.12.2.

In the shaded regions D_0 and $D_1, s \equiv const$. Region D_1, which represents a **finite stabilization time is of** particular importance: for each section $x = x_0 > \delta$ there exists a finite time $t_0 = (x_0 - \delta)q^{-1}$, when $s(x_0, t) \equiv 1$ at $t \geq t_0$.

2.6 AN ANALYTICAL SOLUTION OF FLOW EQUATIONS FOR TWO NONLINEARLY VISCOUS FLUIDS

2.6.1 Introduction

Mathematical modelling of fluid flow through porous media uses the laws of conservation of mass, momentum and empirical Darcy's Law which links velocity vectors with pressure gradients [107]. Many researchers have studied models which do not use Darcy's Law. Let us consider one-dimensional isothermal flow of two viscous fluids in a nondeformable porous medium. In the absence of phase transitions, this model's equations take the form [27, 61].

$$\frac{\partial}{\partial t}(\rho_i s_i) + \frac{\partial}{\partial x}(\rho_i s_i v_i) = 0, s_1 = ms, s_2 = m(1-s),$$

$$\rho_i s_i\left(\frac{\partial v_i}{\partial t} + v_i\frac{\partial v_i}{\partial x}\right) = \frac{\partial s_i\sigma_i}{\partial x} + f_i + \rho_i s_i g, \sigma_i = -p_i + \mu_i\frac{\partial v_i}{\partial x},$$

$$f_i = p_i\frac{\partial s_i}{\partial x} + \varphi_i, p_1 - p_2 = p_c(s) \ \ (i = 1,2),$$

where ρ_i is the density of the ith phase; v_i is its true velocity; m is porosity; s is the phase saturation of the pore space with the first fluid; p_i is phase pressure; $p_c(s)$ is capillary pressure; μ_i is the dynamic viscosity coefficient; f_i is phase interaction force; φ_i is interfacial friction force, $\varphi_1 = -\varphi_2 = K(v_2 - v_1)$; K is the phase interaction coefficient; g is the acceleration of gravity. The condition $\rho_i = const$ leads to a closed system of equations for s_i, v_i, p_i:

$$\frac{\partial s_i}{\partial t} + \frac{\partial}{\partial x}(s_i v_i) = 0, \tag{2.84}$$

$$\rho_i s_i\left(\frac{\partial v_i}{\partial t} + v_i\frac{\partial v_i}{\partial x}\right) - \frac{\partial}{\partial x}\left(\mu_i s_i\frac{\partial v_i}{\partial x}\right) = -s_i\frac{\partial p_i}{\partial x} + \varphi_i + \rho_i s_i g, \tag{2.85}$$

$$s_1 + s_2 = m, \varphi_1 = K(v_2 - v_1) = -\varphi_2, p_1 - p_2 = p_c(s). \tag{2.86}$$

where $K = const > 0, \mu_i = \overline{\mu}_i|\partial v_i/\partial x|^{\alpha}, \overline{\mu}_i = const > 0, \alpha = const \geq 0$. It is also assumed that, $m = 1, g = 0, p_c(s) = 0, p_1 = p_2 \equiv p$.

In this section, we consider a traveling wave analytical solution of the system (2.84)–(2.86). Assuming that all the target functions depend only on the variable $\xi = x - ct$ (where c is a constant parameter), we obtain the following equation system:

$$(\rho_i v_i - c\rho_i)' = 0, \rho_i(v_i v_i' - cv_i') - (\mu_i s_i v_i')' = -s_i p' + \varphi_i \tag{2.87}$$

$(i = 1,2)$, in which the prime indicates differentiation in ξ. We consider system (2.86), (2.87) for $\xi > 0$, augmenting it with the boundary conditions

$$v_i(0) = v_i^0, s_1(0) = s^0, \lim_{\xi\to\infty} v_i(\xi) = u^+, \lim_{\xi\to\infty} s_1(\xi) = s^+ \ (i = 1,2), \tag{2.88}$$

where v_1^0, v_2^0, s^0, s^+ are specified constants which satisfy the conditions $s^0 \neq s^+, v_1^0 \neq v_2^0$. Because it follows from (2.87) that $s_i(v_i - c) = A_i, i = 1, 2,$ then by bringing in (2.88) we can arrive at the following equation system for the unknown constants A_1, A_2, u^+, c:

$$s^0(v_1^0 - c) = A_1, s^+(u^+ - c) = A_1, (1 - s^0)(v_2^0 - c) = A_2, (1 - s^+)(u^+ - c) = A_2.$$

This system is solved using the formulae

$$c = \frac{s^+(1 - s^0)v_2^0 - s^0(1 - s^+)v_1^0}{s^+ - s^0}, u^+ = s^0 v_1^0 + (1 - s^0)v_2^0,$$

$$A_1 = s^0(1 - s^0)(v_1^0 - v_2^0)\frac{s^+}{s^+ - s^0}, A_2 = \frac{s^+}{s^+ - s^0}A_1.$$

Adding the momentum equations for $i = 1$ and 2 to one another, we arrive at the equation for $p(\xi)$:

$$p'(\xi) = \sum_{i=1}^{2}(v_i\rho_i(c - v_i) + \mu_i s_i v_i')'. \tag{2.89}$$

Taking (2.89) into consideration, we can use (2.88) to derive the equation for $s(\xi)$:

$$\lambda^\alpha\left[a_1(s)(|s'|^\alpha s')' + \left(\frac{2\alpha + 1}{2\alpha + 2}\right)|s'|^\alpha|s'|^2\frac{da_1}{ds}\right] + \lambda a_2(s)s' - Ka_3(s)(s - s^+) = 0, \tag{2.90}$$

where

$$a_1(s) = \frac{\overline{\mu}_1(s^+)^{\alpha+1}}{s^{2\alpha+2}} + \frac{\overline{\mu}_2(1-s^+)^{\alpha+1}}{(1-s)^{2\alpha+2}}, \lambda = \frac{s^0(1 - s^0)(v_1^0 - v_2^0)}{s^0 - s^+},$$

$$a_2(s) = \frac{\rho_1(s^+)^2}{s^3} + \frac{\rho_2(1-s^+)^2}{(1-s)^3}, a_3(s) = \frac{1}{s^2(1-s)^2}.$$

Equation (2.90) is the complemented with the relevant conditions for $s(\xi)$ taken from (2.88). Our aim in this section is to prove the existence of a unique solution of this problem for $s(\xi)$.

2.6.2 Problem Solvability

In what follows, we will assume that conditions

$$\lambda > 0, s^+ \neq s^0, (s^+, s^0) \in (0, 1). \tag{2.91}$$

required by these equations have been satisfied.

Our proof of the existence of a solution is based on Schauder's theorem [138] and uses standard auxiliary constructions. Having designated $a(s) = a_1^{(2\alpha+1)/(2\alpha+2)}(s)$, let us rewrite equation (2.90) in a divergent form:

$$\lambda^{\alpha}\left[a(s)|s'|^{\alpha}s'\right]' + \lambda \frac{a_2(s)}{a^{1/(2\alpha+1)}(s)}s' - K\frac{a_3(s)}{a^{1/(2\alpha+1)}(s)}(s - s^+) = 0. \quad (2.92)$$

Assuming that in (2.92) $u(s(\xi)) = \int_{s^+}^{s(\xi)} a^{\chi}(\tau)d\tau$, where $\chi = 1/\alpha + 1 > 0$, we obtain

$$(|u'|^{\alpha}u')' + \lambda_0 b(u)u' - \lambda^{-\alpha}Kd(u)u = 0, u(0) = \int_{s^+}^{s^0} a^{\chi}(\tau)d\tau \equiv u_0, u(\infty) = 0, \quad (2.93)$$

where $\lambda_0 = \lambda^{1-\alpha}, b(u) = \frac{a_2(s(u))}{a^{1/(2\alpha+1)}(s(u))}, d(u) = \frac{a_3(s(u))}{a^{1/(2\alpha+1)}(s(u))} \cdot \frac{s(u) - s^+}{u}$.

It follows from the conditions of the problem that the constants K and λ are positive, while functions $d(s), a(s)$ and $a_3(s)$ are positive for all $s \in (0, 1)$. In whatfollows, we will assume that $s^+ < s^0$ (the case of $s^+ > s^0$ will be considered in a similar way).

Examining the intercept $[0, n]$, let us consider the auxiliary problem for $v(\xi) = \int_{s^+}^{s(\xi)} a^{\chi}(\tau)d\tau$:

$$(|v'|^{\alpha}v')' + \lambda_0 b(v)v' - \lambda^{-\alpha}Kd(v)v = 0, v(0) = u_0, v(n) = 0. \quad (2.94)$$

The maximum principle means that the solutions of this equation satisfy the inequalities $u_0 \geq v(\xi) \geq 0$ for all $\xi \in [0, n)$. Consequently, the functions $b(v)$ and $d(v)$ are strictly positive and bounded. Let us assume that

$$a_0(\tau) = [\overline{\mu}_1(s^+)^{\alpha+1}(1-\tau)^{2(\alpha+1)} + \overline{\mu}_2(1-s^+)^{\alpha+1}\tau^{2(\alpha+1)}]^{(2\alpha+1)/(2\alpha+2)},$$

$$b_0(\tau) = (\rho_1(s^+)^2(1-\tau)^3 + \rho_2(1-s^+)^2\tau^3)/{a_0}^{1/(2\alpha+1)}.$$

In that case

$$a(\tau) = \frac{a_0(\tau)}{[\tau(1-\tau)]^{2\alpha+1}}, b(\tau) = \frac{b_0(\tau)}{\tau^2(1-\tau)^2}, \tau d(\tau) = \frac{\tau - s^+}{a_0^{1/(2\alpha+1)}(\tau)\tau(1-\tau)}.$$

Let α_1, α_2- be the minimum values of the functions $a_0(\tau), b_0(\tau)$, and β_1, β_2 their maximum values at $\tau \in [0, 1]$ (clearly, α_i, β_i are strictly positive and depend only on $\overline{\mu}_i, \rho_i, s^+, i = 1, 2$). This results in

$$\alpha_1^* = \frac{\alpha_1}{[s^0(1-s^+)]^{2\alpha+1}} \leq a(v) \leq \frac{\beta_1}{[s^+(1-s^0)]^{2\alpha+1}} = \beta_1^*, \quad (2.95)$$

$$\alpha_2^* = \frac{\alpha_2}{[s^0(1-s^+)]^2} \leq b(v) \leq \frac{\beta_2}{[s^+(1-s^0)]^2} = \beta_2^*, \tag{2.96}$$

$$\alpha_3^* = \frac{1}{s^0(1-s^+)(\beta_1)^{1/(2\alpha+1)}} \leq \frac{vd(v)}{s(v)-s^+} \leq \frac{1}{s^+(1-s^0)(\alpha_1)^{1/(2\alpha+1)}} = \beta_3^*. \tag{2.97}$$

The derivative of the solution of (2.94) at the point $\xi = n$ is nonpositive, since in view of the boundary condition $v(n) = 0$, the assumption that $v'(n) > 0$, contradicts the non-negativeness of $v(\xi)$. Rewriting (2.94) in the form

$$\varnothing' = \lambda^{-\alpha} Kvd(v) \geq 0, \varnothing = |v'|^\alpha v' + \lambda_0 \int_0^v b(\tau)d\tau, \tag{2.98}$$

we conclude that the monotonically increasing function $\varnothing(\xi)$ is nonpositive. Therefore, $v'(\xi) \leq 0$ for all $\xi \in [0, n]$.

It follows from (2.98) and (2.95)–(2.97) that $|v'|^\alpha v' + \lambda_0 \alpha_2^* v \leq 0$. Multiplying this inequality by $v_0^\alpha > 0$ and taking into account that $v^{\alpha+1} \leq v_0^\alpha v$, we obtain the inequality $v_0^\alpha |v'|^\alpha v' + \lambda_0 \alpha_2^* v^{\alpha+1} \leq 0$. In that case, $v' + \beta_0 v \leq 0$, where $\beta_0 = (\lambda_0 \alpha_2^*/v_0^\alpha)^{1/(\alpha+1)}$.

We can now obtain

$$v(\xi) \leq v_0 \exp(-\beta_0 \xi). \tag{2.99}$$

Let us rewrite equation (2.94) in the form

$$\left(|v'|^\alpha v' + \lambda_0 \int_0^v b(\tau)d\tau\right)' - \lambda^{-\alpha} Kvd(v) = 0$$

and integrate ξ from 0 to the current value ξ:

$$|v'(\xi)|^\alpha v'(\xi) - |v'(0)|^\alpha v'(0) = \lambda_0 \int_{v(\xi)}^{v_0} b(\tau)d\tau + \lambda^{-\alpha} K \int_0^\xi \tau d(\tau)d\tau \equiv \varphi(\xi) \geq 0.$$

This gives $|v'(0)| - \varphi^{1/(\alpha+1)}(\xi) \leq (|v'(0)|^{\alpha+1} - \varphi(\xi))^{1/(\alpha+1)} = -v'(\xi)$. Integrating this inequality over ξ from 0 to n, we obtain

$$|v'(0)| \leq \frac{1}{n}\left(\int_0^n \varphi^{1/(\alpha+1)}(\tau)d\tau + v_0\right) \equiv N < \infty. \tag{2.100}$$

As $\varnothing(0) \le \varnothing(\xi) \le \varnothing(n)$, we have

$$|v'(0)|^{\alpha}v'(0) + \lambda_0 \int_0^{v_0} b(\tau)d\tau \le |v'(\xi)|^{\alpha}v'(\xi) + \lambda_0 \int_0^{v(\xi)} b(\tau)d\tau.$$

From this it follows that for all $\xi \in [0, n]$

$$|v'(\xi)| \le |v'(0)| \le N. \tag{2.101}$$

By virtue of (2.100), (2.101)

$$y \equiv |v'(\xi)|^{\alpha+1} \le |v'(0)|^{\alpha}|v'(\xi)| \le N^{\alpha}y^{1/(\alpha+1)}. \tag{2.102}$$

Let us rewrite equation (2.94) in the form

$$y' + \lambda_0 b(v)y^{1/(\alpha+1)} + \lambda^{-\alpha}Kd(v)v = 0.$$

Taking into account (2.102), we arrive at the inequality $y' + b_0 y \le 0$, from which it follows that

$$|v'(\xi)| \le N\exp(-b_0\xi), \tag{2.103}$$

where $b_0 > 0$ depends on the data specified in the problem, and does not depend on n.

Let us rewrite the solution of (2.95) in the form

$$v(\xi) = v_0 - \int_0^{\xi} (|v'(0)|^{\alpha+1} - \varphi(\tau))^{1/(\alpha+1)}d\tau \equiv T(v), \tag{2.104}$$

where $|v'(0)|$ is calculated from

$$\int_0^{n} (|v'(0)|^{\alpha+1} - \varphi(\xi))^{1/(\alpha+1)}d\xi = v_0.$$

Examining the space of continuous functions $C[0, n]$, let us consider the closed, bounded convex set $M = \{v(\xi)|0 \le v(\xi) \le u_0, \xi \in [0, n]\}$. The operator T is determined on the set M, and the maximum principle leads to the nesting $T(M) \subset M$. The continuity of T can be checked directly by means of (2.104). It follows from estimates (2.99), (2.101), (2.103) that T is compact. Therefore, in accordance with Schauder's theorem, (2.94) has at least one solution in the set M. This solution is unique if $(vd(v))'_v > 0$. Indeed, let $f(\xi)$ be a sufficiently smooth function determined in the interval $[0, n]$ and equal to zero at $\xi = 0$ and $\xi = n$. Let us multiply both parts of equation (2.94) by $f(\xi)$ and integrate the resultant equality over ξ from zero to n, in a process of single integration by parts. This will yield the following integral equality:

$$\int_0^n \left(|v'(\xi)|^\alpha v'(\xi)f' + \lambda_0 f' \int_0^v b(z)dz + \lambda^{-\alpha} Kfvd(v) \right) d\xi = 0. \qquad (2.105)$$

Let v_1 and v_2 be two different solutions of (2.94). Let us assume that $A = |v_1'(\xi)|^\alpha v_1'(\xi) - |v_2'(\xi)|^\alpha v_2'(\xi)$. Hence $A = (v_1 - v_2)' B(v_1', v_2')$, where $B(v_1', v_2') = |v_2'(\xi)|^\alpha + \alpha |v_1'(\xi)| \theta^{\alpha-1}, \theta \in [|v_1'|, |v_2'|] \vee \theta \in [|v_2'|, |v_1'|]$. Let us also assume that $U = v_1 - v_2$, $U(0) = U(n) = 0$. Following integration by parts, (2.105) yields

$$\int_0^n U[(Bf')' - \lambda_0 B_1(\xi)f' - \lambda^{-\alpha} KB_2(\xi)f]d\xi = 0,$$

where

$$B_1(\xi) = U^{-1} \int_{v_1}^{v_2} b(z)dz, \, B_2(\xi) = U^{-1}(v_1 d(v_1) - v_2 d(v_2)) > 0.$$

Let us determine $f(\xi)$ as the solution of the linear problem $(Bf')' - \lambda_0 B_1(\xi)f' - \lambda^{-\alpha} B_2(\xi)f = h(\xi)$, $f(0) = f(n) = 0$, where $h(\xi)$-is an arbitrary continuous function. As stated in [138], and taking into account the boundedness and positive nature of B, this problem is solvable for any continuous right hand side, and therefore $U = 0$.

Let us now solve (2.88), (2.93) for an infinite interval, producing the solution in the form of a limit of the sequence $\{v_n(\xi)\}$ of solutions $v_n(\xi)$ of (2.94) for $n \to \infty$, and using estimates of (2.99), (2.101) and (2.103) which do not depend on n. Because the solutions of (2.94) are unique, the bounded sequence $\{v_n(\xi)\}$ increases monotonically and thus converges to a function $u(\xi)$. By performing passages to the limit on the equalities (2.104) written down for $v_n(\xi)$, we arrive at a similar equality for the limiting function. This means that $u(\xi)$ represents a classical solution of (2.88), (2.93). The asymptotic behaviour of the solution is due to the inequalities (2.99), (2.101), (2.103).

Let us formulate sufficient conditions of uniques of the solution of (2.94) in terms of the initial data of (2.88), (2.92). Condition $(vd(v))' > 0$ is equivalent to

$$\frac{r(s)}{s^{2\alpha+5}(1-s)^{2\alpha+5} a_1^{(2\alpha+3)/(2\alpha+2)}(s)} > 0.$$

In this fraction, the denominator is always positive for $s \in (0, 1)$, and the numerator takes the form $r(s) = \overline{\mu}_1 (s^+)^{\alpha+1} (1-s)^2 g(s, s^+) + \overline{\mu}_2 (1-s^+)^{\alpha+1} s^2 g(1-s, 1-s^+)$, where $g(\tau, \eta) = 2\tau^2 - 3\eta\tau + \eta = 2\tau(\tau - \eta) + \eta(1-\tau)$. Note that $g(\tau, \eta) > 0$ when $\eta < 8/9, \tau \in (0, 1)$. Let $s^0 > s^+$, in which case

$g(s, s^+) > 0$. For $g(1-s, 1-s^+)$ to be positive, it is sufficient to require $1-s^+ < 8/9$, to be satisfied, i.e. $s^+ > 1/9$, in which case $r(s) > 0$. In the same way, when $s^0 < s^+, g(1-s, 1-s^+) > 0$. For $g(s, s^+)$ to be positive, it is sufficient to require that $s^+ < 8/9$ and $r(s) > 0$.

We have therefore proved the following assertion:

Theorem When conditions $s^+ \in \left(\frac{1}{9}, \frac{8}{9}\right)$ and (2.91) are satisfied, there exists a unique classical solution of (2.88), (2.93).

2.7 VALIDATION OF THE USE OF A SPECIFIC APPROXIMATE METHOD IN TWO-PHASE NON-ISOTHERMAL FLOW

2.7.1 A One-dimensional Model

In the case of two-phase non-isothermal flow in a homogeneous medium, the model's one-dimensional equations have the form

$$\begin{aligned} &\theta_t = (\lambda\theta_x - v\theta)_x, \; ms_t = (a_1 s_x + a_2\theta_x - vb)_x, \\ &v_x \equiv -(k(P_x + a_3\theta_x))_x = 0, \end{aligned} \tag{2.106}$$

where x is distance; t is time; θ is temperature; s is the saturation of the aqueous phase of the mixed flow; p is pressure; v is mixed flow rate; $m = const; a_1, a_2, a_3, k, b, \lambda$ are the preset class C^2 functions of θ and s. Furthermore, $|\ln(\lambda, a_1, k)| \leq M = const$, i.e. the case we are considering is nondegenerative. Without restriction of generality, let us assume that $m \equiv 1$.

Antimontsev, Papin and Kruzhkov [6, 76] have examined the application of approximate methods to this model in the case of isothermal flow. Their application to nonisothermal flow is discussed in Section 2.1. In this section, we examine an approximate method of solving the first boundary value problem for (2.106), which is the same as that described in [6, 76], but applies to nonisothermal flow, and assumes unidimensionality and a preset flow rate. These assumptions have enabled us to prove the convergence of the approximate solution to the exact one.

2.7.2 The Analytical Variables

On the assumption that $v \equiv v(t) = \dfrac{q}{\sqrt{t+1}}, q = const$, equations (2.106) permit analytical solutions having the form $s(y), \theta(y), y = x/\sqrt{t+1}$, which satisfy the equation system:

$$(\lambda\theta_y - q\theta)_y + (1/2)y\theta_y = 0, (a_1 s_y + a_2\theta_y - qb)_y + \left(\frac{1}{2}\right)ys_y = 0. \tag{2.107}$$

Assuming that $v \equiv v(t) = \dfrac{q}{\sqrt{t+1}}$, $q = const$, replace the variables in equations (2.106) as follows: $\overline{x} = \dfrac{x}{\sqrt{t+1}}$, $\overline{t} = \ln(t+1)$.

Let us now formulate the initial boundary value problem for the region $R = \{(x,t): x \in [0,X], t \in [0,T]\}$: in new variables (the vinculum over t and x has been omitted):

$$\theta_t = (\lambda\theta_x - q\theta)_x + \left(\frac{1}{2}\right)x\theta_x,\ s_t = (a_1 s_x + a_2\theta_x - qb)_x + (1/2)xs_x,$$

$$\theta(0,t) = \theta_1,\ \theta(X,t) = \theta_2,\ \theta(x,0) = \theta_0(x), \tag{2.108}$$

$$s(0,t) = s_1,\ s(X,t) = s_2,\ s(x,0) = s_0(x).$$

Note that the solutions of (2.107) [analytical solutions of (2.106)] are stationary solutions of (2.108). The theorem of existence of generalized solutions of (2.108) was proved in [15], and the solution was further smoothed for some conditions in [44]. In what follows, we shall assume that these conditions have been satisfied.

In the case of (2.108), we would propose the following approximate method [6, 44]. Let us divide the time interval $[0, T]$ into N sections ($\tau = T/N$) and let us solve the problem with respect to $\theta_{i+1}(x,t)$, $s_{i+1}(x,t)$: separately for each section $I_i \equiv [i\tau, (i+1)\tau](i = 0, \ldots, N-1)$:

$$L_1\theta_{i+1} \equiv -(\theta_{i+1})_t + [(\lambda^{(i)}\theta_{i+1})_x - q\theta_{i+1}]_x + \frac{1}{2}x(\theta_{i+1})_x = 0,$$

$$L_2 s_{i+1} \equiv -(s_{i+1})_t + [a_1^{(i)}(s_{i+1})_x + a_2^{(i)}(\theta_{i+1})_x - qb^{(i)}]_x + \frac{1}{2}x(s_{i+1})_x = 0, \tag{2.109}$$

$$\theta_{i+1}(0,t) = \theta_1,\ \theta_{i+1}(X,t) = \theta_2,\ \theta_{i+1}(x, i\tau) = \theta^{(i)}(x),$$

$$s_{i+1}(0,t) = s_1,\ s_{i+1}(X,t) = s_2,\ s_{i+1}(x, i\tau) = s^{(i)}(x).$$

Here $\theta^{(i)}(x) \equiv \theta_i(x, i\tau)$, $\theta^{(0)}(x) \equiv \theta_0(x)$, $s^{(i)}(x) \equiv s_i(x, i\tau)$, $s^{(0)}(x) \equiv s_0(x)$ and if $a = a(\theta, s)$, then $a^{(i)} = a(\theta^{(i)}(x), s^{(i)}(x))$.

For each section I_i (2.109), there exists a system of linear uniformly parabolic equations with smooth coefficients for which the general theory of parabolic equations postulates the existence of a classical solution. Let us introduce the designations $\Omega = [0, X]$; $I_i^* = [i\tau, t_1]$, t_1-an arbitrary point selected from $I_i/\{i\tau\}$; $\theta_\tau(x, t) = \theta_{i+1}(x, t)$; $s_\tau(x, t) = s_{i+1}(x, t)$ for $x \in \Omega, t \in I_i$ $(i = 0, \ldots, N-1)$. In this way, we have determined functions $\theta_\tau(x, t)$ and $s_\tau(x, t)$ for the whole of the region R.

Let us make use of the designations of functional spaces and norms adopted in [83]. All C, C_k, C_{kl}-type constants are independent of τ.

2.7.3 Convergence of the Family of Functions (θ_τ, s_τ)

Theorem When $\tau \to 0$, the functions $(\theta_\tau(x, t), s_\tau(x, t))$ converge to a classical solution $(\theta(x, t), s(x, t))$ of (2.108),with the rate of convergence estimated as follows:

$$\|\theta - \theta_\tau\|_{V_2} + \|s - s_\tau\|_{V_2} \leq C\tau, \tag{2.110}$$

$$\|\theta - \theta_\tau\|_\infty + s - \|s_\tau\|_\infty \leq C\tau^\beta, \beta \in (0, 1). \tag{2.111}$$

Proof Examining the region $\Omega \times I_i$ let us subtract the first equation in (2.109) from the first equation in (2.108), and also the second equations in each case. Let us designate $\omega = \theta - \theta_\tau$, $u = s - s_\tau$, $\omega_i(x) = \theta(x, i\tau) - \theta^{(i)}(x)$, $u_i(x) = s(x, i\tau) - s^{(i)}(x)$. This will yield the following problem for $\vec{\omega} = (\omega, u), \vec{\omega}_i = (\omega_i, u_i)$, as described in [76, formulae (80), (81)]:

$$-\vec{\omega}_t + (A(x, t)\vec{\omega}_x + B(x, t)\vec{\omega}_{ix} - Q\vec{\omega})_x + 0,5x\vec{\omega}_x + \vec{c}(x, t)\tau = 0,$$

$$\vec{\omega}(0, t) = (0, 0), \vec{\omega}(X, t) = (0, 0), \vec{\omega}(x, 0) = (0, 0), \tag{2.112}$$

$$\vec{\omega}(x, i\tau) = (\theta(x, i\tau) - \theta_\tau(x, i\tau), s(x, i\tau) - s_\tau(x, i\tau)),$$

where the components of matrices A, B and the vector $\vec{c}$ are complex functions of $\theta(x, i\tau), s(x, i\tau), \theta^i(x), s^i(x), \theta(x, t), s(x, t), \theta_x(x, t), s_x(x, t)$ and of the coefficients of (2.108) and (2.109). In the above equation, $A_{12} = Q_{12} = Q_{21} = Q_{22} = 0$, $Q_{11} = q$.

Let us multiply the equation for $\omega(x, t)$ in (2.112) by $\omega(x, t)$ and integrate it first over Ω, and then over I_i^*. Using integration by parts (and taking boundary conditions into account), applying the Cauchy inequality with e and selecting the appropriate constants C_{kl}, we obtain the following inequality [76, formula (87)]:

$$\|\omega(x,t_1)\|_2^2 + C_{11}\int_{I_i^*}\|\omega_x\|_2^2 dt \le (1 + C_{12}\tau)\|\omega_i\|_2^2 + C_{13}\int_{I_i^*}\|u_i\|_2^2 dt + C_{14}\int_{I_i^*}\|\omega\|_2^2 dt + C_{15}\tau^3. \tag{2.113}$$

In the same way, by multiplying the equation for $u(x,t)$ in (2.112) by $u(x,t)$, we obtain the inequality

$$\|u(x,t_1)\|_2^2 + C_{21}\int_{I_i^*}\|u_x\|_2^2 dt \le (1 + C_{22}\tau)\|u_i\|_2^2 + C_{23}\int_{I_i^*}\|\omega_i\|_2^2 dt + C_{24}\int_{I_i^*}\|\omega_i\|_2^2 dt + C_{25}\int_{I_i^*}\|u\|_2^2 dt + C_{26}\tau^3, \tag{2.114}$$

Using (2.113) from inequality (2.114), we obtain

$$\|u(x,t_1)\|_2^2 + C_{31}\int_{I_i^*}\|u_x\|_2^2 dt \le (1 + C_{32}\tau)\|u_i\|_2^2 + (1 + C_{33}\tau)\|\omega_i\|_2^2 + C_{34}\int_{I_i^*}(\|u\|_2^2 + \|\omega\|_2^2)dt + C_{35}\tau^3. \tag{2.115}$$

Adding (2.113) and (2.115), discarding the integrals from the left hand side of the resultant inequality and designating $z_i = z(i\tau), z(t_1) = \|\omega(x,t_1)\|_2^2 + \|u(x,t_1)\|_2^2$, we obtain $z(t_1) \le (1 + C_1\tau)z_i + C_2\int_{I_i^*} z(t)dt + C_3\tau^3$. Assuming that $t_1 = \tau(i+1)$, from Gronwall's lemma we obtain

$z(t) \equiv z_{i+1} \le [(1 + C_1\tau)z_i + C_3\tau^3]\exp(C_2\tau)$, whence $z_{i+1} \le (1 + C_4\tau)z_i + C_5\tau^3$. From there, it is easily demonstrated that $z_{i+1} \le C_5\tau^3[1 + (1 + C_4\tau) + (1+C_4\tau)^2 + \cdots + (1+C_4\tau)^i] = C_5\tau^2(((1+C_4\tau)^{i+1} - 1)/C_4) \le C_6\tau^2$.

Returning to designations (ω, u), we obtain

$$\sup_{0\le t\le T}(\|\omega\|_2^2 + \|u\|_2^2) \le C_6\tau^2. \tag{2.116}$$

From (2.113), (2.115), (2.116) we derive

$$\int_{I_i^*}(\|\omega_x\|_2^2 + \|u_x\|_2^2)dt \le C_7\tau^2. \tag{2.117}$$

From (2.116), (2.117) we obtain an estimate of the convergence rate of (2.110), and from (2.110) and the interpolation inequality $\sup_Q|z| \le C(\beta)(\|z\|_{V_2})^{\beta}(\|z\|_{C^\alpha})^{1-\beta}, \beta\in(0,1)$ we obtain estimate (2.111).

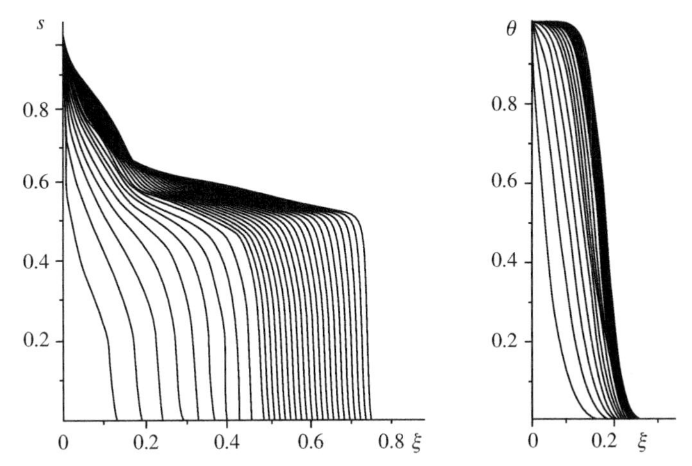

Figure 2.20 The computed distributions of saturation and dimensionless temperature $\overline{\theta} = \frac{(\theta - \theta_1)}{(\theta_2 - \theta_1)}$ during consecutive moments in time, up to the point when the solutions reach stationary profiles (the vinculum over θ has been omitted in the diagrams).

2.7.4 Numerical Calculations

We have carried out numerical calculations for problem (2.108), basing them on the substitution of the time derivative with backward difference and on the existing method of solving the analytical problem (2.108) [93, 99]. Fig. 2.20 shows the computed distributions of saturation s and dimensionless temperature $\overline{\theta} = \frac{(\theta - \theta_1)}{(\theta_2 - \theta_1)}$ during consecutive moments in time, up to the point when the solutions reach stationary profiles (the vinculum over θ has been omitted in the diagrams).

Fig. 2.21 shows analytical solutions of $s(y)$ (Curve 1) and $\overline{\theta}(y)$ (Curve 2) ($y = x/\sqrt{t+1}$) are shown. A comparison of Figs. 2.20 and 2.21 demonstrates that, as was to be expected, stationary solutions of (2.108) coincide with analytical solutions of (2.106). Therefore, while the numerical algorithm of the analytical solution of (2.107) forms an integral part of the method of solving (2.108), the analytical solution itself forms one of the tests of the numerical solution of (2.108).

2.8 COMBINATION OF THE PRINCIPAL MODELS OF TWO-PHASE FLUID FLOW

This section is devoted to the theoretical and numerical analysis of the problem of combining the two principal models of two-phase fluid flow: the Muskat-Leverett model (MLT model), which takes into account the discontinuity between the capillary pressures of the phases, and the

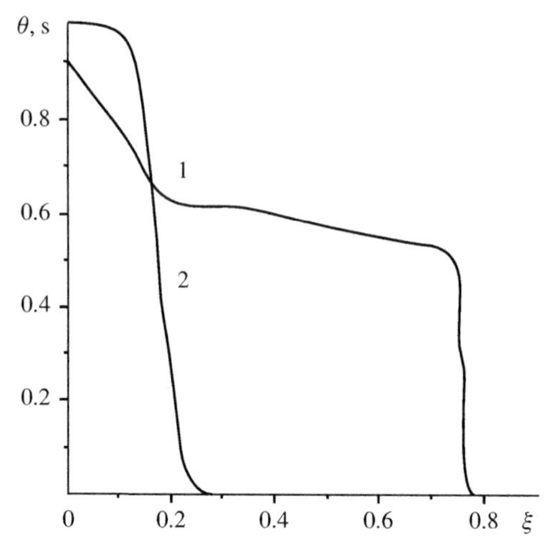

Figure 2.21 Analytical solutions of $s(y)$ (Curve 1) and $\overline{\theta}(y)$ (Curve 2) $(y = x/\sqrt{t+1})$.

Buckley-Leverett model (BL model), in which the phase pressures coincide. Problems of this type need to be examined when modelling the various stages of water drive, each with a different duration and sweep. For instance, in water-encroached parts of the formation, from which virtually all the mobile oil has been displaced, the effect of capillary forces on two-phase fluid flow is slight, and the BL model can be used. Where water encroachment is slight, capillary forces must be taken into account, and therefore the MLT model must be used.

Another situation requiring a combination of the two models is the so-called end effect problem which, in mathematical terms, amounts to the gradient of phase saturation being unlimited in the vicinity of production wells. To obtain boundary conditions capable of numerical implementation for a production well, it is usually assumed that phase pressures coincide in the near-wellbore zone.

Having made this assumption, and regarding the near-wellbore zone as fairly small, we can derive the conditions of proportionality of the phase flow rates and phase "mobilities".

2.8.1 The Combination of MLT and BL Models

In this section, we consider a one-dimensional MLT model of two-phase fluid flow in a homogeneous isotropic porous medium, in the absence of mass forces (Chapter 1, Section 1.1):

$$ms_{\overline{t}} = (Ka(s)s_{\overline{x}} - b(s)Q(\overline{t}))_{\overline{x}}.$$

Here s is the dynamic water saturation; m is porosity; $a = j'(s)k_1(s)k_2(s)(k_1+k_2)^{-1}$, $k_i = \bar{k}_i = \bar{k}_i\mu_i^{-1}$, $\bar{k}_i$ and μ_i are phase permeabilities and viscosities; $j(s)$ is the Leverett J function which reflects the presence of capillary forces; $b(s) = k_1(k_1+k_2)^{-1}$ is the "mobility" coefficient of the displacing phase (water); $K = k\sigma \cos\theta(m/k)^{1/2}$; k is permeability; θ is wetting angle; σ is interfacial tension.

Let us introduce the following new variables: $x = L^{-1}\bar{x}, t = (mL)^{-1}\int_0^{\bar{t}} Q(\xi)d\xi$ (L-characteristic length) and let us rewrite the initial equation in the form

$$s_t = (\lambda(t)a(s)s_x - b(s))_x \equiv -v_x, \tag{2.118}$$

where $\lambda = K(LQ(\bar{t}(t)))^{-1}$; $v = v(x,t)$ is the relative flow velocity (flow rate) of the displacing phase.

Let the line $\Gamma_0 = \{x, t|x = 0, t \geq 0\}$ represent the injection well, $\Gamma_1 = \{x, t|x = 1, t \geq 0\}$ the MLT and BL model combination line, and $\Gamma_l = \{x, t|x = l \geq 1, t \geq 0\}$ - the production well. Let us designate $\Omega_n = \{x, t|0 < x < n, t > 0\}$.

In region Ω_1, the two-phase fluid flow is described by the MLT model and therefore in (2.118) $\lambda \neq 0$, whereas in region $\Omega = (\Omega_l/\Omega_1)$ it is described by the BL model and therefore in (2.118) $\lambda \equiv 0$. To solve equation (2.118) for $s(x,t), (x,t) \in \Omega_l$ and the resultant function $\lambda(t)$ we propose to examine the initial boundary value problem

$$s|_{x=0} = 1; s|_{t=0} = s_0(x); x \in [0, l]; [s]_{x=1} = [v]_{x=1} = 0, \tag{2.119}$$

where $[f]_{x=1} = \{f(1-0,t) - f(1+0,t)\}$ is the discontinuity of the function $f(x,t)$ on $\Gamma_1 = \{x, t|x = 1, t \geq 0\}$.

If the function $s(x,t)$ is smooth, then the equality $as_x|_{x=1-0} = 0$ follows from condition (2.119) along the model combination line Γ_1. As a result, the problem divides into two problems, one in Ω_1 and other in $\Omega = (\Omega_l/\Omega_1)$:

$$Ls \equiv s_t - (\lambda as_x - b)_x = 0; s|_{t=0} = s_0(x), s|_{x=0} = 1, s_x|_{x=1} = 0; \tag{2.120}$$

$$v_x + \varphi_{0t} = 0; v|_{t=0} = v_0, v|_{x=1} = v_1(t). \tag{2.121}$$

Here, $\varphi_0(v) = b^{(-1)}(v), s = b^{(-1)}(v)$ is the inverse of $v = b(s), v_0 = b[s_0(1)], v_1(t) = b[s(t)]$, and the function $s(t) = s(x,t)|_{x=1}$ is determined from the solution of (2.120) for $s(x,t)$. The coefficients $a(s), b(s), \lambda(t)$ satisfy the standard requirements of smoothness and the existence of a fixed sign, which follow from the physical properties of the functional parameters $k_1(s), k_2(s)$ and $j(s)$ which define the MLT and BL models (Chapter 1, 1.0):

i. $(a, b) \in C^{2+\alpha}[0,1], \alpha > 0; (a, b, b_s) > 0, s \in (0,1); a(s)s^{-\gamma_0}(1-s)^{\gamma_1} \leq M < \infty, b(s)s^{-\gamma_0} \leq M, s \in [0,1], 0 \leq \gamma_i \leq 4$, only in isolated points $s_k \in [0,1]$;

ii. $s_0(x) \in C^{2+\alpha}[0, l], s_{0x}(1) = 0, s_0(x) = \delta \geq 0$ at $x \geq 1, 0 < s_0(x) < 1, x \in (0,1); \lambda(t) \in C^{2+\alpha}[0, \infty), \alpha > 0$ and $0 < \delta_0 \leq \lambda(t) \leq M, t \in [0, \infty)$.

2.8.2 Regularization

Let us regularize equations (2.120), replacing the coefficient $a(s)$ with the function $a_\varepsilon(s) = a(s) + \varepsilon, \varepsilon > 0$. Problems (2.120) regularized in this way have a classical solution $s(x,t) \in C^{2+\alpha, 1+\alpha/2}(\Omega_1) \equiv H^{2+\alpha}(\Omega_1), \alpha > 0$ [83]. To obtain estimates of the solution of $s(x,t)$, which do not depend on the regularization parameter ε, let us extend the solution into the region $\Omega_2 = (0,2) \times (0,T)$, assuming that $\bar{s}(x,t) = s(1-x,t)$ when $x \in [1,2]$. The resultant function $s(x,t) = s(1-x,t), x \in [0,2], (s(x,t) = \bar{s}(x,t) x \in [1,2])$ satisfies the initial boundary problem

$$L_0 s \equiv s_t - (\lambda a_\varepsilon s_x)_x + \gamma(x) b_s s_x = 0, (x,t) \in \Omega_2, \tag{2.122}$$

$$s|_{t=0} = s_0(x), x \in [0,2]; s|_{x=0,2} = 1, t \in [0,T], \tag{2.123}$$

where $\gamma(x) = 1$ at $x \in [0,1], s_0 = s_0(2-x), \gamma(x) = -1$, at $x \in [1,2], s_0(x) \in C^{2+\alpha}[0,2]$.

Let us now proceed to the regularization of problem (2.121).

Let us introduce the function $u(x,t) = v(x,t) - v_0, x \in [1, l], v_0 = b(\delta) = const$ and extend it for $t < 0$, assuming that $\overline{u}(x,t) = u(x, -t), t < 0$. Let us consider the following Cauchy problem within a band $\Pi_l = \{x, t | 1 < x < l, |t| < \infty\}$:

$$\Lambda u \equiv u_x + \varphi_t(u) - \varepsilon u_{tt} = 0; u|_{x=1} = u_0(t), |t| < \infty, \tag{2.124}$$

where $u_0 = v_0(t) - \delta$ at $t > 0$ and $u_0 = v_0(-t) - \delta$ at $t < 0, \varphi(u) = \varphi_0(u + \delta)$.

At $\varepsilon = 0$, the function $v(x,t) = u(x,t) + \delta, (x,t) \in \Omega, \Omega = (\Omega_l / \Omega_1)$ satisfies equation (2.121).

Note The problem (2.120), (2.121) and the corresponding regularized problem (2.122)–(2.124) involves the combination of the solutions of

$s = s(x, t), (x, t) \in \Omega_1$ in (2.120), which is evolutionary for t, and of the function $s(x, t) = \varphi[u(x, t)], (x, t) \in \Omega = (\Omega_l/\Omega_1)$, expressed by the solutions of $u(x, t)$ in (2.121) and (2.124), which is evolutionary for x.

The problem of the conjugation of orthogonal flows of the values of $s(x, t)$ as applied to boundary layer equations was first examined by Monakhov and Khusnutdiniova [94].

2.8.3 The Solvability of Problem (2.120)

The maximum principle provides the following values for the solution of (2.122), (2.123):

$$0 < \delta \leq s(x, t) < 1, (x, t) \in \Omega_2, \tag{2.125}$$

so that when $\varepsilon = 0$, (2.122) becomes degenerate (simplified) at only one value, $s = 1$, of the desired solution.

Assuming that $\sigma = 1 - s$, the results described in 2.0 provide us with an estimate which is uniform for ε

$$a|s_x| \leq M, (x, t) \in \overline{\Omega}_2. \tag{2.126}$$

Consequently, $\psi(x, t) = \int_0^{s(x,t)} a_\varepsilon(\xi)d\xi$ is a Holder continuous function, $u \in C^{1,1/2}(\Omega_2)$, as is $s(x, t) \in C^\beta(\overline{\Omega}_2), \beta > 0$ (2.0). Substituting $s(x, t) \in C^\beta(\overline{\Omega}_2)$ into coefficients $a_\varepsilon(s)$ and $b(s)$ in (2.122), we obtain the following estimates (2.0):

$$|s||C^\beta(\overline{\Omega}_2)| \leq M_0; |s_x||C^\beta(\overline{\Omega}_\rho)| \leq M(\rho), \rho \in (0, 1), \beta > 0, \tag{2.127}$$

where $\Omega_\rho = (\rho, 2 - \rho) \times (0, T)$, with the constants M_0, M not dependent on ε. We have thus proved the assertion.

Lemma 1 Problem (2.120) has at least one solution $s(x, t) \in H^{2+\alpha}(\Omega_1), \alpha > 0$ and the estimates (2.125)−(2.127) are true for this solution in regions $\Omega_1 \subset \Omega_2$ and $\Omega_\rho^0 = (\rho, 1) \times (0, T) \subset \Omega_\rho, 0 < \rho < 1$ respectively.

Note 2 Because (2.120) does not degenerate in the neighbourhood of $\Gamma_1 = \{x, t|x = 1, t \geq 0\}$, therefore $s(x, t) \in H^{2+\alpha}(\Omega_1\Gamma_{1\rho}), \Gamma_{1\rho} = \{x, t|x = 1, t \geq \rho > 0\}$. If in addition to assertions (i), (ii) point $x = 1, t = 0$ also satisfies first-order matching conditions, then $s(x, t) \in H^{2+\alpha}(\Omega_1\Gamma_1)$ [83, p. 364].

2.8.4 The Solvability of the Conjugation Problem

On the basis of assumptions (ii), the function $\varphi = \varphi(v), v = u + \delta$ in (2.124) has the properties:

$0 < \frac{d\varphi}{dv}(s) < \infty$ at $s \in (0, 1)$ and function $\frac{d^2\varphi}{dv^2}(s)$ can have only a finite number of zeroes in the interval $s \in (0, 1)$.

Determination Let us define the *generalized solution of the Cauchy problem* (2.124) for $\varepsilon = 0$, as a function $u(x, t)$, bounded in band Π_l which assumes its initial condition at $x = 1$ and has the following properties [67, 76]:

1. It is continuously differentiable everywhere in Π_l except for a finite number of smooth lines, where there exist limiting values $u(x, t)$ located on either side of the line of discontinuity, possibly with the exception of a finite number of points;
2. The inequality

$$\oint_{\Pi_l} (|u - C|f_t + sgn(u - C)[\varphi(u) - \varphi(C)])f_x dtdx \geq 0. \tag{2.128}$$

is satisfied for any constant C and any smooth function $f(x, t) \geq 0$, finite in Π_l.

It clearly follows from the inequality (2.128) at $C = \pm\sup|u(x, t)|$ and due to the arbitrariness of $f(x, t) \geq 0$, that $u(x, t)$ satisfies problem (2.124) for $\varepsilon = 0$, within the meaning of the integral identity

$$\oint_{\Pi_l} (uf_t + \varphi(u)f_{xi})dtdx = 0, \forall f \in \overset{0}{C} \infty(\Pi_l). \tag{2.129}$$

Similarly, the Cauchy problem for the parabolic equation (2.124) at $\varepsilon = 0$ has a single solution $u(x, t) \in H^{2+\alpha}(\Pi_l)$, which satisfies the inequality (2.125):

$$\varphi(\delta) \leq u(x, t) + \delta \leq \varphi(1), (x, t) \in \Pi_l.$$

The truth of this assertion follows from the results obtained in [67; 76] and from the passage to the limit at $\varepsilon \to 0$.

Lemma 2 There exists a generalized solution of the Cauchy problem (2.124) at $\varepsilon \geq 0$.

Theorem 1 The problem of conjugation of (2.120), (2.121) has a generalized solution $s(x,t), (x,t)\in(\Omega_l)$ with the following properties:

a. $s(x,t)\in H^{2+\alpha}(\Omega_1), \alpha>0$;

b. $s(x,t)=b^{-1}[u(x,t)+\delta], (x,t)\in\Omega=(\Omega_l/\Omega_1)$, *where the function* $u(x,t)$ *is a generalized solution of the Cauchy problem (2.124) at* $\varepsilon=0$.

The truth of this assertion follows directly from Lemmas 1 and 2.

2.8.5 Finite-difference Equations

The Muskat-Leverett model (MLT model). Let us introduce a uniform grid E with nodes $x_i=ih, t_n=n\tau, i=1\ldots N, n=0,1\ldots M_1, h$ - space coordinate interval, τ - time interval.

Let us approximate the initial boundary value problem (2.120) containing the regularized operator Les with the equation

$$\frac{s_i^{n+1}-s_i^n}{\tau}=\frac{1}{h^2}\left(\lambda a^n_{\varepsilon i+1/2}(s_{i+1}^{n+1}-s_i^{n+1})-\lambda a^n_{\varepsilon i-1/2}(s_i^{n+1}-s_{i-1}^{n+1})\right)$$
$$-b^n_{si}\left(\frac{s_i^{n+1}-s_{i-1}^{n+1}}{h}\right), i=1\ldots N-1;$$
$$s_0^{n+1}=s_0^n=1; s_i^0=s_{0i}; \frac{s_N^{n+1}-s_N^n}{\tau}=-\frac{2}{h^2}\left(\lambda a^n_{\varepsilon N-1/2}(s_N^{n+1}-s_{N-1}^{n+1})\right)$$
$$-b^n_{sN}\left(\frac{s_N^{n+1}-s_{N-1}^{n+1}}{h}\right).$$
(2.130)

We have taken the relationship between the grid spacings to be $\tau=Kh^2$. The second boundary-value problem [142] was approximated in the same way.

We used regularization to improve convergence of the numerical solution of (2.130).We compared $(\Delta_\varepsilon(E)-\Delta_{\varepsilon^k}(E))\equiv\delta(E)$, where $\delta(E)$ is the regularizer error. We selected the largest ε^k which still retained the order of approximation of $\sim\tau$. In our numerical calculations, we used the regularizer 10^{-8}.

We obtained the numerical solution of (2.130), by the right-hand-side run method [68].

We selected the function $\bar{s} = a_2(1-x)^2 + a_0$, which satisfies the boundary conditions $\bar{s}|_{x=0} = s_0^n = 1, \bar{s}_x|_{x=1} = 0$, as the test function.

The values of $\bar{s}$ served as initial data for (2.130) - $s|_{t=0} = \bar{s}_0(x)$. We then calculated $L_\varepsilon \bar{s} = f(x)$ using the regularized operator L_ε from (2.130). The grid value of $f(E)$ was added to the right hand side of part (2.130) and the resultant solution $s(x, t)$ was compared with $\bar{s}(x)$.

We tested the second boundary value problem in a similar way. The order of magnitude of the error was $O(\tau)$.

We also compared a mixed and a second boundary value problem as part of the same test. The error (at $h = 10^{-2}, K = 1$), $\Delta(0) = \|s_0(t) - \bar{s}(0)\| \approx 10^{-4}, \Delta(E) = \|s^1(x,t) - s^2(x,t)\ \| \approx 10^{-5}$ where $s_0(t) = s|_{x=0}$ was found from the solution of the second boundary-value problem; $s^1(x, t)$ is a grid function which solves the second boundary-value problem with $f(E)$ in its right hand side, and $s^2(x, t)$ is a grid function calculated from (2.130) with $f(E)$ in its right hand side.

The Buckley-Leverett model (BL model). Let us approximate (2.124) by the following finite-difference equation:

$$\frac{u_{i+1}^n - u_i^n}{h} + \varphi_{ui}^n \frac{u_{i+1}^n - u_{i+1}^{n-1}}{\tau} = \varepsilon \frac{u_{i+1}^{n+1} - 2u_{i+1}^n + u_{i+1}^{n-1}}{\tau^2};\ u_i^M = u_{i+1}^{-M},\ u_0^n = u_0^n. \tag{2.131}$$

Note that in (2.131), the evolutionary variable is x, while t plays the role of the space coordinate. Accordingly, we selected the relationship between h and τ on the basis of the condition $h = K\tau^2$. We obtained the numerical solution of (2.131) by the right-hand-side run method [67].

We selected a regularizer for (2.130) as well. In our numerical calculations, we used a regularizer of 10^{-7}.

The test procedure was as described for (2.130). We used the function $\bar{s} = a_2x^2 + a_0$, where a_2 and a_0 were selected from condition $\bar{s}^M = \bar{s}^{-M} = \bar{s}_{i+1}^{-M} = u_{i+1}^M$ as the test function. The deviation from the test solution was of the order of $O(h)$, where at $x \geq 1, h \sim 10^{-4}$.

We examined the regularized problem (2.121) with the initial boundary conditions described below, for small $(l-1)$ values: $v|_{t=0} = b[s_0(1)]$, $v|_{t=1} = b[s(1,1)], v|_{x=1} = b[S(t)]$.

A comparison of the solutions of this equation at $t \in (0,1)$ with the solutions of (2.124) showed the order of the error to be $\sim 10^{-5}$. Since φ_{uu} is sign-variable in the interval $u \in (0,1)$, we usually solved problem (2.124) in our numerical calculations.

2.8.6 Numerical Solution of the Problem of Conjugation of Equations (2.120), (2.121)

Equation (2.130) was selected because of the need to calculate $s|_{x=1}$ without introducing additional grid nodes.

In our numerical experiments, we used the normalized functional parameter values quoted in [142].

When performing numerical calculations, we used a grid E_l for (2.130) with $h=10^{-2}, \tau=10^{-4}, (K=1)$. The $s(t)=s|_{x=1}$ function calculated from (2.130) was monotonic and therefore, when τ was increased by 100 times, the values of $s(t)$ on the new grid E_l were calculated for its corresponding nodes. Since in (2.131) the variable x is evolutionary, we assumed $h=\tau^2=10^{-4}$. We then found the initial data for the calculation of BL $u_i^0=b(s_i)-\delta$, making the calculation in accordance with (2.131). At small values of $(l-1)$, the deviation between the solutions of the conjugation problem and the mixed problem in the segment $[0, l]$ was of the order of $10^{-3}-10^{-4}$. At large values of l, the solutions did not compare well.

2.8.7 Discussion

Low water encroachment levels produce a "blocking effect". In the blocking effect, the wetting phase does not flow out of the formation before the breakthrough (or before the arrival of water at the production well as predicted by the BL model). The effect is due to the change of the sign of b_{ss}. The blocking effect becomes stronger when l increases, but is not much affected by changes of $\lambda(t)$ at $Q=const$. When $\mu=\mu_1/\mu_2$ (the wetting/non-wetting phase viscosity ratio), decreases, so does the blocking effect. When different μ values are considered, the water encroachment of the production well, calculated from the conjugation problem for a moment a time following the breakthrough is different from that calculated from the mixed problem (2.120) for the segment $[0, l]$. For instance, if μ is small, then the water encroachment predicted by the BL model shall be less as compared to problem solution (2.120) on $[0, l]$.

Fig. 2.22 shows the movement of water saturation fronts at $\mu=0,5; \lambda=0,629; s_0=0,041$ (normalized). The solid line shows water saturation calculated from the conjugation problem, and the dotted line the saturation calculated from the mixed problem for $[0, l]$.

For comparison purposes, Fig. 2.23 shows the movement of water saturation fronts calculated using the same parameters from the conjugation

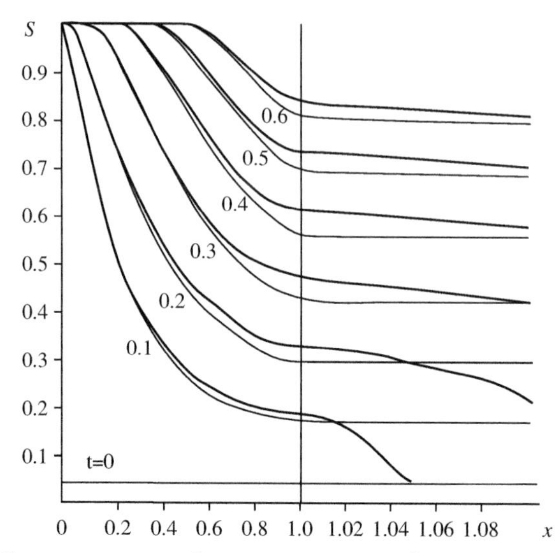

Figure 2.22 The movement of water saturation fronts at $\mu = 0,5$; $\lambda = 0,629$; $s_0 = 0,041$ (normalized).

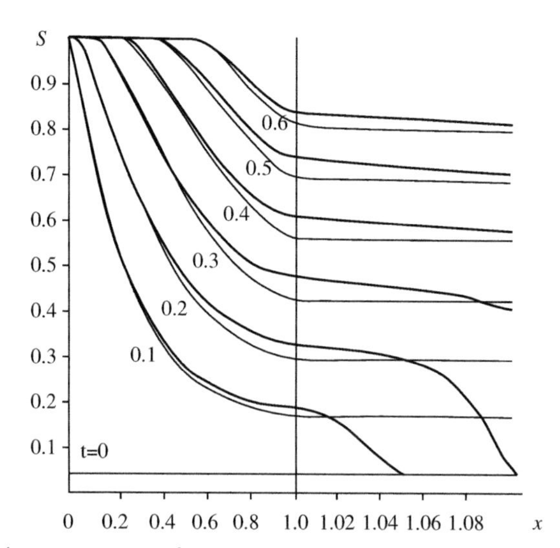

Figure 2.23 The movement of water saturation fronts calculated using the same parameters from the conjugation problem (2.122)–(2.124) (solid line) and from the second boundary value problem on $[0, l]$ (dotted line).

problem (2.122)–(2.124) (solid line) and from the second boundary value problem on $[0, l]$ (dotted line).

2.9 ONE-DIMENSIONAL FLOW OF TWO INTERPENETRATING VISCOUS INCOMPRESSIBLE FLUIDS

In this section, we prove the theorems of the existence and uniqueness of the first boundary value problem "im kleinen" (in small), using the initial data for the equations of flow of two interpenetrating viscous incompressible fluids.

2.9.1 Problem Statement

Below, we consider one-dimensional isothermal two phase flow of viscous incompressible fluids, occurring at the same pressure (Rakhmatulin's hypothesis, [95, 112]) and with no phase transitions. Continuity and momentum equations for each of the phases ($i = 1, 2$) have the form [95, 112]:

$$\frac{\partial \rho_i}{\partial t} + \frac{\partial}{\partial x}(\rho_i v_i) = 0, \rho_i\left(\frac{\partial v_i}{\partial t} + v_i\frac{\partial v_i}{\partial x}\right) = \frac{\partial s_i \sigma_i}{\partial x} + F_i.$$

In these equations, v_i is the relevant phase velocity; ρ_i is reduced density, linked to true density ρ_i^0 and volume concentration s_i by the relationship $\rho_i = s_i \rho_i^0$. The condition $s_1 + s_2 = 1$ arises from the determination of ρ_i. The phase pressure tensor σ_i is an analogue of Stokes' hypothesis [135]: $\sigma_i = -p + \mu_i \frac{\partial v_i}{\partial x}$, where p is pressure (common for the two phases), $\mu_i > 0$ is the dynamic viscosity coefficient of the phases. It is postulated that forces F_i have the form [95, 135]: $F_i = p\frac{\partial s_i}{\partial x} + \varphi_i + \rho_i g$, where $\varphi_1 = K(v_2 - v_1)$, $\varphi_2 = -\varphi_1, K$ - phase interaction coefficient (a present function of their concentrations [95]), and $g = (x, t)$ is a given. The condition $\rho_i^0 = const > 0$ yields the following closed system of equations for $s_i(x, t)$, $v_i(x, t)$ and $p(x, t)$:

$$\frac{\partial s_i}{\partial t} + \frac{\partial}{\partial x}(s_i v_i) = 0, i = 1, 2, \tag{2.132}$$

$$\rho_i\left(\frac{\partial v_i}{\partial t} + v_i\frac{\partial v_i}{\partial x}\right) - \frac{\partial}{\partial x}\left(\mu_i s_i \frac{\partial v_i}{\partial x}\right) = -s_i\frac{\partial p}{\partial x} + \varphi_i + \rho_i g, \tag{2.133}$$

$$s_1 + s_2 = 1; \varphi_1 = K(v_2 - v_1) = -\varphi_2. \tag{2.134}$$

We consider this system for the region $Q_T = \{x|0 < x < 1\} \times (0, T)$, in the following conditions:

$$v_i|_{\partial Q_T} = 0, v_i|_{t=0} = v_i^0(x). \tag{2.135}$$

It is assumed that the initial value of $s^0(x)$ is strictly less than unity and strictly positive:

$$0 < m_0 \leq s^0(x) \leq M_0 < 1, x \in \overline{\Omega}, = \{x|0 \leq x \leq 1\}. \tag{2.136}$$

In what follows, we use the designations adopted in [2; 83] (in particular, $\|u(t)\|$ – is the norm of $u(x, t)$ in $L_2(\Omega)$, $u_x = \frac{\partial u}{\partial x}$).

We will define a generalized solution of (2.132)–(2.135) as consisting of a set of functions (s_i, v_i, p), $i = 1, 2$.

$$s_i(x, t) \in L_\infty(0, T; W_2^2(\Omega)), s_{it} \in L_\infty(0, T; W_2^1(\Omega)),$$

$$v_i(x, t) \in L_\infty(0, T; W_2^1(\Omega)) \cap L_2(0, T; W_2^2(\Omega)), (v_{it}, p_x) \in L_2(Q_T),$$

which satisfy the equations (2.132)–(2.135) at virtually every point in Q_T and which assume specified boundary and initial values in the sense of the traces of the above functions.

Functions (s_i, v_i, p) are termed classical solutions of (2.132)–(2.135), if they have continuous derivatives forming part of (2.132)–(2.134), and satisfy the equations together with the initial and boundary conditions as continuous functions in $\overline{Q}_T = \overline{\Omega} \times [0, T]$.

Theorem 1 *Let the data for the problem (2.132)–(2.135) satisfy the following conditions:* $K(s) \in C^2(0, 1), g \in L_2(0, T; W_2^1(\Omega))$ *and in addition, in the case of (2.136):*

$$v_i^0(x) \in W_2^1(\Omega), s^0(x) \in W_2^2(\Omega), v_i^0(0) = v_i^0(1) = 0, i = 1, 2.$$

In those conditions, there exists a $t_0 > 0, t_0 \in (0, T)$ *such that for all* $t \leq t_0$ *there exists a unique generalized solution* (s_i, v_i, p) *of the problem, and there exist numbers* m, M *such that*

$$0 < m \leq s(x, t) \leq M < 1, (x, t) \in \overline{Q}_{t_0} = \overline{\Omega} \times [0, t_0]. \tag{2.137}$$

If in addition $g(x, t) \in C^{1+\alpha, 1+\alpha/2}(\overline{Q}_T), (s^0(x), u^0(x)) \in C^{2+\alpha}(\overline{\Omega})$ *and the conditions of conformity of initial and boundary data are satisfied, then there exists in* $\overline{Q}_{t_0}$ *a unique classical solution of the problem, which satisfies the inequlity (2.137).*

The proof of Theorem 1 appears in [101] and involves the investigation of the auxiliary problems (2.141), (2.150), (2.156) discussed below (in these classes of functions it is possible to change from the auxiliary functions (s, u, R) to (s_i, v_i, p)).

2.9.2 Continuation of the Solution

In this section, we derive estimates of solutions dependent only on the data of (2.132)–(2.135) and on T, and independent of the interval of existence of the local solution $[0, t_0]$. This makes it possible to extend the local solution to the entire segment $[0, T]$.

We adopt the following expression for the phase interaction coefficient: $K = K_0(s)[s(1-s)]^{-\beta}, s \equiv s_1, \beta \in (-\infty, +\infty)$ and assume the existence of a constant k_0 such that $0 < k_0^{-1} \leq K_0(s) \leq k_0$ for all $s \in [0, 1]$.

2.9.2.1 Velocity and Concentration Energy Inequalities

Lemma 1 The following equations are satisfied for any $t \in [0, T]$:

$$0 \leq s_i(x, t) \leq 1, x \in [0, 1]; \int_0^1 s_i(x, t)dx = \int_0^1 s_i^0(x)dx, i = 1, 2 \tag{2.138}$$

and there exists a bounded measurable function $a(t)$ *such that* $0 \leq a(t) \leq 1$, $s_i(a(t), t) = s_i^0(a(t))$.

The proof of the above is identical to that set out in [2, p. 50].

From (2.132), by virtue of (2.134), (2.135) we obtain

$$sv_1 + (1 - s)v_2 = 0. \tag{2.139}$$

Let us assume $\mu = \mu_1 + \mu_2, \beta_1 = \mu_1/\mu, \beta_2 = \mu_2/\mu$ and introduce a new target function $u(x, t) = \beta_1 v_1(x, t) - \beta_2 v_2(x, t)$. Using (2.139), we obtain

$$v_1 = \frac{1 - s}{a_\mu} u, v_2 = -\frac{s}{a_\mu} u, a_\mu \equiv \beta_1(1 - s) + \beta_2 s. \tag{2.140}$$

Hence, at $i = 1$, equation (2.132) may take the form

$$s_t + (a(s)u)_x = 0, a(s) \equiv \frac{s(1 - s)}{a_\mu}, s|_{t=0} = s^0(x). \tag{2.141}$$

Lemma 2 *The following inequality is true for any* $t\in[0, T]$:

$$\int_0^1 \sum_{i=1}^2 \rho_i^0 s_i v_i^2 dx + 2\int_0^t \int_0^1 \left[\sum_{i=1}^2 \mu_i s_i v_{ix}^2 + K(v_1 - v_2)^2\right] dx d\tau$$

$$\leq \int_0^1 \sum_{i=1}^2 \rho_i^0 s_i^0 (v_i^0)^2 dx + 2\left(\sum_{i=1}^2 \rho_i^0 \int_0^1 s_i^0(x) dx\right)\left(\int_0^1 (|g(x,0)| + |g(x,t)|\right) dx$$

$$+ \int_0^t \int_0^1 \left|g_\tau(x,t)\right| dx d\tau) \equiv N_1(t). \tag{2.142}$$

Proof Let us multiply equation (2.133) by $v_1(x,t)$ and $v_2(x,t)$ respectively, and let us add the resultant equalities. After integrating over Q_t we obtain the following equation (having implicitly assumed summation of i from 1 to 2, over a repeating index i)

$$\int_0^1 \rho_i^0 s_i v_i^2 dx - \int_0^1 \rho_i^0 s_i^0 (v_i^0)^2 dx + 2\int_0^1 \int_0^t (\mu_i s_i v_{ix}^2 + K(v_1 - v_2)^2) dx d\tau$$

$$= 2\int_0^1 \int_0^t \rho_i^0 s_i v_i g dx d\tau \equiv 2I_0(t).$$

Taking into account (2.138) and (2.132), we obtain

$$I_0(t) \leq \left(\sum_{i=1}^2 \rho_i^0 \int_0^1 s_i^0(x) dx\right)\left(\int_0^1 (|g(x,0)| + |g(x,t)|) dx + \int_0^t \int_0^1 |g_\tau(x,\tau)| dx d\tau\right).$$

and thus arrive at (2.142), which proves the lemma.

Let us consider a set of functions $\psi_\beta(s)$, which satisfy the equation

$$\frac{d^2\psi_\beta}{ds^2} = K_0(s)\left(\frac{1}{s(1-s)}\right)^{2+\beta}, s\in(0,1), \beta\in(-\infty, +\infty). \tag{2.143}$$

For $\beta > 0$, let us represent the right hand side of (2.143) in the form $\delta^{(0)}(s) + \delta^{(1)}(s)$, where

$$\delta^{(0)}(s) = \frac{K_0(s)}{s^{2+\beta}(1-s)^q} + \sum_{j=1}^{n+1} c_j \frac{K_0(s)}{s^{j+q}(1-s)^q},$$

$$\delta^{(1)}(s) = \frac{K_0(s)}{(1-s)^{2+\beta} s^q} + \sum_{j=1}^{n+1} b_j \frac{K_0(s)}{(1-s)^{j+q} s^q},$$

n is the whole part of a real number β, $q = \beta - n \in [0, 1]$; c_j, b_j are positive numbers obtained by constructive calculation. In that case

$$\psi_\beta(s) = C_0 + \psi_\beta^{(0)}(s) + \psi_\beta^{(1)}(s), \tag{2.144}$$

where C_0 is an arbitrary constant

$$\psi_\beta^{(0)}(s) = \int_s^1 \left(\int_\xi^1 \delta^{(0)}(y)dy \right) d\xi, \psi_\beta^{(1)}(s) = \int_0^s \left(\int_0^\xi \delta^{(1)}(y)dy \right) d\xi.$$

There exist positive numbers $C_\beta^{(0)}$ and $C_\beta^{(1)}$, dependent only β and k_0, for which the inequalities are true

$$\psi_\beta^{(0)}(s) \geq \frac{k_0^{-1}}{2\beta(1+\beta)s^\beta} - C_\beta^{(0)}, \psi_\beta^{(1)}(s) \geq \frac{k_0^{-1}}{2\beta(1+\beta)(1-s)^\beta} - C_\beta^{(1)}.$$

Having selected $C_0 = C_\beta^{(0)} + C_\beta^{(1)}$ in (2.144), we obtain

$$\psi_\beta(s) \geq \frac{k_0^{-1}}{2\beta(1+\beta)} \left(\frac{1}{s^\beta} + \frac{1}{(1-s)^\beta} \right) > 0.$$

For $\beta = 0$ we have $\psi_0 \equiv \psi_\beta|_{\beta=0} \geq k_0^{-1}\hat{\psi}_0$, where $\hat{\psi}_0 = -\ln s(1-s) + 2s \ln s + 2(1-s)\ln(1-s) + 2 > 0$. For $\beta < 0$ there exists a bounded solution of (2.143) in the form (2.144), which is positive for all $s \in [0,1]$. By virtue of (2.136), we have $\psi_\beta(s)|_{t=0} \leq C(m_0, M_0, \beta, k_0) < \infty$ for all β.

Lemma 3 *The following inequality is true for any* $t \in [0, T]$:

$$\int_0^1 \left(\sum_{i=1}^2 \frac{\mu_i}{s_i(x,t)} (s_{ix}(x,t))^2 + \psi_\beta(s(x,t)) \right) dx \leq$$
$$\int_0^1 \left[\sum_{i=1}^2 \left(3\frac{\mu_i}{s_i^0(x)} (s_{ix}^0(x))^2 + 2\frac{(\rho_i^0)^2}{\mu_i} s_i^0(x)(v_i^0(x))^2 \right) + 4\psi_\beta(s^0(x)) \right] dx$$
$$+ 8\left(\frac{\rho_1^0}{\mu_1} + \frac{\rho_2^0}{\mu_2} \right) N_1 + 12\rho^0 \int_0^t \int_0^1 [|g(x,\tau)| + |g_x(x,\tau)|] dx d\tau \equiv N_2(t).$$

(2.145)

Proof Let us substitute derivatives v_{ix} in equations (2.133), taking them we from the corresponding equations (2.132). Let us now multiply the transformed equations (2.133) by s_{ix}/s_i and add the resultant equalities. Integrating over Q_t and taking into account the estimate $\rho_i^0 v_i s_{ix} \leq \mu_i s_{ix}^2/4s_i + (\rho_i^0)^2 s_i v_i^2/\mu_i$, we obtain the inequality (by summation over the repeating index i from 1 to 2)

$$\frac{1}{4}\int_0^1 \frac{\mu_i}{s_i} s_{ix}^2 dx \leq \int_0^1 \left[\frac{3\mu_i}{4s_i^0}(s_{ix}^0)^2 + \frac{(\rho_i^0)^2}{2\mu_i} s_i^0 (v_i^0)^2 \right] dx + \frac{(\rho_i^0)^2}{\mu_i} \int_0^1 s_i v_i^2 dx +$$
$$\rho_i^0 \int_0^t \int_0^1 s_i v_{ix}^2 dx d\tau + \rho_i^0 \int_0^t \int_0^1 s_{ix} g dx d\tau + I_1(t) + I_2(t),$$

(2.146)

in which

$$I_1(t) = \int_0^t \int_0^1 K\frac{s_x}{s}(v_2 - v_1)dxd\tau,\ I_2(t) = \int_0^t \int_0^1 K\frac{(1-s)_x}{1-s}(v_1 - v_2)dxd\tau.$$

Let us transform the terms $I_1(t)$ and $I_2(t)$, expressing $v_i(x, t)$, in accordance with (2.142), through $u(x, t)$. Then

$$I_1(t) + I_2(t) = -\int_0^t \int_0^1 K\frac{us_x}{s(1-s)a_\mu}dxd\tau. \tag{2.147}$$

Let us then write down equation (2.141) in the form

$$(\psi_\beta(s))_t + (a(s)\psi'_\beta(s)u)_x = a(s)\psi''_\beta(s)us_x.$$

Integrating the above equation over Q_t and taking into account (2.147), we derive

$$I_1(t) + I_2(t) = -\int_0^1 \psi_\beta(s(x,t))dx + \int_0^1 \psi_\beta(s^0(x))dx.$$

Let us now estimate the remaining terms of the right hand side of (2.146). It follows from (2.142) that

$$\sum_{i=1}^2 \frac{\rho_i^0}{\mu_i}\int_0^1 \rho_i^0 s_i v_i^2 dx + \sum_{i=1}^2 \int_0^t \int_0^1 \rho_i^0 s_i v_{xi}^2 dxd\tau \le 2\left(\frac{\rho_1^0}{\mu_1} + \frac{\rho_2^0}{\mu_2}\right)N_1.$$

Let us integrate terms containing $g(x, t)$ by parts and estimate the resultant relation using (2.138) and the multiplicative inequalities [83, p. 79]. We will then obtain

$$\left|\sum_{i=1}^2 \rho_i^0 \int_0^t \int_0^1 g s_{ix} dxd\tau\right| \le 3\rho^0 \int_0^t \int_0^1 [|g(x,t)| + |g_x(x,t)|]dxd\tau$$

and thus arrive at (2.145), which proves the lemma.

In what follows, it is assumed that functions $N_1(t)$ and $N_2(t)$ are bounded for all $t\in[0, T]$. By virtue of Lemma 1 and of equation (2.145), the function

$$F(s) = \int_{s(a(t),t)}^{s(x,t)} \left(\frac{1}{\tau(1-\tau)}\psi_\beta(\tau)\right)^{1/2} d\tau$$

is also bounded for all $(x, t)\in\overline{Q}_T$. Therefore at $\beta \ge 1$ there exist numbers m and M, depending on the data contained in the problem and on T, and such that inequalities (2.137) are true for all $(x,t)\in\overline{Q}_T$. The values m and M may be specified constructively if we include the estimates which follow from (2.144), (2.145) at $\beta\geq 1$:

$$\int_0^1\left[\frac{s_x^2}{s(1-s)}+\frac{1}{s^\beta(1-s)^\beta}\right]dx\leq\left(\frac{\mu}{\mu_1\mu_2}+\frac{k_0\beta(1-\beta)}{2}\right)N_2(t)\equiv N_3(t). \tag{2.148}$$

Lemma 4 If $\beta\geq 1$, then inequality (2.137) in which $m=m_0e^{-N_3}, M=(1-M_0)e^{-N_3}$ at $\beta=1$ is true for any $t\in[0,T], x\in[0,1]$. If $\beta>1$, then $m=km_0, M=1-k(1-M_0), k=\left(1+\frac{1}{2}(\beta-1)N_3m_0^{\frac{\beta-1}{2}}\right)^{\frac{2}{1-\beta}}$.

Lemma 4 yields the estimates

$$\int_0^t(\|u_x(\tau)\|^2+\|u(\tau)\|^2+\|u(\tau)s_x(\tau)\|^2)d\tau+\|u(t)\|^2+\|s_x(\tau)\|^2\leq C_1N_3(t), \tag{2.149}$$

in which the constant C_1 depends only on $\beta,\mu_i,\rho_i,m,M,k_0$.

Note 1 At sufficiently "low" values of $N_2(t)$ estimate (2.137) is valid also for $\beta<1$.

2.9.2.2 Estimation of $\max_{0\leq x\leq 1}|u(x,t)|$

By excluding pressure from system (2.133), we arrive at the following equation for $u(x,t)(u|_{\partial Q_T}=0, u|_{t=0}=\beta_1v_1^0(x)-\beta_2v_2^0(x)\equiv u^0(x))$:

$$u_t-v(b_0(s)u_x)_x+v_1\frac{b_0(s)}{a_\mu^3(s)a(s)}us_x^2+\frac{a_0(s)}{a^{1+\beta}(s)}u=v\left(b_0(s)\frac{a'(s)}{a_\mu^2(s)a(s)}-b_0'(s)\right)$$
$$u_xs_x-a_2(s)uu_x-a_3(s)u^2s_x+b_0(s)g_0, \tag{2.150}$$

where

$$v_1=v\beta_1\beta_2, v=\frac{\mu}{\rho^0}, \rho^0=\rho_1^0+\rho_2^0, a_0(s)=\frac{K_0(s)b_0(s)}{\rho^0a_\mu^{2+\beta}(s)}, b_0(s)=\frac{a_\mu(s)}{a_\rho(s)},$$

$$a_\rho\equiv\alpha_1(1-s)+\alpha_2s, \alpha_i=\frac{\rho_i^0}{\rho^0}, i=1,2, a_2(s)=a_1(s)+\frac{b_0'(s)}{b_0^2(s)}, a_3=\frac{a_1'(s)}{2}+$$
$$\frac{a'(s)b_0'(s)}{b_0^2(s)}, a_1(s)=\frac{(\alpha_1(1-s)^2-\alpha_2s^2)}{a_\mu^2}, a_1'(s)\equiv\frac{da_1}{ds}, g_0\equiv(\alpha_1-\alpha_2)g.$$

Let us assume that

$$C_2 = \min\left\{ \nu \min_{0\leq s\leq 1} b_0(s), \nu_1 \min_{0\leq s\leq 1} \frac{b_0(s)}{a_\mu^3(s)}, \min_{0\leq s\leq 1} a_0(s) \right\},$$

$$C_3 = \max\left\{ \max_{0\leq s\leq 1} |a_2'(s)|, \nu \max_{0\leq s\leq 1} b_0(s) \frac{|a_0'(s)|}{a_\mu^2(s)}, \max_{0\leq s\leq 1} b_0(s), \max_{0\leq s\leq 1} |a_3(s)| \right\}.$$

Lemma 5 In the conditions of Lemma 4, the following estimate is valid for all $t \in [0, T]$:

$$\max_{0\leq x\leq 1} |u(x,t)| \leq \left[\max_{0\leq x\leq 1} |u^0(x)| + C_{03} \int_0^t \max_{0\leq x\leq 1} |g_0(x,\tau)| d\tau \right] \times \exp\left\{ 2C_1 \frac{C_3^2}{C_2} N_3(t) \right\} \equiv N_4(t) \tag{2.151}$$

Proof Let us multiply equation (2.150) by $u^{2l-1}(x,t), l>1$ and integrate the resultant equality over $x \in [0,1]$, integrating term $a_2(s)u^{2l}u_x$ by parts. This yields the inequality

$$\frac{2}{2lC_2}\frac{d}{dt}\int_0^1 u^{2l}dx + (2l-1)\int_0^1 u^{2l-2}u_x^2 dx + \int_0^1 \left[\frac{1}{a(s)} u^{2l} s_x^2 + \frac{1}{a^{1+\beta}(s)} u^{2l} \right] dx$$
$$\leq \frac{C_3}{C_2} \int_0^1 \left[\left(1 + \frac{1}{a(s)} \right) |u_x s_x| + u^2 |s_x| + |g_0| \right] |u|^{2l-1} dx = \sum_{i=1}^3 A_i. \tag{2.152}$$

The following estimates are true for the terms A_i of the right hand side of (2.152):

$$A_1 \leq \frac{(2l-1)}{2} \int_0^1 u^{2l-2} u_x^2 dx + \left(\frac{C_3}{C_2} \right)^2 \frac{1}{(2l-1)} \max_{0\leq x\leq 1} \left(\frac{1}{a(s(x,t))} \right) \int_0^1 \frac{1}{a(s)} u^{2l} s_x^2 dx,$$

$$A_2 \leq \frac{1}{2}\int_0^1 \frac{1}{a(s)} u^{2l} s_x^2 dx + \left(\frac{C_3}{C_2}\right)^2 \int_0^1 a(s) u^{2l+2} dx,$$

$$A_3 \leq \frac{C_3}{C_2}\left(\int_0^1 |g_0|^{2l} dx\right)^{1/2l}\left(\int_0^1 u^{2l} dx\right)^{(2l-1)/2l}.$$

Let us select a number l_0 on the basis of the condition

$$\frac{1}{2} - \left(\frac{C_3}{C_2}\right)^2 \frac{1}{(2l_0 - 1)} \max_{0 \leq x \leq 1}\left(\frac{1}{a(s(x,t))}\right) \geq \frac{1}{4}.$$

In that case, for all $l \geq l_0$ we have

$$\frac{1}{2lC_2}\frac{d}{dt}\int_0^1 u^{2l} dx + \frac{(2l-1)}{2}\int_0^1 u^{2l-1} u_x^2 dx + \int_0^1 \left[\frac{1}{4a(s)} u^{2l} s_x^2 + \frac{1}{a^{1+\beta}(s)} u^{2l}\right] dx$$

$$\leq \frac{C_3}{C_2}\left(\int_0^1 |g_0|^{2l} dx\right)^{1/2l}\left(\int_0^1 u^{2l} dx\right)^{(2l-1)/2l} + \left(\frac{C_2}{C_1}\right)^2 \int_0^1 a(s) u^{2l+2} dx.$$

(2.153)

Let us estimate the last term of the right hand side of (2.153) as follows:

$$\int_0^1 a(s) u^{2l+2} dx \leq \max_{0 \leq x \leq 1} a(s) u^2(x,t) \int_0^1 u^{2l} dx.$$

This yields the following inequality for $y(t) = \left(\int_0^1 u^{2l} dx\right)^{1/2l}$:

$$\frac{dy(t)}{dt} \leq C_3\left(\int_0^1 g_0^{2l} dx\right)^{1/2l} + \frac{C_3^2}{C_2} \max_{0 \leq x \leq 1} a(s) u^2(x,t) \cdot y(t),$$

from which it follows that

$$y(t) \leq \left(y(0) + C_3\left(\int_0^1 g_0^{2l} dx\right)^{1/2l}\right) \exp\left\{\frac{C_3^2}{C_2}\int_0^t \max_{0 \leq x \leq 1} a(s) u^2(x,\tau) d\tau\right\}.$$

(2.154)

Taking into account the estimate which follows from (2.149):

$$\int_0^t \max_{0\leq x\leq 1} a(s)u^2(x,\tau)d\tau \leq C_1(\max_{0\leq s\leq 1} a(s))N_3(t)$$

and passing to the limit in (2.154) at $l \to \infty$, we obtain (2.151).

2.9.2.3 Estimation of Derivatives

Let us express the derivative $\frac{\partial v_i}{\partial x}$ as stated in equation (2.132) and substitute it into equation (2.133). We will then obtain

$$\frac{\partial R_i}{\partial t} + v_i \frac{\partial R_i}{\partial x} = -\frac{\partial p}{\partial x} + \frac{\varphi_i}{s_i} + \rho_i^0 g, R_i \equiv \rho_i^0 v_i + \frac{\mu_i}{s_i}\frac{\partial s_i}{\partial x}, i = 1, 2. \quad (2.155)$$

Let us introduce function $R(x, t)$, assuming

$$R(x,t) = R_1(x,t) - R_2(x,t) = \frac{\mu}{a(s)}\frac{\partial s}{\partial x} + b(s)u, b(s) \equiv \rho^0 \frac{a_\rho}{a_\mu}.$$

By excluding pressure from (2.155) and using (2.140), we arrive at the following equation for $R(x, t)(U \equiv a'(s)u, \delta = \mu_2\rho_1^0 - \mu_1\rho_2^0)$:

$$R_t g + UR_x - \frac{\beta_1\beta_2}{a_\mu^3} uRs_x = -\frac{\delta}{\mu a_\mu^2}\left[a(s)uu_x + \frac{\beta_1(1-s) - \beta_2 s}{a_\mu^2} u^2 s_x\right]$$
$$- \frac{K}{a_\mu^2 a(s)} u + \rho^0 g_0 \equiv f_1, R|_{t=0} = R^0(x). \quad (2.156)$$

Let us assume that $\rho(x, t) = a(s)R^2(x, t)$. By virtue of (2.141), (2.156) we have

$$\rho_t + (U\rho)_x = 2a(s)Rf_1 \equiv f_2. \quad (2.157)$$

The following estimates are true for $\rho(x, t)$ and $f_2(x, t)$:

$$\|\sqrt{\rho(t)}\|^2 \leq C_4 N_3(t), |f_2| \leq C_4(|Ruu_x| + \rho u^2 + |Ru^3| + |uR| + |g_0 R|),$$

where the constant C_4 depends only on $m, M, \mu_i, k_0, \rho_i^0, i = 1, 2$.

Lemma 6 In the conditions of Lemma 5, the following inequalities are true for all $t\in[0, T]$:

$$\|s_x(t)u(t)\|^2 + \|u_x(t)\|^2 + \int_0^t \|u_\tau(t)\|^2 d\tau \le C_5\left(\|s_x^0 u^0\|^2 + \|u_x^0\|^2 + \int_0^t \|g_0(\tau)\|^2 d\tau + K_1 N_3(t)\right) \equiv N_5(t), \tag{2.158}$$

$$\int_0^t \left(\int_0^1 |u_{xx}|dx\right)^2 d\tau \le C_5(N_5 + K_2 N_3(t)), \tag{2.159}$$

where the constant C_5 depends only on $m, M, \mu_i, k_0, \rho_i^0, i = 1, 2,$

$K_1(t) = \max_{0\le x\le 1}\max_{0\le\tau\le t}|g(x,\tau)| + \sum_{i=1}^4 (N_4^t)^i + (N_3^t)^3 + (N_3^t)^{3/2} + (N_3^t)^2 + N_3^t N_4^t + (N_3^t)^2 N_4^t, K_2(t) = 1 + N_3^t + N_4^t, N_3^t = \max_{0\le\tau\le t} N_3(\tau), N_4^t = \max_{0\le\tau\le t} N_4(\tau).$

Proof Let us rewrite equation (2.150) in the form

$$\frac{a_\rho}{a_\mu}u_t - \nu u_{xx} - \nu\frac{a'(s)}{a(s)}; u_x s_x + \frac{\nu\beta_1\beta_2}{\mu^2 a_\mu^3}u\rho = g_0 - \frac{Ku}{\rho^0 a(s) a_\mu^2} - \frac{1}{2}(a_1(s)u^2)_x - \left(\frac{a_\rho}{a_\mu}\right)' u s_t - \frac{\nu\beta_1\beta_2 a(s)}{\mu^2 a_\mu^3}\left(2\rho^0\frac{a_\rho}{a_\mu}uR + \left(\rho^0\frac{a_\rho}{a_\mu}\right)^2 u^2\right)u \equiv f_3. \tag{2.160}$$

Let us now multiply (2.160) by $a(s)u_t$ and integrate the resultant equality over $x\in[0, 1]$. Let us integrate the left hand side of the equation by parts, using (2.157). These transformations yield the equality ($\nu_2 \equiv \nu\beta_1\beta_2/\mu^2$):

$$\frac{1}{2}\frac{d}{dx}\int_0^1\left(\nu a(s)u_x^2 + \nu_2\frac{a(s)}{a_\mu^3}u^2\rho\right)dx + \int_0^1 a(s)\frac{a_\rho}{a_\mu}u_t^2 dx = -\frac{\nu}{2}\int_0^1 a'(s)U s_x u_x^2 dx - \frac{\nu}{2}\int_0^1 a(s)a'(s)u_x^3 dx - \frac{\nu_2}{2}\int_0^1\left(\frac{a(s)}{a_\mu^3}\right)' a(s)u^2\rho u_x dx - \nu_2\int_0^1\frac{a(s)}{a_\mu^3}U u\rho u_x dx + \frac{\nu_2}{2}\int_0^1\frac{a(s)}{a_\mu^3}u^2 f_2 dx + \int_0^1 a(s)f_3 u_t dx \equiv \sum_{i=1}^6 B_i(t). \tag{2.161}$$

Let us estimate terms $B_1, B_3 - B_6$ from the right hand side of (2.161):

$$B_1(t) \leq C\max_{0 \leq x \leq 1}|u(x,t)u_x(x,t)| \cdot \|s_x(t)\| \cdot \|u_x(t)\|,$$

$$|B_3(t)| + |B_4(t)| \leq C\max_{0 \leq x \leq 1}|u^2(x,t)u_x(x,t)| \cdot \|s_x(t)\|^2 \leq$$
$$C\max_{0 \leq x \leq 1}|u_x(x,t)| \cdot \|s_x(t)\|^2 \cdot \|u(t)\| \cdot \|u_x(t)\|,$$

$$|B_5(t)| \leq C\|s_x(t)u(t)\|[(\max_{0 \leq x \leq 1}|u(x,t)|)^2(\|s_x(t)u(t)\| + \|u_x(t)\|) + \|u(t)\|$$
$$((\max_{0 \leq x \leq 1}|u(x,t)|)^3 + \max_{0 \leq x \leq 1}|u(x,t)| + \max_{0 \leq x \leq 1}|g_0(x,t)|)],$$

$$|B_6(t)| \leq \varepsilon_1\|u_t(t)\|^2 + \frac{C}{\varepsilon_1}[\|g_0(t)\|^2 + (\max_{0 \leq x \leq 1}|u(x,t)|)^2(\|s_x(t)u(t)\|^2 +$$
$$\|u_x(t)\|^2 + (\max_{0 \leq x \leq 1}|u(x,t)|)^2\|u(t)\|^2)].$$

Here ε_1 is an arbitrary positive number; and the constant C depends only on $m, M, \mu_i, k_0, \rho_i^0, i = 1, 2$.

Let us rewrite B_2 in the form

$$B_2(t) = \frac{\nu}{2}\int_0^1 (aa')'us_xu_x^2dx + \int_0^1 \frac{a_\rho}{a_\mu}aa'uu_xu_tdx - \nu\int_0^1 (a')^2us_xu_x^2dx +$$
$$\nu_2\int_0^1 \frac{aa'}{a_\mu^3}u^2u_x\rho dx - \int_0^1 aa'uu_xf_3dx \equiv \sum_{i=1}^5 B_2^i(t).$$

$B_2^1(t)$ and $B_2^3(t)$ are estimated in the same way as $B_1(t)$, while $B_2^4(t)$ is estimated in the same way as $B_3(t)$. For $B_2^2(t)$ and $B_2^5(t)$ we have

$$|B_2^2(t)| \leq \varepsilon_2\|u_t(t)\|^2 + \frac{C}{\varepsilon_2}(\max_{0 \leq x \leq 1}|u(x,t)|)^2\|u_x(t)\|^2,$$

$$|B_2^5(t)| \leq C\|u_x(t)\|[\max_{0 \leq x \leq 1}|g_0(x,t)| \cdot \|u(t)\| + \max_{0 \leq x \leq 1}|u(x,t)|(\|u(t)\|$$
$$+ \|u_x(t)\| + \|s_x(t)u(t)\|)].$$

Here ε_2 is an arbitrary positive number and the constant C depends only on $m, M, \mu_i, k_0, \rho_i^0, i = 1, 2$.

Let us introduce the following designations:

$$A(t) = \max_{0\le x\le 1}\max_{0\le \tau\le t}|u(x,\tau)|,\ B(t) = \max_{0\le\tau\le t}\|u(\tau)\|^2,$$
$$D(t) = \max_{0\le\tau\le t}\|s_x(\tau)\|^2,$$

$$Z(t) = \int_0^t (\|u_x(\tau)\|^2 + \|u(\tau)\|^2 + \|s_x(\tau)u(\tau)\|^2)d\tau.$$

It follows from (2.148) and (2.149) that

$$A(t)\le C\max_{0\le\tau\le t}N_4(\tau),\ B(t)+D(t)\le C\max_{0\le\tau\le t}N_3(\tau),\ Z(t)\le CN_3(\tau).$$

Therefore,

$$\int_0^t (|B_5(\tau)| + |B_6(\tau)| + |B_2^2(\tau)| + |B_2^5(\tau)|)d\tau \le (\varepsilon_1+\varepsilon_2)$$
$$\int_0^t \|u_\tau(\tau)^2\|d\tau + \frac{C}{\varepsilon_1}\int_0^t \|g_0(\tau)\|^2 d\tau + C\left[\max_{0\le x\le 1}\max_{0\le\tau\le t}|g_0(x,\tau)|\right.$$
$$\left. + \frac{1}{\varepsilon_2}A^2 + A + A^3 + A^4\right] Z(t).$$

Let us continue the estimation of $B_1(t)$. Since $u(0,t) = u(1,t)$, there exists a point $x_0(t)\in[0,1]$ such that $u_x(x_0(t),t) = 0$. Therefore, from (2.160) we deduce that

$$\max_{0\le x\le 1}|u_x(x,t)| \le \int_0^1 |u_{xx}(x,t)|dt \le C$$
$$[\|u_t(t)\| + \|u_x(t)\|\cdot\|s_x(t)\| + \max_{0\le x\le 1}|u(x,t)|(\|s_x(t)\|^2 + \|u(t)\|^2)$$
$$+ \|g_0(t)\| + \|u(t)\|(\|s_x(t)u(t)\| + \|u_x(t)\| + 1)] \tag{2.162}$$

and therefore

$$\int_0^t |B_1(\tau)|d\tau \le \varepsilon_3\int_0^t \|u_\tau(\tau)\|^2 d\tau + C\int_0^t \|g_0(\tau)\|^2 d\tau$$
$$+ C\left[\frac{1}{\varepsilon_3}A^2 + AD + ADB^{\frac{1}{2}} + A^2DB^{\frac{1}{2}} + A^2D^2\right] Z(t).$$

Here ε_3 is an arbitrary positive number, and the constant C depends only on $m, M, \mu_i, k_0, \rho_i^0$ and $mes\Omega$. In the same way, for $B_3(t), B_4(t)$ and an arbitrary number ε_4 we have

$$\int_0^t |B_3(\tau)| + |B_4(\tau)|d\tau \leq \varepsilon_4 \int_0^t \|u_\tau(\tau)\|^2 d\tau + C\int_0^t \|g_0(\tau)\|^2 d\tau +$$

$$C\left[\frac{1}{\varepsilon_4} BD^2 + B^{1/2}D + BD + ADB\right] Z(t).$$

Combining the estimates for $B_1(t) - B_6(t)$, let us select ε_i on the basis of the condition $\sum_{i=1}^4 \varepsilon_i \leq \frac{1}{2}\min_{m\leq s\leq M} a(s)\frac{a_\rho(s)}{a_\mu(s)}$. Integrating (2.161) over time and taking into account (2.148), (2.149), we arrive at the inequality

$$\|s_x(t)u(t)\|^2 + \|u_x(t)\|^2 + \int_0^t \|u_\tau(\tau)\|^2 d\tau \leq C\Bigg(\|s_x^0 u^0\|^2 + \|u_x^0\|^2$$

$$+ \int_0^t \|g_0(\tau)\|^2 d\tau + \Bigg(\max_{0\leq x 1}\max_{0\leq\tau\leq t}|g_0(x,\tau)| + \sum_{i=1}^4 A^i + AD + A^2D^2$$

$$+ ADB^{1/2} + A^2DB^{1/2} + BD^2 + B^{1/2}D + BD + ABD\Bigg) Z(t)\Bigg),$$

which leads to (2.158). By squaring the inequality (2.162) and integrating it over time, we arrive at estimate (2.159).

Note 2 If the function $g(x,t)$ satisfies the conditions

$$sup_{0\leq t<\infty}\max_{0\leq x\leq 1}|g(x,t)| + \int_0^\infty \|g(t)\|^2 dt + \int_0^\infty \int_0^1 [|g_t(x,t)| + |g_x(x,t)|$$
$$+ |g(x,t)|]dxdt \leq N_6 < \infty, \tag{2.163}$$

then the estimations of Lemmas 1-6 are uniform over t.

Let us obtain an estimate of the derivative $s_x(x,t)$. For simplicity, we will assume that condition (2.163) has been satisfied, and that will allow us to regard the constants $N_1 - N_5$ as independent of time (the case of N_i being dependent on t does not introduce major difficulties into the equation).

Let us multiply equation (2.156) by function $a(s)R^{2p-1}(x,t), p>0$ and integrate over $x\in[0,1]$. Let us integrate the left hand side of the equation by parts, using equation (2.141). Using the Holder inequality to estimate the right hand side of the resultant inequality, we obtain

$$\frac{1}{2p}\frac{d}{dt}\int_0^1 a(s)R^{2p}dx \le C_6\left[\frac{(p-1)}{p}\left(\int_0^1 a(s)R^{2p}dx\right)^{\frac{2p-1}{2p}}\left(\int_0^1 (a(s)uR^2)^{2p}dx\right)^{\frac{1}{2p}} + \left(\int_0^1 a(s)R^{2p}dx\right)^{\frac{2p-1}{2p}}\left(\int_0^1 (f_1)^{2p}dx\right)^{\frac{1}{2p}}\right],$$

where the positive constant C_6 depends on $\mu_1, \mu_2, \rho_1^0, \rho_2^0, m, M, k_0$. Taking into account $|f_1| \le C_6(|uu_x| + |R|u^2 + |u|^3 + |u| + |g_0|)$, we arrive at the inequality

$$\left(y(t) = \left(\int_0^1 a(s)R^{2p}dx\right)^{1/2p}\right)$$

$$\frac{dy}{dt} \le C_6\left(\int_0^1 (uR^2)^{2p}dx\right)^{\frac{1}{2p}} + (\max_{0\le x\le 1}|u(x,t)|)^2 y(t) + V_p(t),$$

$$V_p = \left(\int_0^1 g_0^{2p}dx\right)^{\frac{1}{2p}} + \left(\int_0^1 (uu_x)^{2p}dx\int_0^1\right)^{\frac{1}{2p}} + \left(\int_0^1 u^{6p}dx\right)^{\frac{1}{2p}},$$

from which it follows that

$$\begin{aligned}&\max_{0\le x\le 1}|R(x,t)|\\ &\le C_6\left[sup_{0\le t<\infty}\max_{0\le x\le 1}|u(x,t)|\int_0^t (\max_{0\le x\le 1}|R(x,\tau)|)^2 d\tau\right.\\ &\left.+\max_{0\le x\le 1}|R^0(x)| + \int_0^t V_\infty(\tau)d\tau\right]e^{C_6\int_0^t (\max_{0\le x\le 1}|u(x,\tau)|)^2 d\tau},\end{aligned} \tag{2.164}$$

$$V_\infty(t) = \max_{0\le x\le 1}|g_0(x,t)| + \max_{0\le x\le 1}|u(x,\tau)u_x(x,t)| + (\max_{0\le x\le 1}|u(x,t)|)^3.$$

By virtue of (2.158), (2.159), (2.163) and (2.148), (2.149) we conclude that

$$\int_0^t \left(\max_{0\le x\le 1}\left|u(x,\tau)\right|\right)^2 d\tau + \sup_{0\le t<\infty}\int_0^t V_\infty(\tau)d\tau \le C_6(N_2,N_3,N_4,N_5,N_6)$$

is uniform over t.

Let functions $u^0(x)$ and $g_0(x,t)$ satisfy both (2.163) and the additional conditions

$$\max_{0\le x\le 1}|u^0(x)| \le \Delta_1,\ sup_{0\le t<\infty}\int_0^t \max_{0\le x\le 1}|g_0(x,\tau)|d\tau \le \Delta_2, \tag{2.165}$$

$$\Delta = \max\{\Delta_1,\Delta_2\} \le \frac{1}{2C_6^2T}\left[\max 0\le x\le 1|R^0(x)| + \sup_{0\le t<\infty}\int_0^t V_\infty(\tau)d\tau\right]^{-1}, \tag{2.166}$$

where the constant C_6 depends only on the data specified in the problem. By virtue of Lemma 5 we have $sup_{0\le t<\infty}\max_{0\le x\le 1}|u(x,t)| \le 2\Delta$, while the inequality shown below follows from (2.164)

$$\max_{0\le x\le 1}|R(x,t)| \le C_6\left(\max_{0\le x\le 1}|R^0(x)| + sup_{0\le t<\infty}\int_0^t V_\infty(\tau)d\tau + \Delta\int_0^t (\max_{0\le x\le 1}|R(x,t)|)^2 d\tau\right),$$

and provided that conditions (2.165), (2.166) are satisfied for all $t\in[0,T]$, this yields the estimate

$$\max_{0\le x\le 1}|R(x,t)| \le 2C_6\left(\max_{0\le x\le 1}|R^0(x)| + sup_{0\le t<\infty}\int_0^t V_\infty(\tau)d\tau\right) \equiv N_7.$$

Thus, if conditions (2.163), (2.166) as well as $u^0(x)\in W_2^1(\Omega)$, $s_x^0\in L_\infty(\Omega)$ are satisfied, then the estimate $\max_{0\le x\le 1}|s_x(x,t)| \le C_7(N_1,N_2,N_3,N_4,N_5,N_6,N_7,T)$ is true for all $t\in[0,T]$. In the same initial data conditions, Lemma 6 leads to the nesting of $u_t\in L_2(Q_T)$ and therefore of $u_{xx} \in L_2(Q_T)$. Let in addition $(s^0(x), u^0(x)) \in W_2^2(\Omega), g(x, t) \in L_2(0, T; W_2^1(\Omega))$.

Differentiating equation (2.156) over x, we arrive at an equation for $z(x, t) \equiv R_x(x, t)$, which takes the form $(z|_{t=0} = R_x^0(x))$

$$z_t + Uz_x + U_x z = f_4, \; |f_4| \leq C_7(1 + |z| + |u_{xx}| + |g_{0x}|).$$

It will be easily seen that $z(x, t) \in L_2(Q)$ and therefore $R_t \in L_2(Q)$. After this step, it follows from system (2.133) that $v_i \in L_2(0, T; W_2^2(\Omega))$, $p_x \in L_2(Q_T)$, and that the coefficients of (2.160) belong to space $C^{\alpha,\alpha/2}(Q_T)$ in which $\alpha \in (0, 1)$. If in addition $g(x, t) \in C^{1+\alpha,\alpha/2}(Q_T)$, $u^0(x) \in C^{2+\alpha}(\Omega)$ and the appropriate data fitting conditions are satisfied, then $u(x, t) \in C^{2+\alpha,1+\alpha/2}(Q_T)$. Returning to the equations for $R(x, t)$ and $z(x, t)$, we obtain $z(x, t), R_t(x, t) \in C^{\alpha,\alpha/2}(Q_T)$, and proceeding to functions, we conclude that $v_i(x, t) \in C^{2+\alpha,1+\alpha/2}(Q_T), p_x(x, t) \in C^{\alpha,\alpha/2}(Q_T)$.

Theorem 2 The assertions of theorem 1 are true for all $t \in [0, T]$, where T satisfies the inequality (2.166).

CHAPTER 3

Multidimensional Numerical Models of Subsurface Fluid Dynamics

3.1 INTRODUCTION

Currently there is significant progress in the study of complex problems involving subsurface fluid dynamics. These advances are mainly due to the wide implementation of computing simulation into the practice of applied science. The most impressive results have been obtained using the finite difference and finite element methods.

Due to nonlinearity and variability of boundary conditions in the problems addressed by subsurface fluid dynamics and restrictions imposed by hardware (calculation rate, memory limits, etc.) it is necessary to develop efficient numerical models that are easy to run on computers.

The main models used in this chapter involve the ML-model, Navier-Stokes and Zhukovsky models describing fluid flow in porous media.

Nonlinear Navier-Stokes equations are numerically solved for conventional variables and for the variables: flow function – turbulence of velocity. There is no apparent advantage of using one method relative to the other. Each of them has its own advantage in specific situations. For instance, the variables flow function – turbulence, velocity – flow function are widely used to solve linear problems in fluid dynamics.

However, one is faced with the problem of calculating the turbulence value at the surface of solid body using the boundary conditions for velocities. If we consider problems of fluid flow inside a channel having given tangent velocity components and 4 pressure at the inlet and outlet then additional complications might arise in association with determining the turbulence function for given boundary conditions. In this chapter we consider the method of numerical integration of nonlinear equations to describe the flow of liquid in porous media under specified boundary conditions. This is followed by the development of a numerical algorithm to solve a problem of liquid flow in formation adjacent to a well under specified pressure at the inlet and outlet for an unknown formation pressure.

Numerical simulation of the Navier-Stokes model for variables velocity – pressure brings about a difficulty with fixing the boundary conditions for pressure. Therefore, the numerical solution of these equations was made for the variables velocity – flow function. Some sections below give results of numerical calculation for problems of classical fluid dynamics using the above variables.

3.2 CONVERGENCE OF FINITE-DIFFERENCE SCHEMES FOR A NAVIER-STOKES MODEL WITH VELOCITY – PRESSURE VARIABLES

3.2.1 Introduction

In this paragraph we address the issues of stability and convergence of implicit finite-difference splitting schemes developed for approximated Navier-Stokes equations of the second order with space variables. The limits and convergence of solutions of schemes considered for the above problems in an incompressible fluid using Navier-Stokes equations are proved by methods of a priori estimations using a periodic flow approach under specified pressure drop inside a planar channel. Numerous studies were conducted using the difference technique applied to solve problems involving Navier-Stokes equations for variables (u, v, p) with initial-boundary conditions. The system of Navier-Stokes equations describes the flow of incompressible fluid, which is known to differ from the Cauchy-Kovalevskaya system. Consequently, in many papers the efficient algorithm of numerical solution in variables (u, v, p) is developed to approximate Navier-Stokes equations through an evolving system approach. The numerical solution of regularized system of Navier-Stokes equations, examination of mathematical problems of stability and convergence of proposed finite-difference schemes are considered, for example, in papers [29, 70, 71, 77–79, 82, 98, 130]. It should be noted that the concept to approximate Navier-Stokes equations using equations of an evolving type was formulated in a pioneering paper by N.N. Vladimirova, B.G. Kuznetsov and N.N. Yanenko [29]. R. Temam [130] proposed another approach to e-approximation of the Navier-Stokes equations while considering the behaviour of a regularized solution when $\varepsilon \to 0$. The difference scheme examined proved that under certain conditions imposed on τ, h, ε the solution of the difference problem converges to a solution of the equations of Navier-Stokes. O.A.Ladyzhenskaya and

V. Y. Rivkind [82] considered different schemes to split regularized equations and problems of stability and convergence.

In paper [98] A.P. Oskolkov analysed the parabolic approximation of Navier-Stokes equations.

Some splitting difference schemes and iteration approaches are studied in papers by G.M. Kobelkov [70, 71]. The convergence of a solution of some difference splitting schemes and difference properties of the auxiliary regularized problem were studied in papers [77, 79]. The above papers mainly considered schemes of the first approximation order using a conventional difference grid template having arbitrary convergence.

Another approach to numerical solution of the two-dimensional Navier-Stokes equations in variables (u, v, p) is provided by methods to develop difference schemes using various grids for respective dependent variables. This approach combines advantages of reliable methods of finding solutions to equations in compressible viscous gas, hereinafter referred to as the method of "coarse particles" for incompressible fluid. Various modifications of the "coarse particles" method have been described in the book by O.M. Belotserkovsky [10]. Several papers, among them those by G.M. Kobelkov [70], G.I. Timukhin and M.M. Timukhina [131, 132], N. Danayev, B. Zhumagulov, B.G. Kuznetsov, Sh. Smagulov [38] have addressed the issue of stability and convergence of difference solutions of "coarse particles" methods applied to Navier-Stokes equations in an incompressible fluid. In [34] the stability and convergence are analysed using a heuristic approach based on the use of a type of parabolic difference approximation.

For practical calculations the explicit method of "coarse particles" [10] is frequently used both in our country and abroad. A downside of using the above method is the arbitrary stability of difference solutions.

Papers [7, 11, 33–35] have considered implicit difference schemes:

$$\frac{\vec{u}^{n+1/2} - \vec{u}^{n}}{\tau} + (\vec{u}^{n} \cdot \nabla_n)\vec{u}^{n+1/2} + \nabla_h p^n = \frac{1}{Re}\Delta_h \vec{u}^{n+1/2},$$

$$\frac{\vec{u}^{n+1} - \vec{u}^{n+1/2}}{\tau} + \nabla_h(p^{n+1} - p^n) = 0, \; div_h \vec{u}^{n+1} = 0. \qquad (*)$$

One can easily deduce that the use of implicit calculation at the initial stage of a difference scheme allows to avoid a rigid constraint imposed on parameters of the difference grid. The difference scheme (*) requires the transformation of two-dimensional operators of the type $(E - \tau\nu\Delta_h)$ in each time step. Paper [132] made use of an efficient difference splitting scheme of a stabilizing correction type. Numerical calculations were made over a wide range of parameters considered, yet it was pointed out that that substantiation of stability and convergence of difference solutions was not verified.

The current section examines the use of implicit difference schemes to split approximations of the second order against space variables to regularize the system of Navier-Stokes equations and to substantiate the stability and convergence of implicit splitting "coarse particle" schemes using an example of flow of fluid having a known pressure drop inside a planar channel.

3.2.2 Formulation of Problem Having Initial-boundary Conditions

We assume that area $\Omega \in R^2$ is a square. Inside Ω we consider the system of Navier-Stokes equations

$$\frac{\partial \vec{u}}{\partial t} + (\vec{v} \cdot \nabla)\vec{v} + \nabla p = \nu \Delta \vec{v} + \alpha(t)\nabla x_1, div\vec{v} = 0 \tag{3.1}$$

having the following initial and boundary conditions:

$$\begin{aligned} \vec{v}(x_1, x_2, 0) = \vec{v}_0(x_1, x_2), \left.\frac{\partial^k \vec{v}}{\partial x_1^k}\right|_{x=0} = \left.\frac{\partial^k \vec{v}}{\partial x_1^k}\right|_{x_1=1}, \left.\frac{\partial^k p}{\partial x_1}\right|_{x_1=0} \\ = \left.\frac{\partial^k p}{\partial x_1^k}\right|_{x_1=1}, k = 0, 1, 2, \ldots \end{aligned} \tag{3.2}$$

$\vec{v} = 0$ at $x_2 = 0, x_2 = 1$, that is the components of velocity vector $\vec{v}$ are equal to zero at the lower and upper boundaries of the calculation domain. In this case coefficient $\alpha(t)$ in equation (3.1) characterizes the known pressure drop.

3.2.3 Convergence of Fractional Step Difference Schemes to Regularize a Problem

Difference properties of the problem (3.2) having initial-boundary conditions for regularized system of equations (3.1) have been studied in paper [53]. A priori estimations for solutions uniform in regularization parameters have been obtained. Here and hereinafter we will assume domain Ω being covered by a rectangular grid Ω_h. Let's designate the sets of boundary and boundary-adjacent nodes as follows:

$$\Gamma_1 = \left\{ (x_{1i}, x_{2j}) \middle| i = 0, \frac{1}{2}, 1, \frac{3}{2}; j = 0, \frac{1}{2}, 1, \frac{3}{2}, \ldots, N; x_{1i} = ih, x_{2j} = jh, h = 1/N \right\},$$

$$\Gamma_2 = \left\{ (x_{1i}, x_{2j}) \middle| i = N - \frac{1}{2}, N, N + \frac{3}{2}; j = 0, \frac{1}{2}, 1, \frac{3}{2}, \ldots, N; x_{1i} = ih, x_{2j} = jh, h = 1/N \right\},$$

$$\Gamma_3 = \left\{ (x_{1i}, x_{2j}) \middle| i = 0, \frac{1}{2}, 1, \frac{3}{2}, \ldots, N; j = 0, \frac{1}{2}; x_{1i} = ih, x_{2j} = jh, h = 1/N \right\},$$

$$\Gamma_4 = \left\{ (x_{1i}, x_{2j}) \middle| i = 0, \frac{1}{2}, 1, \frac{3}{2}; j = N - \frac{1}{2}, N; x_{1i} = ih, x_{2j} = jh, h = 1/N \right\}.$$

Similar to the method of cell particles [10], we can interpret each cell as a volume element having a pressure that can be calculated in the cell's centre. Consider the following auxiliary difference scheme:

$$(u_{i+1/2j})_{x_1\bar{x}_1} + (u_{i+1/2j})_{x_2\bar{x}_2} - (p_{ij})_{x_1} = 0, i = \overline{1, N-1}, \; j = \overline{1, N-1},$$

$$(v_{ij+1/2})_{x_1\bar{x}_1} + (v_{ij+1/2})_{x_2\bar{x}_2} - (p_{ij})_{x_2} = 0, i = \overline{1, N-1}, \; j = \overline{1, N-2}, \quad (3.3)$$

$$(u_{i+1/2j})_{\bar{x}_1} + (v_{ij+1/2})_{\bar{x}_2} = \rho_h, i = \overline{2, N+1}, \; j = \overline{1, N-1}, \; or \; i = \overline{1, N},$$

having boundary conditions

$$\vec{u}\big|_{\Gamma_1} = \vec{u}\big|_{\Gamma_2}; \vec{p}\big|_{\Gamma_1} = \vec{p}\big|_{\Gamma_2}; \vec{u}\big|_{\Gamma_3 \cup \Gamma_4} = 0, \quad (3.4)$$

accounting for the fact that $\sum_{i=1}^{N} \sum_{j=1}^{N-1} p_{ij} h^2 = 0$.

3.2.3.1 Estimates of Difference Solutions

Lemma 1 *Solution of problems (3.3),(3.4) can be estimated as*

$$\left\| v_{x_1\bar{x}_1} \right\|^2 + \left\| v_{x_1\bar{x}_2} \right\|^2 \leq c \left\| p_{h,\bar{x}_1} \right\|^2. \quad (3.5)$$

Estimation (3.5) is proved just in the same way as lemma 1 in paper [54].

Further we consider the system of equations (3.3) and (3.4) with inhomogeneous boundary conditions, that is

$$v_{i,1/2} = \psi_1(x_1), v_{i,N-1/2} = \psi_2(x_1), \tag{3.6}$$

assuming that $\rho_h = 0$.

We assume that ψ_1, ψ_2- are sufficiently smooth functions which are periodic in x_1 and that the corresponding conformity conditions are met.

Lemma 2 *Solution of problems (3.3),(3.4),(3.5) can be estimated as follows*

$$\begin{aligned} &\left\|\vec{v}_{x_1\bar{x}_1}\right\|^2 + \left\|v_{x_1\bar{x}_2}\right\|^2 + \left\|p_{x_1}\right\|^2 \\ &\leq c\left\{\frac{1}{h}\sum_{i=1}^{N}\left[((\psi_{1,i})_{x_1})^2 + (\psi_{2,i,x_1})^2\right]h + \left[(\psi_{1,i,x_1,\bar{x}_1})^2 + (\psi_{2,i,x_1,\bar{x}_1})^2\right]h\right\}, \end{aligned} \tag{3.7}$$

where c is a constant independent of grid spacing h_0.

Proof Let's introduce an auxiliary function

$$\xi_{ij+1/2} = \begin{cases} \psi_1(x_1), 0 \leq x_1 \leq 1, \; j = 0 \\ 0, 0 \leq x_1 \leq 1, \; j = \overline{1, N-1} \\ \psi_2(x_1), 0 \leq x_1 \leq 1, \; j = N. \end{cases}$$

Multiply the first equation (3.3) by $h^2 u_{i+1/2j}$, the second equation (3.3) by $h^2(v_{ij+1/2} - \xi_{ij+1/2})$.

Using formula of summation by parts we have

$$\begin{aligned} &\left\|\vec{v}_{x_1}\right\|^2 + \left\|\vec{v}_{x_2}\right\|^2 + \sum_{j=1}^{N-1}\sum_{i=2}^{N+1} p_{ij}(u_{i+1/2j})_{\bar{x}_1} h^2 \\ &+ \sum_{j=1}^{N}\sum_{i=2}^{N} p_{ij}(u_{ij+1/2} - \xi_{ij+1/2})_{\bar{x}_2} + (v_{\bar{x}_1}, \xi_{\bar{x}_1}] + (v_{\bar{x}_2}, \xi_{\bar{x}_2}] = 0. \end{aligned}$$

This brings us to expression

$$\begin{aligned} &1/2(\left\|\vec{v}_{\bar{x}_1}\right\|^2 + \left\|\vec{v}_{\bar{x}_2}\right\|^2) + \sum_{i=2}^{N+1}\sum_{j=1}^{N} p_{ij}(\xi_{ij+1/2})_{\bar{x}_2} h^2 \\ &+ \sum_{j=1}^{N}\sum_{i=2}^{N+1} p_{ij}(u_{i+1/2,\, j,\bar{x}_1} + v_{ij+1/2,\bar{x}_2})h^2 \leq c(\left\|\vec{\xi}_{\bar{x}_1}\right\|^2 + \left\|\vec{\xi}_{\bar{x}_2}\right\|^2). \end{aligned}$$

that yields the estimation

$$\left\|\vec{v}_{\bar{x}_1}\right\|^2_{L_2(\Omega_n)} + \left\|\vec{v}_{\bar{x}_2}\right\|^2_{L_2(\Omega_n)} + \left\|p\right\|^2_{L_2(\Omega_n)} \le c(\left\|\vec{\xi}_{\bar{x}_1}\right\|^2_{L_2(\Omega_n)} + \left\|\vec{\xi}_{\bar{x}_2}\right\|^2_{L_2(\Omega_n)}). \quad (3.8)$$

We note that

$$\left\|\vec{\xi}_{\bar{x}_1}\right\|^2 + \left\|\vec{\xi}_{\bar{x}_2}\right\|^2 \le c\sum_{i=1}^{N}\{1/2[(\psi_{i,1})^2 + (\psi_{2,i})^2] + (\psi_{1,i,\bar{x}_1})^2 + (\psi_{2,i,\bar{x}_1})^2\}h. \quad (3.9)$$

If we differentiate equation (3.3) against x_1 and introduce designations $p_{x_1} = q, u_{x_1} = \tilde{u}, v_{x_1} = \tilde{v}, \psi_{1,x_1} = \tilde{\psi}_1, \psi_{2,x_1} = \tilde{\psi}_2$, then we can formulate problem for $q, \tilde{u}$, and $\tilde{v}$

$$(\tilde{u}_{i+1/2j})_{x_1\bar{x}_1} + (\tilde{u}_{i+1/2j})_{x_2\bar{x}_2} - (q_{ij})_{x_1} = 0,$$

$$(\tilde{v}_{ij+1/2})_{x_1\bar{x}_2} + (\tilde{v}_{ij+1/2})_{x_2\bar{x}_2} - (q_{ij})_{x_2} = 0, i = \overline{1,N},\ j = \overline{1,N-1},$$

$$(\tilde{u}_{i+1/2j})_{\bar{x}_1} + (\tilde{v}_{ij+1/2})_{\bar{x}_2} = 0, i = \overline{1,N},\ j = \overline{1,N-1}\ or\ i = \overline{2,N+1}, \quad (3.10)$$

$$\sum_{i=1}^{N}\sum_{j=1}^{N-1} q_{ij}h^2 = 0.$$

The system of equations (3.10) is solved with the following conditions

$$\vec{\tilde{v}}\big|_{\Gamma_1} = \vec{\tilde{v}}\big|_{\Gamma_2}; \vec{\tilde{v}}\big|_{\Gamma_3} = (0, \tilde{\psi}_1); q\big|_{\Gamma_1} = q\big|_{\Gamma_2}; \vec{\tilde{v}}\big|_{\Gamma_4} = (0, \tilde{\psi}_2). \quad (3.11)$$

$$\left\|\vec{v}_{\bar{x}_1}\right\|^2_{L_2(\Omega_n)} + \left\|\vec{v}_{\bar{x}_2}\right\|^2_{L_2(\Omega_n)} + \left\|q\right\|^2_{L_2(\Omega_n)}$$
$$\le \frac{c}{h}\left\{\sum_{i=1}^{N}\left[(\psi_{1,i,\bar{x}_1})^2 + (\psi_{2,i,\bar{x}_1})^2 + (\psi_{1,i,x_1,\bar{x}_1})^2 + (\psi_{2,i,x_1,\bar{x}_2})^2\right]h\right\}.$$

Lemma 2 is proved.

3.2.3.2 Linear Non-stationary Problem ($\lambda = 0$)

$$\frac{\partial\vec{v}}{\partial t} + \lambda(\vec{v}\cdot\nabla)\vec{v} + \nabla p = \nu\Delta\vec{v} + \alpha(t)\nabla x_1, div\vec{v} = 0, \quad (3.12)$$

$$\vec{v}(x_1, x_2, 0) = \vec{v}_0(x_1, x_2), \vec{v}\big|_\gamma = 0, \int_\Omega p dx_1 dx_2 = 0,$$

$$\left.\frac{\partial^k \vec{v}}{\partial x_1^k}\right|_{x_1=0} = \left.\frac{\partial^k \vec{v}}{\partial x_1^k}\right|_{x_1=1}, \left.\frac{\partial^k p}{\partial x_1^k}\right|_{x_1=0} = \left.\frac{\partial^k p}{\partial x_1^k}\right|_{x_1=1}, k = 0, 1, 2, \ldots \tag{3.13}$$

In order to obtain the numerical solution for problems (3.12), (3.13) we will use difference schemes developed for regularized systems of the following shape:

$$A\vec{v}_{\bar{t}}^{n+1} + \nabla_h p^{n+1} = \nu \Delta_h \vec{v}^{n+1} + \alpha(t_{n+1})\nabla_h x_1 - \Delta t \cdot R(\vec{v}^{n+1}, \vec{v}^n), \tag{3.14}$$

$$-\varepsilon p_{\bar{t}}^{n+1} + \underline{div}_h \vec{v}^{n+1} = 0. \tag{3.15}$$

Here $\vec{v}$ is a vector having the components $(u_{i+1/2j}, v_{ij+1/2})$, $i = \overline{1, N}$, $j = \overline{1, N-1}$, so equation (3.14) is calculated at points $i = \overline{2, N+1}$ *or* $i = 1, \ldots, N$, $j = \overline{1, N-1}$, where p- is the grid pressure in point (i, j). Operators $\underline{div}_h$ and ∇_h have the form:

$$\underline{div}_h \vec{v}^{n+1} = \frac{u_{i+1/2j} - u_{i-1/2j}}{h} + \frac{v_{ij+1/2} - v_{ij-1/2}}{h},$$

$$\nabla_h p^{n+1} = \left(\frac{p_{i+1j} - p_{ij}}{h}, \frac{p_{ij+1} - p_{ij}}{h}\right) = (p_{x_1}, p_{x_2}).$$

Let $A\vec{v}_{\bar{t}}^{n+1} = (Au_{\bar{t}}^{n+1}, Av_{\bar{t}}^{n+1})$. R, A – are positively determined, self-adjoint operators. In addition, the condition $A > E$ is satisfied,

$$A = \begin{pmatrix} E - \Delta t \alpha_0 \Delta_h + \Delta t \chi \Lambda_{11} & \Delta t \chi \Lambda_{12} \\ \Delta t \chi \Lambda_{21} & E - \Delta t \alpha_0 \Delta_h + \Delta t \chi \Lambda_{22} \end{pmatrix}, \chi = \frac{\Delta t}{\varepsilon}. \tag{3.16}$$

Difference operators Λ_{km}, $k, m \in [1, 2]$, are such that

$$\begin{aligned} &\Lambda_{kk}\vec{v}^{n+1} = \vec{v}^{n+1}_{x_k \bar{x}_k}, k = 1, 2, \Lambda_{12} v^{n+1} = \Lambda_{12} v^{n+1}_{ij+1/2} = (v^{n+1}_{ij+1/2})_{\bar{x}_1 x_2}, \\ &\Lambda_{21} u^{n+1} = \Lambda_{21} u^{n+1}_{i+1/2j} = (u^{n+1}_{i+1/2j})_{x_1 \bar{x}_2}. \end{aligned} \tag{3.17}$$

As to $R(\vec{v}^{n+1}, \vec{v}^n) = (R_1(u^{n+1}, u^n), R_2(v^{n+1}, v^n))$ let us assume that conditions specified in paper [54] are fulfilled.

Let's write down components in (3.14), (3.15):

$$(E - \Delta t \alpha_0 \Delta_h + \Delta t \chi \Lambda_{11}) \cdot \frac{u^{n+1}_{i+1/2j} - u^n_{i+1/2j}}{\Delta t} + \Delta t \chi \Lambda_{12} \cdot \frac{v^{n+1}_{ij+1/2} - v^n_{ij+1/2}}{\Delta t}$$
$$= \nu \Delta_h u^{n+1}_{i+1/2j} + \alpha(t_{n+1}) + \chi(\underline{div}_h \vec{v}^{n+1})_{x_1} - p^n_{x_1} - \Delta t R_1(u^{n+1}_{i+1/2j}, u^n_{i+1/2j}), \tag{3.18}$$

$$(E - \Delta t \alpha_0 \Delta_h + \Delta t \chi \Lambda_{22}) \cdot \frac{v^{n+1}_{ij+1/2} - v^n_{ij+1/2}}{\Delta t} + \Delta t \chi \Lambda_{21} \cdot \frac{u^{n+1}_{i+1/2j} - u^n_{i+1/2j}}{\Delta t}$$
$$= \nu \Delta_h u^{n+1}_{ij+1/2} + \chi(\underline{div}_h \vec{v}^{n+1})_{\bar{x}_2} - p^n_{x_2} - \Delta t R_2(v^{n+1}_{ij+1/2}, v^n_{ij+1/2}), \tag{3.19}$$

$$\varepsilon p^{n+1}_{\bar{t}} + \underline{div}_h \vec{v}^{n+1} = 0. \tag{3.20}$$

From (3.18) and (3.19) it follows that

$$\frac{\vec{v}^{n+1} - \vec{v}^n}{\Delta t} = (\alpha_0 + \nu)\Delta_h \vec{v}^{n+1} - \alpha_0 \Delta_h \vec{v}^n - \nabla_h p^n + \chi \nabla_h \underline{div}_h \vec{v}^n$$
$$+ \alpha(t_{n+1}) \nabla_h x_1 - \Delta t R(\vec{v}^{n+1}, \vec{v}^n). \tag{3.21}$$

Using equation in (3.21)

$$R(\vec{v}^{n+1}, \vec{v}^n) = (\alpha_0 + \nu)^2 \Lambda_{11} \Lambda_{22} (\vec{v}^{n+1} - \vec{v}^n),$$

we obtain a stabilizing correction scheme:

$$\frac{\vec{v}^{n+1/2} - \vec{v}^n}{\Delta t} = (\alpha_0 + \nu)\Lambda_{11} \vec{v}^{n+1/2} + (\alpha_0 + \nu)\Lambda_{22} \vec{v}^n - \alpha_0 \Delta_h \vec{v}^n$$
$$+ \chi \nabla_h \underline{div}_h \vec{v}^n - \nabla_h p^n + \alpha(t_{n+1}) \nabla_h x_1, \tag{3.22}$$

$$\frac{\vec{v}^{n+1} - \vec{v}^{n+1/2}}{\Delta t} = (\alpha_0 + \nu)\Lambda_{22}(\vec{v}^{n+1} - \vec{v}^n), p^{n+1} = p^n - \chi div_h \vec{v}^{n+1}.$$

Schemes (3.18)–(3.20), and (3.22) have the boundary conditions:

$$\vec{v}\big|_{t=0} = \vec{v}^{(0)}; p\big|_{t=0} = 0; \vec{v}^{n+1}\big|_{\Gamma_1} = \vec{v}^{n+1}\big|_{\Gamma_2}; \vec{v}^{n+1}\big|_{\Gamma_3 \cup \Gamma_4} = 0. \tag{3.23}$$

3.2.3.3 Stability and Convergence of Difference Schemes of a Stabilizing Correction Type (3.22), (3.23)

We assume that solution of problem (3.12), (3.13) is sufficiently smooth to ensure conformity conditions. Condition of unique determination of pressure is provided by $\sum_{\omega_h} ph^2 = 0$.

Theorem 1 *Let solution of problem (3.12),(3.13) be sufficiently smooth. Then the solution of problems (3.14),(3.15) is stable and the following estimations are valid:*

$$\max_m \left\| A^{1/2}\vec{v}^m \right\|^2_{L_2(\Omega_h)} + \sum_{m=1}^{M} \left\| \Delta_h^{1/2}\vec{v}^m \right\|^2_{L_2(\Omega_h)}\Delta t + \varepsilon \left\| p^{n+1} \right\|^2_{L_2(\Omega_h)} \leq c, \tag{3.24}$$

$$\left\| A^{1/2}\vec{v}^m - A^{1/2}\vec{v}_h^m \right\|^2_{L_2(\Omega_h)} + \sum_{m=1}^{M} \left\| \Delta_h^{1/2}\vec{v}^m - \Delta_h^{1/2}\vec{v}_h^m \right\|^2_{L_2(\Omega_h)} \Delta t \leq c((\Delta t)^2 + h^3) \tag{3.25}$$

at $\Delta t, h \to 0$.

Proof Let us multiply (3.14), (3.15) by $2\Delta t\vec{v}^{n+1}h^2, 2p^{n+1}\Delta th^2$ and sum with respect to Ω_h, using the Green difference formula and Cauchy inequality. As a result we obtain estimate (3.24). Estimate (3.24) provides the stability of the difference solution and proves that difference schemes (3.14), (3.15) have solutions. Now we are left with proving estimate (3.25). Let $\vec{v}_h^{n+1} = (u_{h,i+1/2j}^{n+1}, v_{h,ij+1/2}^{n+1}), p_{ij,h}^{n+1}$ – is the value of exact solution of system (3.12), (3.13) in the corresponding node. Let us consider the deficiency

$$A\vec{v}_{h,t}^{n+1} - \nu\Delta_h\vec{v}_h^{n+1} + \nabla_h p_h^{n+1} - \alpha(t_{n+1})\nabla_h x_1 + \Delta t R(\vec{v}_h^{n+1}, \vec{v}_h^n) = \vec{r}_h^n, \tag{3.26}$$

$$\varepsilon p_{h,\bar{t}}^{n+1} + \underline{div}_h \vec{v}_h^{n+1} = \rho_h^n. \tag{3.27}$$

Let's denote

$$\vec{\omega}^{n+1} = \vec{v}_h^{n+1} - \vec{v}^{n+1}, \pi^{n+1} = p_h^{n+1} - p^{n+1}. \tag{3.28}$$

Then for $\vec{\omega}^{n+1}, \pi^{n+1}$ we obtain equation

$$A\vec{\omega}_{\bar{t}}^{n+1} = \nu\Delta_h\vec{\omega}^{n+1} - \nabla_h\pi^{n+1} - \Delta t R(\vec{\omega}^{n+1}, \vec{\omega}^n) + \vec{r}_h^n, \varepsilon\pi_{\bar{t}}^{n+1} + \underline{div}_h\vec{\omega}^{n+1} = \rho_h^n,$$

$$\vec{\omega}^{(0)} = 0, \pi^{(0)} = 0, \tag{3.29}$$

$$\begin{aligned} &\vec{\omega}^{(n+1)}\big|_{\Gamma_1} = \vec{\omega}^{(n+1)}\big|_{\Gamma_2}, \pi^{(n+1)}\big|_{\Gamma_1} = \pi^{(n+1)}\big|_{\Gamma_2}, \\ &\vec{\omega}^{(n+1)}\big|_{\Gamma_3} = (0, \nu^{(n+1)})\big|_{\Gamma_3}, \vec{\omega}^{(n+1)}\big|_{\Gamma_4} = (0, \nu^{(n+1)})\big|_{\Gamma_4}. \end{aligned} \tag{3.30}$$

Let us consider the auxiliary problems:

$$1.\ \nu\Delta_h\vec{\varphi}^m - \nabla_h q^m = 0, \underline{div}_h\vec{\varphi}^m = \rho_h^m, m = 0, 1, 2, \ldots, M, \tag{3.31}$$

$$\vec{\varphi}^{(m)}\big|_{\Gamma_1} = \vec{\varphi}^{(m)}\big|_{\Gamma_2}, q^{(m)}\big|_{\Gamma_1} = q^{(m)}\big|_{\Gamma_2}, \vec{\varphi}^{(m)}\big|_{\Gamma_3 \cup \Gamma_4} = 0, \sum\nolimits_{\omega_h} q_{ij}^{(m)} h^2 = 0. \tag{3.32}$$

$$2.\ \nu\Delta_h\vec{\psi}^m - \Delta_h q^m = 0, \underline{div}_h\vec{\psi}^m = 0, m = 0, 1, 2, \ldots, M, \tag{3.33}$$

$$\vec{\psi}^{(m)}\big|_{\Gamma_1} = \vec{\psi}^{(m)}\big|_{\Gamma_2}, \vec{\psi}^{(m)}\big|_{\Gamma_3} = (0, \nu^{(m)})\big|_{\Gamma_3}, \vec{\psi}^{(m)}\big|_{\Gamma_4} = (0, \nu^{(m)})\big|_{\Gamma_4}. \tag{3.34}$$

Owing to lemmas 1, 2 solution of problems (3.31–3.34) can be estimated as:

$$\begin{aligned} &\left\|\vec{\varphi}_{\bar{x}_1}^m\right\|_{L_2(\Omega_h)}^2 + \left\|\vec{\varphi}_{\bar{x}_2}^m\right\|_{L_2(\Omega_h)}^2 + \left\|q^m\right\|_{L_2(\Omega_h)}^2 \le c\left\|p_h^m\right\|_{L_2(\Omega_h)}^2, \\ &\left\|\vec{\varphi}_{\bar{t}\bar{x}_1}^{m+1}\right\|_{L_2(\Omega_h)}^2 + \left\|\vec{\varphi}_{\bar{t}\bar{x}_2}^{m+1}\right\|_{L_2(\Omega_h)}^2 + \left\|q_{\bar{t}}^{m+1}\right\|_{L_2(\Omega_h)}^2 \le c\left\|p_{h,\bar{t}}^{m+1}\right\|_{L_2(\Omega_h)}^2, \end{aligned} \tag{3.35}$$

$$\left\|\vec{\varphi}_{\bar{x}_1\bar{x}_1}\right\|_{L_2(\Omega_h)}^2 + \left\|\vec{\varphi}_{\bar{x}_1\bar{x}_2}\right\|_{L_2(\Omega_h)}^2 \le c\left\|p_{h,\bar{x}_1}\right\|_{L_2(\Omega_h)}^2, \tag{3.36}$$

$$\begin{aligned} &\left\|\vec{\psi}_{\bar{x}_1}^m\right\|_{L_2(\Omega_h)}^2 + \left\|\vec{\psi}_{\bar{x}_2}^m\right\|_{L_2(\Omega_h)}^2 + \left\|q^m\right\|_{L_2(\Omega_h)}^2 \\ &\le c\sum\nolimits_{i=1}^N \left\{\frac{1}{h}\left[\left(\nu_{h,i,1/2}^m\right)^2 + \left(\nu_{h,i,N+1/2}^m\right)^2\right] + \left[\left(\nu_{h,i,1/2}\right)_{\bar{x}_1}^2 + \left(\nu_{h,i,N+1/2}\right)_{\bar{x}_1}^2\right]h \right. \\ &\qquad \left. + \left[\left(\nu_{h,i,1/2}^m\right)_{\bar{x}_1}^2 + \left(\nu_{h,i,N-1/2}^m\right)_{\bar{x}_1}^2\right]h\right\}, \end{aligned} \tag{3.37}$$

$$\left\|\vec{\psi}_{\bar{t}\bar{x}_1}^{m+1}\right\|_{L_2(\Omega_h)}^2 + \left\|\vec{\psi}_{\bar{t}\bar{x}_2}^{m+1}\right\|_{L_2(\Omega_h)}^2 + \left\|q_{\bar{t}}^{m+1}\right\|_{L_2(\Omega_h)}^2$$
$$\leq \frac{c}{h}\left\{\sum_{i=1}^{N}\left[\left(v_{h,i,1/2\bar{t}}^{m+1}\right)^2 + \left(v_{h,i,N+1/2,\bar{t}}^{m+1}\right)^2\right]h\right\} \tag{3.38}$$
$$+ c\sum_{i=1}^{N}\left[\left(v_{h,i,1/2\bar{t}\bar{x}_1}^{m+1}\right)^2 + \left(v_{h,i,N+1/2,\bar{t},\bar{x}_1}^{m+1}\right)^2\right]h,$$

$$\left\|\vec{\psi}_{\bar{x}_1\bar{x}_1}^{m}\right\|_{L_2(\Omega_h)}^2 + \left\|\vec{\psi}_{\bar{x}_1\bar{x}_2}^{m}\right\|_{L_2(\Omega_h)}^2 \leq \frac{c}{h}\left\{\sum_{i=1}^{N}\left[\left(v_{h,i,1/2\bar{x}_1}^{m}\right)^2 + \left(v_{h,i,N-1/2,\bar{x}_1}^{m}\right)^2\right]h\right\}$$
$$+ c\sum_{i=1}^{N}\left[\left(v_{h,i,1/2\bar{x}_1x_1}^{m}\right)^2 + \left(v_{h,i,N+1/2,\bar{x}_1x_1}^{m}\right)^2\right]h. \tag{3.39}$$

Let us make a substitution

$$\vec{\omega}^{n+1} = \vec{\omega}^{n+1} + \vec{\varphi}^{n+1} + \vec{\psi}^{n+1}. \tag{3.40}$$

Using equations (3.28), (3.40), (3.29)–(3.34) we obtain the following problem for $\vec{\omega}^{n+1}$:

$$A\vec{\omega}_{\bar{t}}^{n+1} = \nu\Delta_h\vec{\omega}^{n+1} - \nabla_h\pi^{n+1} - \Delta t R(\vec{\omega}^{n+1},\vec{\omega}^n) + \vec{f}^{n+1} + \vec{r}_h^{n}, \varepsilon\pi_{\bar{t}}^{n+1}$$
$$+ \underline{div}_h\vec{\omega}^{n+1} = 0. \tag{3.41}$$

Here

$$\vec{f}^{n+1} = \nu\Delta_h(\vec{\varphi}^{n+1} + \vec{\psi}^{n+1}) - A(\vec{\varphi}_{\bar{t}}^{n+1} + \vec{\psi}_{\bar{t}}^{n+1}) - \Delta t R(\vec{\varphi}^{n+1} + \vec{\psi}^{n+1}, \vec{\varphi}^n + \vec{\psi}^n)$$

with the following conditions for $\vec{\omega} = (\omega_{1,i+1/2j}, \omega_{2,\,j+1/2i})$:

$$\vec{\omega}^{(0)} = \vec{\varphi}^{(0)} + \vec{\psi}^{(0)}$$
$$\vec{\omega}^{(n+1)}\Big|_{\Gamma_1} = \vec{\omega}^{(n+1)}\Big|_{\Gamma_2}, \vec{\omega}^{(n+1)}\Big|_{\Gamma_3\cup\Gamma_4} = 0, \pi^{(n+1)}\Big|_{\Gamma_1} \tag{3.42}$$
$$= \pi^{(n+1)}\Big|_{\Gamma_2}, \sum_{\omega_h}\pi^{(n+1)}h^2 = 0$$

Now let us multiply (3.41) by $2\Delta t h^2 \vec{\omega}^{(n+1)}, 2\Delta t \pi^{(n+1)} h^2$ and sum in Ω_h. As a result we obtain

$$\begin{aligned}
&\left\|A^{1/2}\vec{\omega}^{n+1}\right\|^2 - \left\|A^{1/2}\vec{\omega}^{n}\right\|^2 + 2\Delta t\nu\left\|\Delta^{1/2}\vec{\omega}^{n+1}\right\|^2 + \varepsilon\left(\left\|\pi^{n+1}\right\|^2 - \left\|\pi^{n}\right\|^2\right)\\
&\quad \le c\left\|\Delta^{1/2}\left(\vec{\varphi}^{n+1} + \vec{\psi}^{n+1}\right)\right\| \cdot \left\|\Delta^{1/2}\vec{\omega}^{n+1}\right\|\\
&\qquad + \left\|A^{1/2}\left(\vec{\varphi}_{\bar{t}}^{n+1} + \vec{\psi}_{\bar{t}}^{n+1}\right)\right\| \cdot \left\|A^{1/2}\vec{\omega}^{n+1}\right\|\Delta t\\
&\qquad + \Delta t\left\|\vec{\omega}^{n+1}\right\| \cdot \left\|\vec{r}_h^{n}\right\| + 2(\Delta t)^3\left(\vec{\varphi}_{\bar{t}\bar{x}_1\bar{x}_2}^{n+1} + \vec{\psi}_{\bar{t}\bar{x}_1\bar{x}_2}^{n+1}, \vec{\omega}_{\bar{x}_1\bar{x}_2}^{n+1}\right)\\
&\qquad - (\Delta t)^2\left(\left\|\vec{\omega}_{\bar{x}_1\bar{x}_2}^{n+1}\right\|^2 - \left\|\vec{\omega}_{\bar{x}_1\bar{x}_2}^{n}\right\|^2\right).
\end{aligned} \tag{3.43}$$

From estimate (3.43) we obtain the following inequality:

$$\begin{aligned}
&\left\|A^{1/2}\vec{\omega}^{n+1}\right\|^2 - \left\|A^{1/2}\vec{\omega}^{n}\right\|^2 + 2\nu\left\|\Delta^{1/2}\vec{\omega}^{n+1}\right\|^2\Delta t\\
&+(\Delta t)^2\left(\left\|\vec{\omega}_{\bar{x}_1\bar{x}_2}^{n+1}\right\|^2 - \left\|\vec{\omega}_{\bar{x}_1\bar{x}_2}^{n}\right\|^2\right)\\
&+\varepsilon\left(\left\|\pi^{n+1}\right\|^2 - \left\|\pi^{n}\right\|^2\right) \le c\left\|\Delta^{1/2}\vec{\omega}^{n+1}\right\|\Delta t\\
&+c\left(\left\|\Delta^{1/2}\left(\vec{\varphi}^{n+1} + \vec{\psi}^{n+1}\right)\right\|\Delta t + \left\|A^{1/2}\left(\vec{\varphi}_{\bar{t}}^{n+1} + \vec{\psi}_{\bar{t}}^{n+1}\right)\right\|^2\Delta t + \left\|\Delta^{1/2}\vec{\omega}^{n+1}\right\|^2\Delta t\right)\\
&+(\Delta t)^3\left(\left\|\vec{\varphi}_{\bar{t}\bar{x}_1\bar{x}_2}^{n+1}\right\|^2 + \left\|\vec{\psi}_{\bar{t}\bar{x}_1\bar{x}_2}^{n+1}\right\|^2 + \left\|\vec{\omega}_{x_1x_2}^{n+1}\right\|^2\right) + \Delta t\left\|r_h^n\right\|^2 + \Delta t\left\|\vec{\omega}^{n+1}\right\|^2.
\end{aligned} \tag{3.44}$$

Using the difference analogy of the Gronuoll lemma (3.44) we obtain

$$\begin{aligned}
&\max_m\left\|\vec{\omega}^{m}\right\|^2 + \sum_{m=1}^{M}\left\|\Delta^{1/2}\vec{\omega}^{m}\right\|^2\Delta t \le c\sum_{m=0}^{M}\left\|r_n^m\right\|^2\Delta t\\
&\quad + \sum_{m=1}^{M}\left\|A^{1/2}\left(\vec{\varphi}_{\bar{t}}^{m} + \vec{\psi}_{\bar{t}}^{m}\right)\right\|^2\Delta t + (\Delta t)^2\sum_{m=1}^{M}\left\|\vec{\varphi}_{\bar{t}\bar{x}_1\bar{x}_2}^{m} + \vec{\psi}_{\bar{t}\bar{x}_1\bar{x}_2}^{m}\right\|^2\Delta t\\
&\quad + \left\|A^{1/2}\vec{\omega}^{0}\right\|^2.
\end{aligned} \tag{3.45}$$

It is obvious that $\vec{r}_h^n \sim O(\Delta t + h^2)$. Let's consider estimates (3.29)–(3.39). We assume that $\varepsilon = \Delta t$, which allows us to write $\rho_h^n \sim O(\Delta t + h^2)$. Then

$$\left\| A^{1/2}\vec{\varphi}^m \right\|, \left\| A^{1/2}\vec{\varphi}_{\bar{t}}^{m+1} \right\| \sim O(\Delta t + h^2), v_{h,i,1/2}^m \sim O(h^2),$$

$$\begin{aligned}
v_{h,i,1/2x_1}^m &= \frac{v_{h,i+1,1/2}^m - v_{n,i,1/2}^m}{h} \\
&= \frac{v_{h,i,1/2}^m + h(v_{h,i,1/2}^m)_{x_1} + 0,5h^2(v_{h,i,1/2}^m)_{x_1x_2} + O(h^3) - v_{h,i,1/2}^m}{h} \\
&= (v_{h,i,1/2}^m)_{x_1} + \frac{h}{2}(v_{h,i,1/2}^m)_{x_1x_2} + O(h^2) \\
&= (v_{h,i,0})_{x_1} + \frac{h}{2}(v_{i,0})_{x_1x_2} + \frac{h^2}{8}(v_{i,0})_{x_1x_2x_2} + \frac{h}{2}(v_{i,0})_{x_1x_2} + O(h^2).
\end{aligned}$$

Consequently,

$$\begin{aligned}
&(v_{h,i,1/2}^m)_{(x_1)} \sim O(h^2), (v_{h,i,1/2}^m)_{(x_1x_1)} \sim O(h^2), (v_{h,i,1/2}^m)_{(x_1t)} \sim O(h^2), \\
&(v_{h,i,1/2}^m)_{tx_1x_1)} \sim O(h^2).
\end{aligned} \tag{3.46}$$

Owing to validity of estimates (3.38), (3.39) and (3.46) we have

$$\left\| A^{1/2}\vec{\psi}_{\bar{t}}^m \right\|^2, \left\| A^{1/2}\vec{\psi}^m \right\|^2 \sim O(h^3), \left\| \vec{\varphi}_{\bar{t}\bar{x}_1\bar{x}_2}^m \right\|^2, \left\| \vec{\psi}_{\bar{t}\bar{x}_1\bar{x}_2}^m \right\|^2 \sim O(h). \tag{3.47}$$

Using (3.45)–(3.47), we find

$$\max_m \left\| \vec{\omega}^m \right\|^2 + \sum_{m=1}^{M} \left\| \Delta^{1/2}\vec{\omega}^m \right\|^2 \Delta t \le c((\Delta t)^2 + h^3), \tag{3.48}$$

from which accounting for (3.48) and formula (3.41) we finally obtain

$$\max_m \left\| \vec{\omega}^{m+1} \right\|^2 + \sum_{m=0}^{M} \left\| \vec{\omega}_{\bar{x}}^{m+1} \right\|^2 \Delta t \le O((\Delta t)^2 + h^3).$$

This proves theorem 1.

3.2.3.4 Difference Scheme for Nonlinear Equation

We will examine the following difference scheme that approximates equation (3.1):

$$A\vec{v}_{\bar{t}}^{n+1} = \nu\Delta_h\vec{v}^{n+1} - \nabla_n p^{n+1} + \alpha(t_{n+1})\nabla_n x_1 - L_h(\vec{v}^n) - \Delta t R(\vec{v}^{n+1}, \vec{v}^n),$$
$$\varepsilon p_{\bar{t}}^{n+1} + \underline{div}_h\vec{v}^{n+1} = 0, \tag{3.49}$$

with initial-boundary conditions in the form of (3.24). $L_h(\vec{v})$is the difference operator approximating the convective terms in equation (3.1) to the second order of accuracy [54].

We will analyse the convergence of a difference scheme solution (3.49), which is linear relative to $\vec{v}^{n+1}$. Solvability and convergence of (3.49) are proved using method proposed for linear equations. We will consider inversion of the nonlinear term in more detail. Let $\vec{v}_h^{n+1} = (u_{i+1/2j}^{n+1}, v_{ij+1/2}^{n+1}), p_{h,ij}$ be the values of the exact solution. Let us consider the discrepancy

$$A\vec{v}_{h,\bar{t}}^{n+1} + L_h(\vec{v}_h^n) + \nabla_h p_h^{n+1} - \nu\Delta_h\vec{v}^{n+1} + \Delta t \cdot R(\vec{v}_h^{n+1}, \vec{v}_h^n) - \alpha(t_{n+1})\nabla_h x_1 = \vec{r}_h^n,$$
$$\varepsilon p_{h,\bar{t}}^{n+1} + div_h\vec{v}_h^{n+1} = s_h^n. \tag{3.50}$$

Using designations (3.29), we come to the solution of a problem

$$A\vec{\omega}_{\bar{t}}^{n+1} = \nu\Delta_h\vec{\omega}^{n+1} - \nabla_h\pi^{n+1} - \Delta t \cdot R(\vec{\omega}^{n+1}, \vec{\omega}^n) + L_h(\vec{v}_h^n) - L_h(\vec{v}^n) + \vec{r}_h^n,$$
$$\varepsilon\pi_{\bar{t}}^{n+1} + \underline{div}_h\vec{v}_h^{n+1} = s_h^n \tag{3.51}$$

having the initial-boundary conditions (3.30).

We will dwell on the nonlinear terms in more detail $L_h(\vec{v}_h^n) - L_h(\vec{v}^n)$. We have here

$$\begin{aligned} &L_h(v_h^n) - L_n(v^n) \\ &= \left[u_{h,i+1/2j}^n u_{h,i+1/2j\overset{0}{x}_1}^n + \frac{1}{4}(v_{h,i+1j+1/2}^n + v_{h_1,i+1j-1/2}^n + v_{hij+1/2}^n + v_{hij-1/2}^n) u_{h,i+1/2j\overset{0}{x}_1}^n \right] \\ &- \left[u_{i+1/2j}^n u_{i+1/2j\overset{0}{x}_1}^n + \frac{1}{4}(v_{i+1j+1/2}^n + v_{i+1j-1/2}^n + v_{ij+1/2}^n + v_{ij-1/2}^n) u_{i+1/2j,\overset{0}{x}_2}^n \right] \\ &= \sum_{k=1}^4 I_k. \end{aligned} \tag{3.52}$$

$$
\begin{aligned}
I_1 + I_3 &= u^n_{h,i+1/2j} u^n_{h,i+1/2j\overset{0}{x}_1} - u^n_{i+1/2,j} u^n_{i+1/2j\overset{0}{x}_1} \\
&= u^n_{h,i+1/2j} u^n_{h,i+1/2j\overset{0}{x}_1} - \left(u^n_{h,i+1/2j} + \omega^n_{1i+1/2j} \right)\left(u^n_{h,i+1/2j\overset{0}{x}_1} + \omega^n_{1i+1/2j\overset{0}{x}_1} \right) \\
&= -\omega^n_{1i+1/2j} u^n_{h,i+1/2j\overset{0}{x}_1} - \omega^n_{1i+1/2j}\omega^n_{1i+1/2j\overset{0}{x}_1} - u^n_{h,i+1/2,j}\omega^n_{1i+1/2j\overset{0}{x}_1},
\end{aligned}
\tag{3.53}
$$

$$
\begin{aligned}
I_2 + I_4 = &-\frac{1}{4}(\omega^n_{2i+1j+1/2} + \omega^n_{2i+1j-1/2} + \omega^n_{2ij+1/2} + \omega^n_{2ij-1/2}) u^n_{h,i+1/2j\overset{0}{x}_2} \\
&-\frac{1}{4}(v^n_{hi+1j+1/2} + v^n_{hi+1j-1/2} + v^n_{hij+1/2} + v^n_{hij-1/2})\omega^n_{1i+1/2j\overset{0}{x}_2} \\
&-\frac{1}{4}(\omega^n_{2i+1j+1/2} + \omega^n_{2i+1j-1/2} + \omega^n_{2ij+1/2} + \omega^n_{2ij-1/2})\omega^n_{1i+1/2j\overset{0}{x}_2}.
\end{aligned}
\tag{3.54}
$$

Let $\vec{\varphi}^m, \vec{\psi}^m$ be solution of problems (3.31)−(3.34). Using substitution (3.40) we can write equations:

$$
\begin{aligned}
A\vec{\omega}^{n+1}_{\bar{t}} &= \nu\Delta_h\vec{\omega}^{n+1} + \vec{r}^{\,n}_h - \nabla_h\pi^{n+1} - \Delta t\cdot R(\vec{\omega}^{n+1}, \vec{\omega}^n) + \sum\nolimits_{k=1}^{4} I_k \\
&= \nu(\Delta_h\vec{\varphi}^{n+1} + \Delta_h\vec{\psi}^{n+1}) - A(\varphi^{n+1}_{\bar{t}} + \psi^{n+1}_{\bar{t}}) \\
&\quad - \Delta t R(\vec{\varphi}^{n+1} + \vec{\psi}^{n+1}, \vec{\varphi}^n + \vec{\psi}^n),
\end{aligned}
\tag{3.55}
$$

$$
\varepsilon\pi^{n+1}_{\bar{t}} + \underline{div}_h\vec{\omega}^{n+1} = 0
$$

having initial-boundary conditions (3.42).

We can multiply (3.55) by $2\Delta t\vec{\omega}^{n+1}h^2, 2\Delta t\pi^{n+1}h^2$ and take sum in Ω_h. Then using an embedding inequality we obtain

$$
\begin{aligned}
&\left\|A^{1/2}\vec{\omega}^{n+1}\right\|^2 - \left\|A^{1/2}\vec{\omega}^n\right\|^2 + 2\nu\left\|\Delta^{1/2}\vec{\omega}^{n+1}\right\|^2\Delta t \\
&+ (\Delta t)^2\left(\left\|\vec{\omega}^{n+1}_{\bar{x}_1\bar{x}_2}\right\|^2 - \left\|\vec{\omega}^n_{\bar{x}_1\bar{x}_2}\right\|^2\right) + \varepsilon\left(\left\|\pi^{n+1}\right\|^2 - \left\|\pi^n\right\|^2\right) \le \delta\left\|\Delta^{1/2}\vec{\omega}^{n+1}\right\|^2\Delta t \\
&+ \Delta t c\left\|\Delta^{1/2}(\vec{\varphi}^{n+1} + \vec{\psi}^{n+1})\right\|^2 + \left\|A^{1/2}(\varphi^{n+1}_{\bar{t}} + \psi^{n+1}_{\bar{t}})\right\|^2 \\
&\Delta t + c\left\|A^{1/2}\vec{\omega}^{n+1}\right\|\Delta t + (\Delta t)^3\left(\left\|\vec{\varphi}^{n+1}_{\bar{t}\bar{x}_1\bar{x}_2} + \vec{\psi}^{n+1}_{\bar{t}\bar{x}_1\bar{x}_2}\right\|^2 + \left\|\vec{\omega}^{n+1}_{\bar{x}_1\bar{x}_2}\right\|^2\right) \\
&+ \left\|\vec{r}^{\,n}_h\right\|^2\Delta t + 2\Delta t\left|(L_h(\vec{v}^{\,n}_h) - L_h(\vec{v}^{\,n}), \vec{\omega}^{n+1})\right|.
\end{aligned}
\tag{3.56}
$$

Let us estimate nonlinear terms in (3.56):

$$\left(\omega_1^n u^n_{\underset{h,x_1}{0}}, \omega_1^{n+1}\right) = \left((\omega_1^n + \varphi_1^n + \psi_1^n) u^n_{\underset{h,x_1}{0}}, \omega_1^{n+1}\right)$$
$$\le \max_{\Omega h}\left|u^n_{\underset{h,x_1}{0}}\right| \left(\left\|\vec{\omega}^n\right\|_{L_2(\Omega_n)} + \left\|\vec{\varphi}^n\right\|_{L_2(\Omega_n)} + \left\|\vec{\psi}^n\right\|_{L_2(\Omega_n)}\right) \left\|\vec{\omega}^{n+1}\right\|_{L_2(\Omega_n)}$$
$$\le c\left(\left\|\vec{\omega}^{n+1}\right\|^2_{L_2(\Omega_n)} + \left\|\vec{\varphi}^n\right\|_{L_2(\Omega_n)} + \left\|\vec{\psi}^n\right\|^2_{L_2(\Omega_n)} + \left\|\vec{\omega}^n\right\|^2_{L_2(\Omega_n)}\right).$$

$$\left|\left(u_h^n \omega^n_{\underset{1x}{0}}, \omega_1^{n+1}\right)\right| = \left|u_h^n \left(\omega^n_{\underset{1,x}{0}} + \varphi^n_{\underset{1,x}{0}} + \psi^n_{\underset{1,x}{0}}\right)\omega_1^{n+1}\right|$$
$$\le \max_{\Omega h}\left|u_h^n\right|\left(\left\|\vec{\omega}^n_{\bar{x}}\right\| \cdot \left\|\vec{\omega}^{n+1}\right\| + \left(\left\|\vec{\varphi}^n_{\bar{x}}\right\| + \left\|\vec{\psi}^n_{\bar{x}}\right\|\right) \cdot \left\|\vec{\omega}^{n+1}\right\|\right) \le \delta\left\|\vec{\omega}^n_{\bar{x}}\right\|^2$$
$$+ c\left(\left\|\vec{\omega}^{n+1}\right\|^2 + \left\|\vec{\varphi}^n_{\bar{x}}\right\|^2 + \left\|\vec{\psi}^n_{\bar{x}}\right\|^2\right). \tag{3.57}$$

The remaining nonlinear terms are evaluated in the same manner. Assume that

$$\nu - c\left\|\vec{\omega}^n_x\right\| \cdot \left\|\vec{\omega}^n\right\| \ge 0. \tag{3.58}$$

Then from (3.49)–(3.51) we find

$$\max_m \left\|A^{1/2}\vec{\omega}^m\right\|^2 + \sum_{m=1}^{M} \left\|\Delta^{1/2}\vec{\omega}^m\right\|^2 \Delta t \le c((\Delta t)^2 + h^3). \tag{3.59}$$

Let us determine the sufficiency condition for inequality (3.58) to take place. From (3.58) accounting for (3.59) we obtain

$$\nu - c\left\|\vec{\omega}_x\right\| \cdot \left\|\vec{\omega}^n\right\| \ge \nu - \frac{c}{h} \cdot \left\|\vec{\omega}^n\right\|^2 = \nu - \frac{c_0}{h}((\Delta t)^2 + h^3) \ge 0. \tag{3.60}$$

One can easily see that when (3.60) is valid estimate (3.59) takes place.

This provides a proof to the following theorem.

Theorem 2 *Let solution of problems (3.1),(3.2) be sufficiently smooth and conditions (3.60) are fulfilled. Then solutions of problems (3.23),(3.49) converge to solution of problems (3.1),(3.2) in energy metrics and*

$$\max_m \left\|\vec{\omega}^m\right\|_{L_2(\Omega_h)} + \sum_{m=1}^{M} \left\|\omega^m_{\bar{x}}\right\|^2 \Delta t \le c((\Delta t)^2 + h^3).$$

Note For various choice in $R(\vec{v}^{n+1}, \vec{v}^n)$ one has different types of fractional steps bringing about different rates of solution convergence. These issues can be analysed using above approach.

3.2.4 Convergence of Finite-difference Schemes of "Coarse Particles" Type

3.2.4.1 Linear Case

For linear Navier-Stokes equations (3.12) having initial-boundary conditions (3.13) we can develop a difference scheme in stages.

In the first stage we solve equations:

$$\frac{u_{i+1/2j}^{n+1/2}-u_{i+1/2j}^{n}}{\Delta t}=\nu((u_{i+1/2j}^{n+1/2})_{x_1\bar{x}_1}+(u_{i+1/2j}^{n})_{x_2\bar{x}_2})-(p_{ij}^{n})_{x_1}+\alpha(t_n), \quad (3.61)$$

$$\frac{v_{ij+1/2}^{n+1/2}-v_{ij+1/2}^{n}}{\Delta t}=\nu((v_{ij+1/2}^{n+1/2})_{x_1\bar{x}_1}+(v_{ij+1/2}^{n})_{\bar{x}_2\bar{x}_2})-(p_{ij}^{n})_{x_2}, i=\overline{1,N}, j=\overline{1,N-1}.$$

In the second stage we do the same for equations:

$$\frac{\bar{u}_{i+1/2j}^{n+1/2}-u_{i+1/2j}^{n}}{\Delta t}=\nu(\bar{u}_{i+1/2j}^{n+1/2})_{\bar{x}_2\bar{x}_2}, \quad (3.62)$$

$$\frac{\bar{v}_{ij+1/2}^{n+1/2}-v_{ij+1/2}^{n}}{\Delta t}=\nu(\bar{v}_{ij+1/2}^{n+1/2})_{x_2\bar{x}_2}, i=\overline{1,N}, j=\overline{1,N-1}.$$

In the third stage equations

$$\frac{u_{i+1/2j}^{n+1}-\bar{u}_{i+1/2j}^{n+1/2}}{\Delta t}-(p_{ij}^{n+1}-p_{ij}^{n})_{x_1}=0,$$

$$\frac{v_{ij+1/2}^{n+1}-\bar{v}_{ij+1/2}^{n+1/2}}{\Delta t}-(p_{ij}^{n+1}-p_{ij}^{n})_{x_2}=0, \quad (3.63)$$

$$\underline{div}_h\vec{v}^{n+1}=(u_{i+1/2j}^{n+1})_{\bar{x}_1}+(v_{ij+1/2}^{n+1})_{\bar{x}_2}=0.$$

are solved. Solutions (3.61)–(3.63) should meet boundary conditions of the form (3.4).

Theorem 3 *Let solution of the difference problem (3.12),(3.13) be given by a sufficiently smooth function Then the solutions of difference equations (3.4),(3.61)–(3.63) converge to solution of initial problem at a rate of* $O(\Delta t+h^{3/2})$ *in energy metrics.*

Proof From (3.61), (3.62) we have

$$\frac{\overline{u}^{n+1/2}_{i+1/2j} - u^{n}_{i+1/2j}}{\Delta t} = \nu\left(\left(\overline{u}^{n+1/2}_{i+1/2j}\right)_{x_1\overline{x}_1} + \left(\overline{u}^{n+1/2}_{i+1/2j}\right)_{x_2\overline{x}_2}\right) - \Delta t\left(\overline{u}^{n+1/2}_{i+1/2j}\right)_{x_1\overline{x}_1x_2\overline{x}_2}$$

$$+ \alpha(t_n) - (p^{n}_{ij})_{x_1},$$

$$\frac{\overline{v}^{n+1/2}_{ij+1/2} - v^{n}_{ij+1/2}}{\Delta t} = \nu\left(\left(\overline{v}^{n+1/2}_{ij+1/2}\right)_{x_1\overline{x}_1} + \left(\overline{v}^{n+1/2}_{ij+1/2}\right)_{x_2\overline{x}_2}\right) - \Delta t\left(\overline{v}^{n+1/2}_{ij+1/2j}\right)_{x_1\overline{x}_1x_2\overline{x}_2}$$

$$- (p^{n}_{ij})_{x_2}. \tag{3.64}$$

Consider (3.63) and (3.64) with initial-boundary conditions (3.4). Hereinafter we drop the "bar symbol" in (3.63), (3.64). Let us multiply (3.63) and (3.64) by $2\Delta t u^{n+1/2}_{i+1/2j}h^2, 2\Delta t v^{n+1/2}_{ij+1/2}h^2$ and $2\Delta t u^{n+1}_{i+1/2j}h^2, 2\Delta t v^{n+1}_{i+1/2j}h^2$ and summing on Ω_h. As a result we obtain

$$\left\|\vec{v}^{n+1/2}\right\|^2 - \left\|\vec{v}^{n}\right\|^2 - \left\|\vec{v}^{n+1/2} - \vec{v}^{n}\right\|^2 + 2\nu\left\|\vec{v}^{n+1/2}_{x}\right\|^2\Delta t$$
$$+ \Delta t\left(\left\|\vec{v}^{n+1/2}_{\overline{x}_1\overline{x}_2}\right\|^2 - \left\|\vec{v}^{n}_{\overline{x}_1\overline{x}_2}\right\|^2\right) = 2\Delta t\left(\alpha(t_n), \vec{v}^{n+1/2}\right) + 2\Delta t\left(p^n, \underline{div}_h\vec{v}^{n+1/2}\right), \tag{3.65}$$

$$\left\|\vec{v}^{n+1}\right\|^2 - \left\|\vec{v}^{n+1/2}\right\|^2 + \left\|\vec{v}^{n+1} - \vec{v}^{n+1/2}\right\|^2$$
$$= -2\Delta t\left(\Delta_h\left(p^{n+1} - p^n\right), \vec{v}^{n+1}\right) \times 2\Delta t\left(\alpha(t_n), \vec{v}^{n+1/2}\right)$$
$$= 2\Delta t(\vec{v}^{n}, \vec{\alpha}(t_n)) + 2\Delta t\left(\vec{v}^{n+1/2} - \vec{v}^{n}, \vec{\alpha}(t_n)\right)$$
$$\leq 2\Delta t\left\|\vec{v}^{n}\right\| \cdot \left\|\vec{\alpha}(t_n) + 2\Delta t\right\|\vec{v}^{n+1/2} - \vec{v}^{n}\right\| \cdot \left\|\vec{\alpha}(t_n)\right\|$$
$$+ 2\Delta t\left(\vec{v}^{n+1} - \vec{v}^{n+1/2}, qrad_h p^n\right)$$
$$= 2(\Delta t)\left(\vec{v}^{n+1} - \vec{v}^{n+1/2}, \nabla_h(p^{n+1} - p^n), + \nabla_h p^{n+1}\right) = 2\left\|\vec{v}^{n+1} - \vec{v}^{n+1/2}\right\|^2$$
$$- 2(\Delta t)^2\left(\nabla_h(p^{n+1} - p^n), \nabla_h p^{n+1}\right) = 2\left\|\vec{v}^{n+1} - \vec{v}^{n+1/2}\right\|^2$$
$$- (\nabla t)^2\left(\left\|\nabla_h p^{n+1}\right\|^2 - \left\|\nabla_h p^{n}\right\|^2 + \left\|qrad_h(p^{n+1} - p^n)\right\|^2\right)$$
$$= \vec{v}^{n+1} - \vec{v}^{n+1/2}\right\|^2 - (\Delta t)^2\left(\left\|\nabla_h p^{n+1}\right\|^2 - \left\|\nabla_h p^{n}\right\|^2\right). \tag{3.66}$$

As a result of summation in (3.65), (3.66) we obtain the estimate

$$\max_m \left\| \vec{v}^m \right\|^2 + 2\nu\Delta t \sum_{m=1}^{M} \left\| \vec{v}_x^{m+1/2} \right\|^2 + \Delta t \max_m \left\| \vec{v}_{\bar{x}_1\bar{x}_2}^{m+1/2} \right\|^2 + (\Delta t)^2 \left\| \nabla_h p^{n+1} \right\|^2 \leq \left\| \vec{v}^0 \right\|^2 + \Delta t \left\| \vec{v}_{\bar{x}_1\bar{x}_2}^0 \right\|^2 + (\Delta t)^2 \left\| \nabla_h p^0 \right\|^2, \tag{3.67}$$

from which due to (3.65) we have

$$\max_m \left\| \vec{v}^{m+1/2} \right\|^2 \leq c < \infty. \tag{3.68}$$

Estimates (3.67), (3.68) provide the stability of difference schemes (3.61)–(3.63). We will determine the convergence of solutions for scheme (3.4), (3.61)–(3.63).

Let $u_{h,i+1/2j}^{n+1/2}, v_{h,ij+1/2}^{n+1/2}, u_{h,i+1/2j}^{n+1}, v_{h,i+1/2j}^{n+1}, p_{h,i,j}^{n}$ – be the values of exact solution of difference problems (3.12), (3.13) in corresponding grid nodes. Assume that

$$\vec{v}_{h,\bar{t}}^{n+1/2} = \nu\Delta_h \vec{v}_h^{n+1/2} - \nabla_h p_h^n + \alpha(t_n)\nabla_h x_1 - \Delta t(\vec{v}_h^{n+1/2})_{x_1\bar{x}_1x_2\bar{x}_2} + \vec{r}_h^n,$$

$$\frac{\vec{v}_h^{n+1} - \vec{v}_h^{n+1/2}}{\Delta t} = -\Delta_h(p_h^{n+1} - p_h^n) + \vec{r}_{1,h}^n, \underline{div}_h \vec{v}_h^{n+1} = \rho_h^{n+1}.$$

Let $\vec{\omega}^{n+1} = \vec{v}_h^{n+1} - \vec{v}^{n+1}, \vec{\omega}^{n+1/2} = \vec{v}_h^{n+1/2} - \vec{v}^{n+1/2}$. Owing to (3.63), (3.64) we obtain the following equation for $\vec{\omega}$

$$\frac{\vec{\omega}^{n+1/2} - \vec{\omega}^n}{\Delta t} = \nu\Delta_h \vec{\omega}^{n+1/2} - grad_h \pi^n - \Delta t(\vec{\omega}^{n+1/2})_{x_1\bar{x}_1x_2\bar{x}_2} + \vec{r}_h^n,$$

$$\frac{\vec{\omega}^{n+1} - \vec{\omega}^{n+1/2}}{\Delta t} + \nabla_h(\pi^{n+1} - \pi^n) = \vec{r}_{1,h}^n, \underline{div}_h \vec{\omega}^{n+1} = \rho_h^{n+1},$$

having conditions

$$\vec{\omega}^{(0)} = 0, \vec{\omega}^{(n+1/2)}\Big|_{\Gamma_3\cup\Gamma_4} = (0, \nu^{(n+1)})\Big|_{\Gamma_3\cup\Gamma_4}, \vec{\omega}^{(n+1)}\Big|_{\Gamma_3\cup\Gamma_4} = (0, \nu^{(n+1)})\Big|_{\Gamma_3\cup\Gamma_4}$$

for $\vec{\omega}^{n+1}, \vec{\omega}^{n+1/2}$and requirement to be periodic in x_1.

Let $\vec{\varphi}^m, \vec{\psi}^m$ be solution to problems (3.31)–(3.34). We introduce designations

$$\vec{\omega}^{m+1/2} = \vec{\omega}^{m+1/2} + \vec{\varphi}^{m+1/2} + \vec{\psi}^{m+1/2}, \vec{\omega}^{m+1} = \vec{\omega}^{m+1} + \vec{\varphi}^{m+1} + \vec{\psi}^{m+1}. \quad (3.69)$$

As a result we obtain the following difference equation for $\vec{\omega}$:

$$\frac{\vec{\omega}^{n+1/2} - \vec{\omega}^n}{\Delta t} = \nu\Delta_h\vec{\omega}^{n+1/2} - \Delta_h\pi^n - \Delta t(\vec{\omega}^{n+1/2})_{x_1\bar{x}_1x_2\bar{x}_2}$$
$$-(\vec{\varphi}_{\bar{t}}^{n+1} + \vec{\psi}_{\bar{t}}^{n+1} - \nu\Delta_h(\vec{\varphi}^{n+1} + \vec{\psi}^{n+1})) + \vec{r}_h^n, \quad (3.70)$$

$$\frac{\vec{\omega}^{n+1} - \vec{\omega}^{n+1/2}}{\Delta t} = -\Delta_h(\pi^{n+1} - \pi^n) + \vec{r}_{1,h}^n, \underline{div}_h\vec{\omega}^{n+1}$$
$$= 0, \vec{\omega} = (\omega_{1,i+1/2j}, \omega_{2,i+1/2j}).$$

The boundary conditions are met for $\vec{\omega}$

$$\vec{\omega}^{(n+1/2)}\big|_{\Gamma_3\cup\Gamma_4} = 0, \vec{\omega}^{(n+1)}\big|_{\Gamma_3\cup\Gamma_4} = 0 \quad (3.71)$$

and periodicity conditions are held for x_1.

Further we multiply (3.70) by $2\Delta th^2\vec{\omega}^{n+1/2}$ and sum up on Ω_h, using Cauchy inequality and integrating by parts. As a result we obtain

$$\left\|\vec{\omega}^{n+1/2}\right\|^2 - \left\|\vec{\omega}^n\right\|^2 + \left\|\vec{\omega}^{n+1/2} - \vec{\omega}^n\right\|^2 + 2\Delta t\nu\left\|\vec{\omega}_x^{n+1/2}\right\|^2$$
$$+ 2(\Delta t)^2\left\|\vec{\omega}_{\bar{x}_1\bar{x}_2}^{n+1/2}\right\|^2 \le \delta\Delta t(\left\|\vec{\omega}_x^{n+1/2}\right\|^2 + \left\|\vec{\omega}_{x_1x_2}^{n+1/2}\right\|^2\Delta t)$$
$$+ \Delta t(\left\|\nabla_h(\vec{\varphi}^{n+1} + \vec{\psi}^{n+1})\right\|^2 + \Delta t\left\|\vec{\varphi}_{\bar{x}_1\bar{x}_2}^{n+1} + \vec{\psi}_{\bar{x}_1\bar{x}_2}^{n+1}\right\|^2)$$
$$+ \Delta tc\left\|\vec{\varphi}_{\bar{t}}^{n+1} + \vec{\psi}_{\bar{t}}^{n+1}\right\|^2 + 2\Delta t(\pi^n, \underline{div}_h\vec{\omega}^{n+1/2}). \quad (3.72)$$

Multiply (3.70) by $2\Delta th^2\vec{\omega}^{n+1}$ and sum up over the calculation domain:

$$\left\|\vec{\omega}^{n+1}\right\|^2 - \left\|\vec{\omega}^{n+1/2}\right\|^2 + \left\|\vec{\omega}^{n+1} - \vec{\omega}^{n+1/2}\right\|^2 = (\vec{r}_{1,h}^n, \vec{\omega}^{n+1}). \quad (3.73)$$

3.2.4.2 Nonlinear Case

We will call problem (3.62), (3.63), (3.74) with conditions (3.4) problem 2. Operators $L_{h,1}(u^n, v^n), L_{h,2}(u^n, v^n)$ are approximations of convective terms in Navier-Stokes equation.

Problem 2 is equivalent to the following:

$$u^{n+1/2}_{\bar{t},i+1/2j} = \nu\Delta_h u^{n+1/2}_{i+1/2j} - \Delta t\nu\left(u^{n+1/2}_{i+1/2j}\right)_{x_1\bar{x}_1x_2\bar{x}_2} - (p^n_{ij})_{x_1} + \alpha(t_{n+1}) + L_{h,1}(u^n, v^n),$$

$$v^{n+1/2}_{\bar{t},ij+1/2} = \nu\Delta_h v^{n+1/2}_{ij+1/2} - \Delta t\nu\left(v^{n+1/2}_{ij+1/2}\right)_{x_1\bar{x}_1x_2\bar{x}_2} - (p^n_{ij})_{x_1} = \nu L_{h,2}(u^n, v^n),$$

$$\frac{u^{n+1}_{i+1/2j} - u^{n+1/2}_{i+1/2j}}{\Delta t} + (p^{n+1}_{ij} - p^n_{ij})_{x_1} = 0,$$

$$\frac{v^{n+1}_{ij+1/2} - v^{n+1/2}_{ij+1/2}}{\Delta t} + (p^{n+1}_{ij} - p^n_{ij})_{x_2} = 0, \underline{div}_h \vec{v}^{n+1} = 0, i = 1, 2, \ldots, N, j = 1, \ldots, N-1 \tag{3.74}$$

with initial-boundary conditions (3.4).

Let $u^{n+1}_h, v^{n+1}_h, p^{n+1}_h$ be exact solution of problems (3.1), (3.2), $\vec{r}^n_h$– discrepancy,

$$\vec{L}_h(u^n, v^n) = (L_{h,1}(u^n, v^n), L_{h,2}(u^n, v^n)),$$

$$v^{n+1/2}_{h,\bar{t}} = \nu\Delta_h v^{n+1/2}_h - \nabla_h p^n_h + \vec{L}_h(u^n_h, v^n_h) + \vec{r}^n_h + \alpha(t_{n+1})\nabla_h x_1$$
$$- \Delta t\nu(v^{n+1/2}_h)_{x_1\bar{x}_1x_2\bar{x}_2},$$

$$\frac{\vec{v}^{n+1}_h - v^{n+1/2}_h}{\Delta t} + \nabla_h(p^{n+1}_h - p^n_h) = \vec{r}^n_{1,h}, \underline{div}_h \vec{v}^{n+1}_h = \rho^n_h.$$

Assume that

$$\vec{\omega}^{n+1} = \vec{v}^{n+1}_h - \vec{v}^{n+1}, \vec{\omega}^{n+1/2} = \vec{v}^{n+1/2}_h - \vec{v}^{n+1/2},$$

$$\vec{\omega}^{n+1/2} = \vec{\omega}^{n+1/2} + \vec{\varphi}^{n+1/2} + \vec{\psi}^{n+1/2},$$

$$\vec{\omega}^{n+1} = \vec{\omega}^{n+1} + \vec{\varphi}^{n+1} + \vec{\psi}^{n+1}.$$

Then for $\vec{\omega}^{n+1}$ we obtain an equation

$$\frac{\vec{\omega}^{n+1/2}-\vec{\omega}^{n}}{\Delta t}=\nu\Delta_h\vec{\omega}^{n+1/2}-\nabla_h\pi^n-\vec{L}_h(v_h^n,u_h^n)$$
$$-\vec{L}_h(u_h^n,v_h^n)-(\vec{\varphi}^{n+1}+\vec{\psi}^{n+1})_{\bar{t}}-\Delta t\nu\vec{\omega}^{n+1/2}_{x_1\bar{x}_1x_2\bar{x}_2}+\vec{r}_h^n \tag{3.75}$$
$$-\Delta t\nu(\varphi^{n+1}+\psi^{n+1})_{x_1\bar{x}_1x_2\bar{x}_2}+\nu\Delta_h(\vec{\varphi}^{n+1}+\vec{\psi}^{n+1}),$$

$$\frac{\vec{\omega}^{n+1}-\vec{\omega}^{n+1/2}}{\Delta t}+\Delta_h(\pi^{n+1}-\pi^n)=\vec{r}_h^n,\ \underline{div}_h\vec{\omega}^{n+1}=0,$$

where $\vec{\omega}^{n+1/2},\vec{\omega}^{n+1}$ are the homogeneous boundary conditions in x_1 and periodicity conditions in x_2. Let us multiply (3.75) by $2\Delta t\vec{\omega}^{n+1/2}h^2$, $2\Delta t\vec{\omega}^{n+1}h^2$ and sum up on Ω_h. As a result we find

$$\left\|\vec{\omega}^{n+1/2}\right\|^2-\left\|\vec{\omega}^{n}\right\|^2+\left\|\vec{\omega}^{n+1/2}-\vec{\omega}^{n}\right\|^2+2\Delta t\nu\left\|\vec{\omega}_x^{n+1/2}\right\|^2$$
$$=2\Delta t(\pi^n\underline{div}_h\vec{\omega}^{n+1/2})+2\Delta t(\vec{L}_h(u_h^n,v_h^n)-\vec{L}_h(u^n,v^n)+\vec{r}_h^n,\vec{\omega}^{n+1/2})$$
$$-2(\Delta t)^2\nu\left\|\vec{\omega}^{n+1/2}_{\bar{x}_1\bar{x}_2}\right\|^2-2\Delta t(\vec{\varphi}_{\bar{t}}^{n+1}+\vec{\psi}_{\bar{t}}^{n+1}-\nu\nabla_h(\vec{\varphi}^{n+1}+\vec{\psi}^{n+1}),\vec{\omega}^{n+1/2})$$
$$+2(\Delta t)^2\nu((\vec{\psi}^{n+1}+\vec{\varphi}^{n+1})_{\bar{x}_1\bar{x}_2},\vec{\omega}^{n+1/2}_{\bar{x}_1\bar{x}_2}), \tag{3.76}$$

$$\left\|\vec{\omega}^{n+1}\right\|-\left\|\vec{\omega}^{n+1/2}\right\|+\left\|\vec{\omega}^{n+1}-\vec{\omega}^{n+1/2}\right\|^2=2\Delta t(\vec{r}_{1,h}^n,\vec{\omega}^{n+1}). \tag{3.77}$$

We will dwell on a nonlinear term in more detail

$$\left(((u_h^n)^2-(u^n)^2)_{\overset{0}{x_1}},\omega_1^{n+1/2}\right)=\left((u_h^n)^2-(u^n)^2,\omega^{n+1/2}_{1,\overset{0}{x_1}}\right)$$
$$=\left(\omega_1^n(u_h^n+u^n),\omega^{n+1/2}_{1,\overset{0}{x_1}}\right)=\left(\omega_1^n(2u_h^n+\omega_1^n),\omega^{n+1/2}_{1,\overset{0}{x_1}}\right) \tag{3.78}$$
$$=\left((\omega_1^n+\varphi_1^n+\psi_1^n)(2u_h^n+\omega_1^n+\varphi_1^n+\psi_1^n),\omega^{n+1/2}_{1,\overset{0}{x_1}}\right).$$

We can estimate (3.78) using Helder inequality:

$$2\left(\omega_1^nu_h^n,\omega^{n+1/2}_{1,\overset{0}{x_1}}\right)\le c\max\left|\vec{u}_h^n\right|\cdot\left\|\vec{\omega}^n\right\|\cdot\left\|\vec{\omega}_x^{n+1/2}\right\|\le\delta\left\|\vec{\omega}_x^{n+1/2}\right\|^2+c\left\|\vec{\omega}^n\right\|^2,$$

$$\left((\omega_1^n), \omega^{n+1/2}_{\substack{0\\1,x_1}}\right) \le \left\|\omega_x^{n+1/2}\right\| \cdot \left\|\omega^n\right\|_{L_4(\Omega_n)}$$

$$\le c\left\|\omega_x^{n+1/2}\right\| \cdot \left\|\omega^n\right\| \cdot \left\|\omega_x^n\right\| \cdot \left(\vec{\omega}_1^n \vec{\varphi}_1^n, \vec{\omega}^n_{\substack{0\\1,x_1}}\right)$$

$$\le \left\|\vec{\omega}^n\right\|_{L_4(\Omega_n)} \left\|\vec{\varphi}^n\right\|_{L_4(\Omega_n)} \left\|\vec{\omega}_{x_1}^{n+1/2}\right\| \le \delta \frac{\left\|\vec{\varphi}_x^n\right\|^2}{h^2} \left\|\vec{\omega}_x^{n+1/2}\right\|^2 + c\left\|\vec{\omega}^n\right\|^2, \tag{3.79}$$

$$\left((\vec{\omega}^n)^2, \vec{\omega}^{n+1/2}_{\substack{0\\x}}\right) \le \left\|\vec{\omega}_x^{n+1/2}\right\| \cdot \left\|\vec{\omega}^n\right\|^2_{L_4(\Omega_n)} \le c\left(\frac{1}{h^2}\left\|\vec{\omega}_x^{n+1/2}\right\| \cdot \left\|\vec{\omega}^n\right\|^2 + \left\|\vec{\omega}^n\right\|^2\right).$$

The remaining terms are estimated in the same manner. Due to (3.76)–(3.79), we obtain the inequality

$$\left\|\vec{\omega}^{n+1}\right\|^2 - \left\|\vec{\omega}^n\right\|^2 + \sum_{m=1}^M\left(\left\|\vec{\omega}_x^{m+1/2}\right\|^2 + \Delta t\left\|\vec{\omega}_{\bar{x}_1\bar{x}_2}^{m+1/2}\right\|^2\right)\Delta t$$

$$\le c\sum_{n=1}^M\left(\left\|\vec{\varphi}_{\bar{t}}^{n+1}\right\|^2 + \left\|\vec{\psi}_{\bar{t}}\right\| + \left\|\nabla_h(\vec{\varphi}^{n+1} + \vec{\psi}^{n+1})\right\|\right.$$

$$\left. + \Delta t\left\|(\vec{\varphi}^{n+1} + \vec{\psi}^{n+1})_{\bar{x}_1 x_2}\right\|\right)\Delta t + \sum_{m=1}^M\left(\left\|\vec{r}_h^m\right\|^2 + \left\|\vec{r}_{1,h}^m\right\|^2\right)\Delta t.$$

So, the assertion is proved:

Theorem 4 *Let the solution of problems (3.1),(3.2) be sufficiently smooth and conditions* $\frac{\Delta t}{h} \le \chi_0 = const$ *apply. Then the solution of difference problem (3.4), (3.61),(3.64) converges to solution of problems (3.1),(3.2) in energy metrics of the order* $O(\Delta t + h^{3/2})$.

3.3 NUMERICAL REALIZATION OF NAVIER-STOKES MODEL IN MULTIPLY CONNECTED DOMAIN IN VELOCITY – FLOW FUNCTION VARIABLES

3.3.1 Introduction

Here we consider numerical calculation of a flow of viscous incompressible fluid inside a planar channel having periodically located barriers in the form of rectangular plates orthogonal to the channel walls. A similar problem was considered earlier by authors [36, 39], but these papers failed to account for unique condition to determine pressure. In the case of multiply connected domains the requirement of unique pressure determination is necessary to formulate the equivalent problem. In papers [121, 125] a unique pressure condition is used to calculate numerically the flow of a viscous incompressible liquid in double connected domains. Paper [121] deals with the numerical solution of a periodic problem of viscous incompressible fluid flow around a plate inside a planar channel where the condition of unique pressure is formulated as follows

$$\oint\left[\frac{1}{Re}(\omega_y+\psi_{ty}+\omega\psi_x)dx+\left(-\frac{1}{Re}\omega_x-\psi_{tx}+\omega\psi_y\right)dy\right]=0. \qquad (*)$$

Integrals in (*) are calculated using the Newton-Cotes formula with an accuracy order of $O(h^6)$. Derivatives are substituted by a finite-difference relation of the fourth order ofaccuracy, that is, calculations are conducted with an error and it is natural that the integral (*), as is mentioned in the above article, is accounted for in an approximate manner. In these papers the problem is reduced to a solution of nonlinear difference equations of the fourth order with non-local boundary conditions and no substantiation is given in the use of the finite-difference schemes.

In this section we propose a stable numerical algorithm to solve the Navier-Stokes equation in a double connected region at a given flow rate. The condition of unique pressure, which is exactly satisfied under numerical calculation, is formulated. The problem is reduced to a solution of the finite-difference equation of the second order with non-local boundary conditions. In contrast to other papers, the algorithm can be easily implemented. The problem is solved through reducing it to the underlying physical mechanisms. A grid flow function is introduced and determined through solution of the Poisson equation with given boundary conditions. Accurate meeting of the requirement of unique pressure is achieved. Numerical calculations are made according to the schemes proposed.

3.3.2 Formulation of the Problem

Navier-Stokes equations are considered inside the channel

$$\Omega=\left\{x,y\middle|0<x<1,\ -\frac{1}{4}<y<\frac{1}{4}\right\}:$$

$$\vec{u}_t+(\vec{u},\nabla)\vec{u}+\nabla p=\frac{1}{Re}\vec{u},\ div\vec{u}=0 \qquad (3.80)$$

They satisfy the following conditions:

a. at the walls of the planar channel Ω and on barrier $u = v = 0(\vec{u} = (u, v))$;

b. periodicity conditions are met at the channel's inlet and outlet, that is

$$\left.\frac{\partial^k \vec{u}}{\partial x^k}\right|_{x=0} = \left.\frac{\partial^k \vec{u}}{\partial x^k}\right|_{x=1}, \left.\frac{\partial^{k+1} p}{\partial x^{k+1}}\right|_{x=0} = \left.\frac{\partial^{k+1} p}{\partial x^{k+1}}\right|_{x=1}, k = 0, 1, \ldots, m. \quad (3.81)$$

In this case it is assumed that the flow rate is given

$$\int_{-0,25}^{0,25} u dy = \psi_0 = 1. \quad (3.82)$$

For pressure to be determined in unique manner it is required that

$$\int_{\gamma} (p_x dx + p_y dy) = 0 \quad (3.83)$$

for any contour of γ around the plate.

3.3.3 Description of Difference Scheme

To write down the difference scheme to solve problem (3.80)−(3.83) in domain Ω let us consider the finite-difference grid using the splitting method [10] in the following form:

$$\Omega_h = \begin{cases} x_{i+1/2} = \left(i + \frac{1}{2}\right)\Delta x = \left(i + \frac{1}{2}\right) h_1, i = 0, 1, \ldots, N; (N+1)h_1 = X_{max}, \\ y_{j+1/2} = \left(j + \frac{1}{2}\right)\Delta y = \left(j + \frac{1}{2}\right) h_2, j = 0, 1, \ldots, M; (M+1)h_2 = Y_{max}, \end{cases}$$

where $\Delta x = h_1, \Delta y = h_2$- grid spacing. N, M- corresponding number of grid cells in directions x, y [a point with coordinates (i, j) coincides with the cell centre].

Let us consider the following difference scheme for equations of the form (3.80) in the finite-difference scheme Ω_h:

$$\frac{u_{i+1/2,\,j}^{(n+1/2)} - u_{i+1/2,\,j}^{(n)}}{\Delta t} + L_{1h}u_{i+1/2,\,j}^{(n)} = \frac{1}{Re}\Delta_h u_{i+1/2,\,j}^{(n)},$$
$$\frac{v_{i,\,j+1/2}^{(n+1/2)} - v_{i,\,j+1/2}^{(n)}}{\Delta t} + L_{2h}v_{i,\,j+1/2}^{(n)} = \frac{1}{Re}\Delta_h v_{i,\,j+1/2}^{(n)}, \tag{3.84}$$

$$\frac{u_{i+1/2,\,j}^{(n+1)} - u_{i+1/2,\,j}^{(n+1/2)}}{\Delta t} + \frac{p_{i+1,\,j}^{(n+1)} - p_{i,\,j}^{(n+1)}}{h_1} = 0,$$
$$\frac{v_{i,\,j+1/2}^{(n+1)} - v_{i,\,j+1/2}^{(n+1/2)}}{\Delta t} + \frac{p_{i,\,j+1}^{(n+1)} - p_{i,\,j}^{(n+1)}}{h_2} = 0, \tag{3.85}$$

$$\underline{div}_h \vec{u}^{(n+1)} = \frac{u_{i+1/2,\,j}^{(n+1)} - u_{i-1/2,\,j}^{(n+1)}}{h_1} + \frac{v_{i,\,j+1/2}^{(n+1)} - v_{i,\,j-1/2}^{(n+1)}}{h_2} = 0;$$

$$L_{1h}u_{i+1/2,j}^{(n)}$$
$$= \frac{1}{2}\left(a_{i+1,j}u_{x_1 i+1/2,j}^{(n)} + a_{ij}u_{\bar{x}_1 i+1/2,j}^{(n)} + b_{i+\frac{1}{2},j+1/2}u_{x_2 i+1/2,j}^{(n)} + b_{i+\frac{1}{2},j-1/2}u_{\bar{x}_2 i+1/2,j}^{(n)}\right);$$

$$L_{2h}v_{i,j+1/2}^{(n)}$$
$$= \frac{1}{2}\left(a_{i+\frac{1}{2},j+1/2}v_{x_1 i,j+1/2}^{(n)} + a_{i-\frac{1}{2},j-1/2}v_{\bar{x}_1 i,j+1/2}^{(n)} + b_{i,j+1}v_{x_2 i,j+1/2}^{(n)} + b_{ij}u_{\bar{x}_2 i,j+1/2}^{(n)}\right);$$

$$a_{ij} = \frac{1}{2}\left(u_{i+\frac{1}{2},j+1}^{(n)} + u_{i-1/2,j}^{(n)}\right),\ b_{i+\frac{1}{2},j+1/2} = \frac{1}{2}\left(v_{i+1,j+1/2}^{(n)} + v_{i,j+1/2}^{(n)}\right);$$

$$a_{i+\frac{1}{2},j+1/2} = \frac{1}{2}\left(u_{i+\frac{1}{2},j+1}^{(n)} + u_{i+1/2,j}^{(n)}\right),\ b_{ij} = \frac{1}{2}\left(v_{i,j+1/2}^{(n)} + v_{i,j-1/2}^{(n)}\right).$$

Here we make use of conventional designations taken from the difference scheme theory [116], corresponding to difference approximation of convective terms in equations (3.80); Δ_h- is the Laplace difference operator:

$$\Delta_h u_{i+1/2,j} = \frac{u_{i+3/2,j} - 2u_{i+1/2,j} + u_{i-1/2,j}}{h_1^2} + \frac{u_{i+\frac{1}{2},j+1} - 2u_{i+1/2,j} + u_{i+\frac{1}{2},j-1}}{h_2^2}.$$

At the first stage of scheme implementation (3.84), (3.85) we can use the explicit formulas tofind the auxiliary value $u^{(n+1/2)}_{i+1/2,\, j}, v^{(n+1/2)}_{i,\, j+1/2}$. Then from relation (3.85) we have $(u^{(n+1)}_{i+1/2,\, j})_{x_2} - (v^{(n+1)}_{i,\, j+1/2})_{x_1} = rot_h \vec{u}^{(n+1/2)} = (u^{(n+1/?)}_{i+1/2,\, j})_{x_2} - (v^{(n+1/2)}_{i,\, j+1/2})_{x_1}$.

Taking into account relation (3.82) and introducing the flow functions we obtain the Poisson equation to determine $\psi^{(n+1)}_{i+1/2,\, j+1/2}$:

$$\Delta_h \psi^{(n+1)}_{i+1/2,\, j+1/2} = \frac{u^{(n+1/2)}_{i+1/2,\, j+1} - u^{(n+1/2)}_{i+1/2,\, j}}{h_2} - \frac{v^{(n+1/2)}_{i+1,\, j+1/2} - v^{(n+1/2)}_{i,\, j+1/2}}{h_1}, \quad (3.86)$$

The values to be found u^{n+1}, v^{n+1}, are known to be determined through relationships

$$u^{(n+1)}_{i+1/2,\, j} = \frac{\psi^{(n+1)}_{i+1/2,\, j+1/2} - \psi^{(n+1)}_{i+1/2,\, j-1/2}}{h_2},\ v^{(n+1)}_{i,\, j+1/2} = \frac{\psi^{(n+1)}_{i+1/2,\, j+1/2} - \psi^{(n+1)}_{i-1/2,\, j+1/2}}{h_1}. \quad (3.87)$$

3.3.4 Conditions Under Which Pressure is Unique

It should be noted that for finding a solution such as (3.84), (3.85) it is not required to solve the Poisson equation for pressure with complex boundary conditions (which have been examined in papers [7, 11, 13, 33–35]). The conditions of pressure uniqueness (3.83) are automatically met.

Indeed, let the contour be circled in a standard manner. Taking the sum in the second relationship (3.85) at $i = i_0$ and spanning j from j_0 to $j_1 - 1$ we obtain:

$$\sum_{j=j_0}^{j_1-1} (p^{(n+1)}_{i_0,\, j+1} - p^{(n+1)}_{i_0,\, j}) = -\frac{h_2}{\Delta t} \sum_{j=j_0}^{j_1-1} (v^{(n+1)}_{i_0,\, j+1/2} - v^{(n+1/2)}_{i_0,\, j+1/2}),$$

$$p_{i_0 j_1} = p_{i_0 j_0} - \frac{h_2}{\Delta t} \sum_{j=j_0}^{j_1 - 1} (v^{(n+1)}_{i_0,\, j+1/2} - v^{(n+1/2)}_{i_0,\, j+1/2}).$$

Then we sum up the first relationship (3.85) at $j = j_1$ varying i from i_0 to $i_1 = 1$, the second relationship (3.85) is summed up at $i = i_1$ running j from $j_1 - 1$ to j_0, in the first we take $j = j_0$ and alter i from i_1 to i_0.

As a result for an unique pressure determination we find that the following equality should be valid

$$\begin{aligned} I = h_1 \sum\nolimits_{i=i_0}^{i_1 - 1} (u^{(n+1)}_{i+1/2, j_0} - u^{(n+1)}_{i+1/2, j_1} + u^{(n+1/2)}_{i+1/2, j_0} - u^{(n+1/2)}_{i+1/2, j_1}) \\ - h_2 \sum\nolimits_{j=j_0}^{j_1 - 1} \left(v^{(n+1)}_{i_0,\, j+1/2} + v^{(n+1/2)}_{i_0,\, j+1/2} - v^{(n+1)}_{i_1,\, j+1/2} + v^{(n+1/2)}_{i_1,\, j+1/2} \right) = 0. \end{aligned} \tag{3.88}$$

Equality (3.88) can be easily transformed into the following difference analogue of an integral pressure condition:

$$I = h_1 h_2 \sum\nolimits_{i=i_0}^{i_1 - 1} \sum\nolimits_{j=j_0}^{j_1 - 1} \left[\Delta_h \psi^{(n+1)}_{i+1/2, j+1/2} - \left(\left(u^{(n+1/2)}_{i+1/2, j} \right)_{x_2} - \left(v^{(n+1/2)}_{i, j+1/2} \right)_{x_1} \right) \right] = 0. \tag{3.89}$$

3.3.5 Boundary Conditions for Flow Function

When numerical type of equations (3.86) is applied to calculate the flow around a flat barrier inside a channel it is necessary to fix the boundary value of the flow function $\psi^{(n+1)}$ at the boundary. This condition is missing in the general formulation of the problem. To find value of $\psi^{(n+1)}$ on the plate surface for every time step let's make use of relationship (3.88). This can be done in the following manner. Represent $\psi^{(n+1)}$ in the form

$$\psi^{(n+1)} = \psi^{(n+1)}_{(2)} + C_0 \psi_{(1)}, \tag{3.90}$$

where $\psi_{(1)}$is the grid function that is a solution of a difference problem

$$\Delta_h \psi_{(1)} = 0 \tag{3.91}$$

with the following boundary conditions:

a. periodicity conditions are satisfied at the boundary of calculation domain, that is at the inlet and outlet;

b. at the solid channel wall $\psi_{(1)} = 0$;

c. at the boundary $\psi_{(1)} = 1$, and the grid function $\psi_{(2)}^{(n+1)}$ satisfies the relationship.

$$\Delta_h \psi_{(2)}^{(n+1)} = rot_h \vec{u}^{(n+1/2)}. \tag{3.92}$$

The following boundary conditions for $\psi_{(2)}^{(n+1)}$ are set:

a. periodic conditions are observed at inlet and outlet;

b. for channel walls $\psi_{(2)}^{(n+2)} = \psi^{(n+1)}$;

c. on the surface of the plate $\psi_{(2)}^{(n+1)} = 0$.

Note that equation (3.91) with corresponding boundary conditions is solved only once, and the solution is entered into the computer's memory.

The unknown value of coefficient C_0 equal to the value $\psi^{(n+1)}$ at the boundary can be found from relationship (3.89).

Indeed, substituting expression (3.90) into identity (3.89) for any contour drawn around the plate we find

$$C_0 = \frac{D_0 - D_1(\psi_{(2)}^{(n+1)}) - D_2(\psi_{(2)}^{(n+1)})}{D_1(\psi_{(1)}) + D_2(\psi_{(1)})}, \tag{3.93}$$

where

$$D_1(\psi) = \frac{h_1}{h_2} \sum_{i=i_0}^{i_1-1} \left(\psi_{i+\frac{1}{2},j_0+1/2} - \psi_{i+\frac{1}{2},j_0-1/2} - \psi_{i+\frac{1}{2},j_1+1/2} + \psi_{i+\frac{1}{2},j_0-1/2} \right);$$

$$D_2(\psi) = \frac{h_2}{h_1} \sum_{j=j_0}^{j_1-1} \left(\psi_{i_0+\frac{1}{2},j+1/2} - \psi_{i_0-\frac{1}{2},j+1/2} - \psi_{i_1+\frac{1}{2},j+1/2} + \psi_{i_1-\frac{1}{2},j+1/2} \right);$$

$$D_0 = h_2 \sum_{j=j_0}^{j_1-1} \left(v_{i_0,j+1/2}^{(n+1/2)} - v_{i_1,j+1/2}^{(n+1/2)} \right) - h_1 \sum_{i=i_0}^{i_1-1} \left(u_{i+1/2,j_0}^{(n+1/2)} - u_{i+1/2,j_1}^{(n+1/2)} \right).$$

Making use of the properties of a solution of the Laplace equation one can readily see that denominator is different from zero in formula (3.93).

For the sake of convenience, calculations of C_0 were made assuming the contour being the same as the outer boundaries of the calculation domain. To conclude description of this algorithm, we will list the sequence of steps to make computer calculations.

1. From equation (3.91) we find a corresponding value $\psi_{(1)}$ and store it in the computer memory.
2. For given initial values explicit formulae (3.85) yield auxiliary values $v^{(n+1/2)}, u^{(n+1/2)}$.
3. Filling the right hand side by known values $u^{(n+1/2)}, v^{(n+1/2)}$ we solve the problem for equation (3.92).
4. Using formula (3.93) and known values $u^{(n+1/2)}, v^{(n+1/2)}, \psi_{(1)}, \psi_{(2)}^{(n+1)}$ we can find coefficient C_0.
5. Formula (3.90) can be used to find the value of flow function for a new time layer.
6. Formulae (3.87) and known values of $\Psi^{(n+1)}$ can be used to find values $u_{i+1/2j}^{(n+1)}, v_{i,\,j+1/2}^{(n+1)}$ within the calculation domain.

In these calculations the criterion for reaching the steady-state regime was provided by condition

$$\left\| \frac{\vec{u}^{(n+1)} - \vec{u}^{n}}{\Delta t} \right\|_{C(\Omega_h)} \leq \varepsilon. \tag{3.94}$$

3.3.6 Numerical Solutions of Equations (3.91), (3.92)

Implementation of a numerical algorithm to solve the periodic problem is provided by solving equations (3.91), (3.92) with corresponding boundary conditions. For example, for equation (3.92) we have a problem with periodic conditions at the inlet and outlet, that is

$$\psi_{-1/2,\,j+1/2} = \psi_{N-1/2,\,j+1/2}, \psi_{1/2,\,j+1/2} = \psi_{N+1/2,\,j+1/2}, \tag{3.95}$$

and known conditions at solid boundaries of the calculation domain. Taking account these peculiarities in boundary conditions the Poisson equation (3.92) is solved using a stabilizing correction method of the following type:

$$\frac{\psi^{(n+1/2)} - \psi^{(n)}}{\tau_0} = L_{11}\psi^{(n+1/2)} + L_{22}\psi^{(n)} - f(x). \tag{3.96}$$

$$\frac{\psi^{(n+1)} - \psi^{(n+1/2)}}{\tau_0} = L_{11}(\psi^{(n+1)} - \psi^{(n)}), \tag{3.97}$$

where τ_0 is the internal iteration parameter; L_{11}, L_{22} are difference operators, corresponding to approximation of second derivatives in x_1 and x_2.

To find $\psi^{(n+1/2)}$ from relationship (3.96) a scalar run is used. As soon as the domain of the problem considered is not regular, i.e. it is impossible to make a direct calculation; the run is conducted in specific parts of the calculation domain.

Let the boundary be confined between the lines corresponding to values x_1 and x_2 having indices $i = i_0, i = i_1, j = j_0, j = j_1$. Then according to the above comment, the first vertical runs are made at $j = \overline{1, N_2 - 1}$ for $i = \overline{0, i_0 - 1}$, then for $j = \overline{1, j_0 - 1}$ at $i = \overline{i_0, i_1}$, and finally for $j = j_1 - 1, N_2 - 1$. The final vertical runs are made at $i = \overline{i_1, N_1}$ for $j = \overline{1, N_2 - 1}$. The calculation formulas for scalar vertical runs are the same.

Now let us describe implementation of the second stage of stabilization scheme, that is solution of equations (3.97). At this stage, calculations are made separately for each subarea of the calculation grid. At $i = \overline{0, N}$ for $j = j_1 - 1,\ N_2 - 1.$ and $j = \overline{j_1 + 1, N_2 - 1}$equation (3.97) is realized through cyclic run for fixed j values.

It is most interesting to implement solution of (3.97) before and after the plate, i.e. it is reduced to the following model equations:

$$u_0 = u_N, u_1 = u_{N_1+1},$$

$$a_n u_{n+1} - b_n u_n + c_n u_{n-1} = -\varphi_n, n = \overline{1, n_0 - 1}, n = \overline{n_1 + 1, N_1}, \tag{3.98}$$

$$u_{n_0} = \alpha, u_{n_1} = \beta.$$

First let us consider equation (3.98) for $n = \overline{n_1 + 1, N_1}$. The solution is represented as a conventional run and the run coefficients can be found in the same manner as in [116]. Taking into account the initial two equations from system (3.98) we can calculate values $u_0, u_1, u_{N_1}, u_{N_1+1}$.

Indeed, in order to determine the above values we can write down the system of four equations

$$u_0 - u_{N_1} = 0, u_1 - u_{N_1+1} = 0, u_1 - X_1 u_0 = Y_1, u_{N_1} - X_{N_1} u_{N_1+1} = Y_{N_1}, \tag{3.99}$$

with denominator differing from zero, as $|X_n| \leq 1$.

For example, from these equations u_0 is calculated using the formula

$$u_0 = \frac{Y_{N_1} + Y_1 X_{N_1}}{1 - X_1 X_{N_1}}.$$

This is followed by determining all u_n values. Cycle calculation formulae, as was mentioned above, are given in [116].

3.3.7 Numerical Calculations and their Comparison with Results of Previous Studies

The method used to calculate the Navier-Stokes equations described here differs from the earlier published splitting approach [7, 11, 13, 33–35]. First of all we take into account the boundary conditions and use a different method of implementing the solution. As is known, in the past, the laminar flow condition was accounted for in a line drawn through the centre of the grid cell contacting the solid surface, whereas the no flow conditions were held for the whole cell. The above condition required solving the Poisson equation for the pressure using complex boundary conditions. Here we are considering the opposite case. In our situation the laminar flow condition is exact, whereas the no flow condition has an order of approximation $O(h^2)$ at a line $h/2$ away from the surface of the solid. This enables us to introduce the flow function. In this case the pressure can be calculated explicitly. Consequently the comparison with results of previous studies can illustrate the potential of the approach proposed as well as guarantee the accuracy of calculations.

As an example, we will consider a well-known problem of the flow of a viscous incompressible fluid in a cylindrical container having a sliding piston at the top. The calculations were made using grids (41, 41), (21, 21).

The simulation was conducted over a wide range of Reynolds numbers. The results indicate that in every situation the steady-state flow regime was reached. As a practical criterion of convergence and stability of the difference scheme used we required that condition

$$\frac{8\tau\nu}{h^2} < 1, h = \min(h_1, h_2)$$

is satisfied.

As a rule, to satisfy condition (3.94) it took $\approx 300-500$ iterations. The data obtained are in agreement with results of the paper [102]. At $Re = 500$ the turbulence zone develops in the left and right bottom corners. However, this flow is absent when $Re = 100$.

The coordinates of the centre of the principal turbulence obtained for $Re = 100, 500$, differ from those of paper [102] by 2%.

3.3.8 Numerical Calculations for Periodic Problem of Streamlines Around Plates

The above algorithm was used in a simulation of the grid (41, 41) having spacing $h_1 = 0,025$ and $h_2 = 0,0125$. A plate having length of 0,15 was located symmetrically across the channel at $x = 0,3$. The fluid flow rate inside the channel was assumed equal to 1, the Re number varied from 100 to 250, and the time interval was selected within the $\tau = 0,0005 \div 0,005$ range.

The initial velocity distribution was assumed to be symmetrical relative to the channel axis. The optimum iteration parameters were found experimentally: $\tau_0 = 0,002, \tau_0 = 0,005$. For $Re = 100 \div 250$ we observed, that the steady-state regime had reverse flow behind the plate. The typical streamline pattern is shown in Fig. 3.1 for $Re = 100$. With increasing Re the size of the turbulence zone expanded and the centre of turbulence shifted towards the plate.

Here we will describe the numerical experiments for an asymmetrical initial velocity distribution in more detail. In order to set u^0, v^0 the Laplace equation was solved with $\Delta_h \psi = 0$, the boundary conditions being: $\psi = 0,1$ on the plate, $\psi = \pm 0,5$ at the corresponding walls. The periodicity conditions applied to the outer regions of the calculation domain. Then the values u^0, v^0 were found using formulae (3.87). Calculations were made for $Re = 100$ and 250 on the surface of the walls (41, 41) and (41, 21).

The calculation results show that the difference in the value of the flow function at the line $x = 0,3$ (where the plate is fixed) is of the order of (5 ± 10) % for the grids specified. When flow develops (at $n \to \infty$) a stable pulsating regime develops in streamlining flow forming vortices next to the plate. The turbulence gets separated at a later stage.

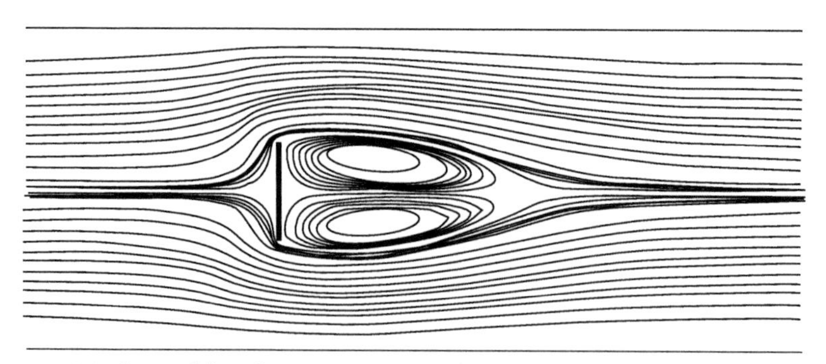

Figure 3.1 Isolines of flow function.

Fig. 3.2 shows the values of flow function at the plate for $Re = 250$.

Also, the value of the flow function is shown together with v behind the plate for the different times on the channel's axis of symmetry. As is seen from the figure, after $n = 1630$ the periodicity of motion is clearly visible.

Figs. 3.3–3.5 depict the isolines of the flow functions showing sequential formation of vortices and their separation followed by the dissipation characteristics of decaying potential vortices.

A similar behaviour of vortices was described in paper [85].

Fig. 3.6 shows the value of velocity profile for $x = 0,3$ and at the inlet for $n = 1630, n = 2100$, when flow function isolines are the closest to each other. We will note that the relative error of inlet velocity in this case is 2%, the profiles of velocity above the plate are almost identical.

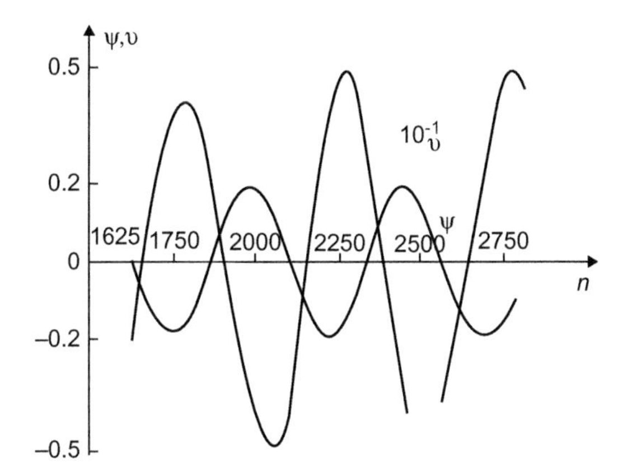

Figure 3.2 The values of flow function at the plate for $Re = 250$.

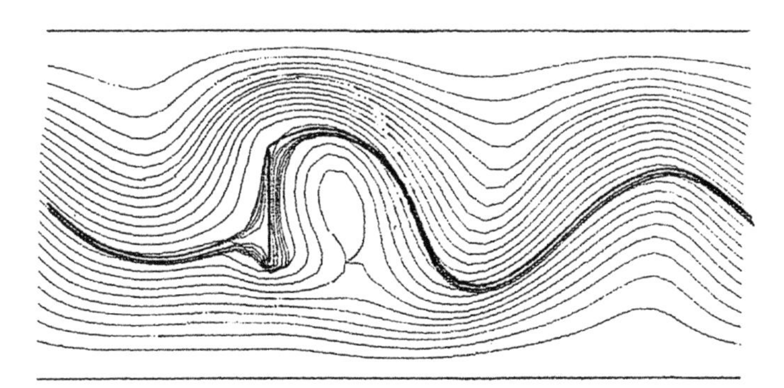

Figure 3.3 Flow function isolines plotted for $Re = 100, T = 2, 10$.

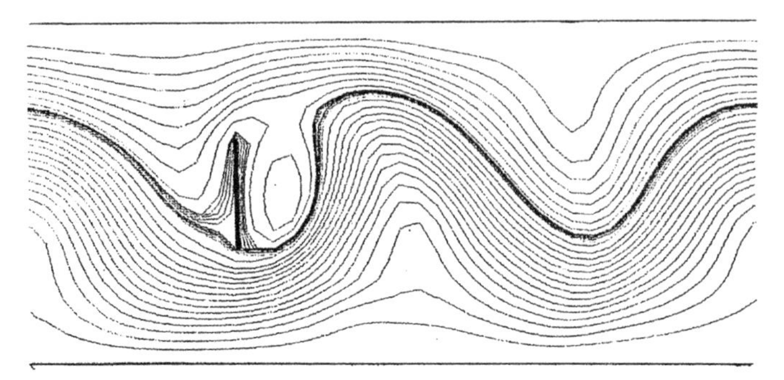

Figure 3.4 Flow function isolines plotted for $Re = 250, T = 1,392$.

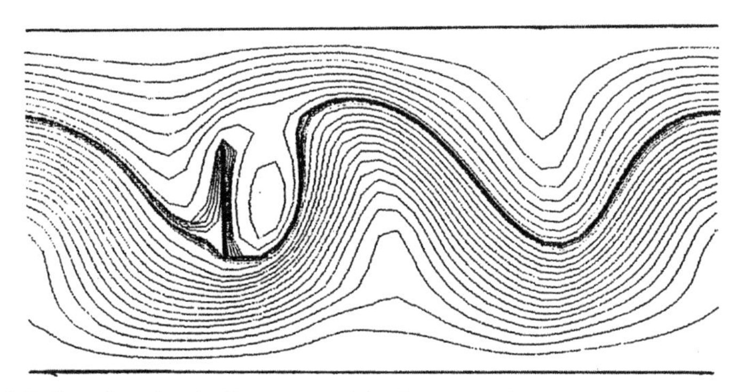

Figure 3.5 Flow function isolines plotted for $Re = 250, T = 1,420$.

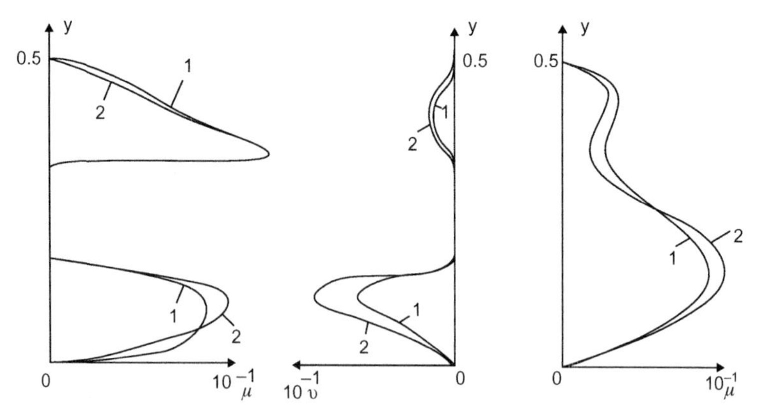

Figure 3.6 $1 - n = 1630$; $2 - n = 2100$.

On average depending on *Re* it took 20÷35 iterations to satisfy the steady state criterion

$$\left\| \Delta_h \psi_{(2)}^{n+1} - rot_h \vec{u}^{n+1/2} \right\|_{L_2(\Omega_h)} \leq 10^{-4}$$

with stabilizing corrections having the form of (3.96), (3.97) for each time step.

3.4 IMPLEMENTING THE METHOD OF SPLITTING INTO UNDERLYING PHYSICAL PROCESSES IN COMPLEX GEOMETRICAL REGIONS

3.4.1 Introduction

The idea of region regularization was put forward by Teachmarch [133] for the problem of determining. However the idea of the method of virtual domains was used for the first time in the papers by V.K. Saulyev [120].

The method of virtual domains involves the consideration of an auxiliary problem with a small parameter in a larger domain. The solution of the auxiliary problem should be close to that of a solution of the initial problem. In this section we will make a numerical calculation of the flow of a viscous incompressible fluid inside a planar channel having periodically arranged barriers and an arbitrary curved boundary. These problems cannot be solved by using the regular grid and homogeneous difference method.

A similar problem was considered earlier in the papers of B.T. Zhumagulov, Sh.S. Smagulov, M.K. Orunkhanov, N.M. Temirbekov and M.I. Iztleuov. However, these studies approached the problem from the standpoint of variables: flow function – turbulence. Numerical algorithms were developed in papers [116], [118]. Their solution convergence was weakly dependent on a small parameter.

It is known that quite often the flow symmetry is broken in viscous flow streamlining symmetric bodies even for relatively small Reynolds numbers. Therefore, the numerical simulation of this problem is interesting both from the theoretical and practical point of view.

V.A. Gushchin [33] and O. M. Belotserkovsky [10] illustrated the efficiency of splitting up the underlying physical processes using an example of transverse streamlining a cylinder by the homogeneous (infinite) flow of an incompressible fluid.

In paper [97] the method of virtual domains is used to analyse the problem of a streamlining a pair of cylinders situated inside a planar channel with periodic boundary conditions in the cross sections. The resultant Navier-Stokes equation is developed in flow function – turbulence variables.

In this section the problem of streamlining a cylindrical body inside a planar channel having periodic boundary conditions is examined numerically. The method of virtual domains is used in combination with the algorithm for splitting up the underlying physical processes.

After finding a steady-state solution with an accuracy of $\varepsilon = 10^{-4}$ the flow is perturbed through equating to zero the velocity component of the symmetrical half. Even at these conditions the numerical iterations result in attaining a steady-state regime.

3.4.2 Formulation of the Problem

Let us consider flow in a planar channel having regular barriers (see Fig. 3.7).

We solve the Navier-Stokes equation in a double connected domain Ω_1, limited by straight lines $x = 0, X$ and solid walls γ and γ_0:

$$\frac{\partial \vec{v}}{\partial t} + (\vec{v}\nabla)\vec{v} = \nu\Delta\vec{v} - \nabla p, div\vec{v} = 0; \tag{3.100}$$

$$\vec{v}\Big|_{t=0} = v_0(x), \frac{\partial^k \vec{v}}{\partial x_1^k}\Big|_{x=0} = \frac{\partial^k \vec{v}}{\partial x_1^k}\Big|_{x_1=X};$$

$$\vec{v}\Big|_{\gamma} = 0, \vec{v}\Big|_{\gamma_0} = 0, \int_0^Y udy = Q, \vec{v} = (u, v). \tag{3.101}$$

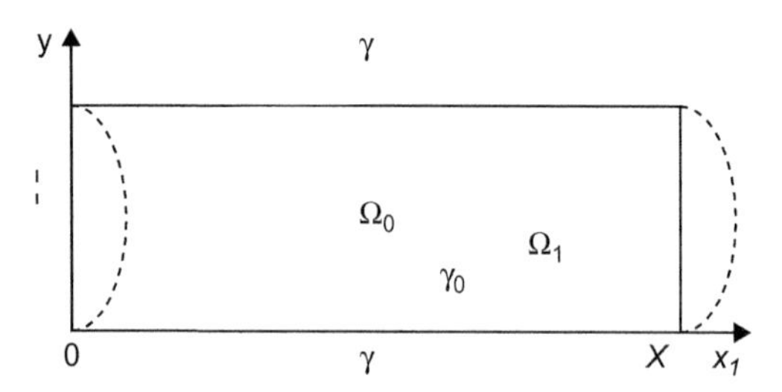

Figure 3.7 Flow in a planar channel with regular barriers.

Problems (3.100), (3.101) are peculiar as they have a curvilinear boundary γ_0. In order to solve the problems using homogeneous difference schemes we will consider an auxiliary problem [22] in line with the method of virtual domains:

$$\frac{\partial \vec{v}^{\varepsilon}}{\partial t} + (\vec{v}^{\varepsilon}\nabla)\vec{v}^{\varepsilon} = \nu\Delta\vec{v}^{\varepsilon} - \frac{\xi(x)}{\varepsilon}\vec{v}^{\varepsilon} - \nabla p^{\varepsilon}, div\vec{v}^{(\varepsilon)} = 0; \qquad (3.100*)$$

$$\vec{v}^{\varepsilon}\Big|_{t=0} = v_0(x), \frac{\partial^k \vec{v}^{\varepsilon}}{\partial x_1^k}\Big|_{x_1=0} = \frac{\partial^k \vec{v}^{\varepsilon}}{\partial x_1^k}\Big|_{x_1=X}, \vec{v}^{\varepsilon}\Big|_{\gamma} = 0, \int_0^y u dy = Q, \vec{v}^{\varepsilon} = (u^{\varepsilon}, v^{\varepsilon}), \qquad (3.101*)$$

where $\xi(x) = 1$ at $x \in \Omega_0$ and $\xi(x) = 0$ at $x \in \Omega_1$.

The paper [22] proved the theorem of existence of generalized solution to problem (3.100*), (3.101*). It was established that the solution of problems (3.100*), (3.101*) converges to the solution of problems (3.100), (3.101) at a rate of ε^{α}:

$$\left\|\vec{v}^{\varepsilon} - \vec{v}\right\|_{L_\infty(0,T;L_2(\Omega_1))} + \int_0^T \left\|\vec{v}^{\varepsilon} - \vec{v}\right\|^2_{W_2^1(\Omega_1)} \leq C\varepsilon^{\alpha}, \qquad (3.102*)$$

where α is a positive constant.

The steady-state Navier-Stokes equations were examined using the methods of virtual domains in paper [122].

3.4.3 Numerical Solution Algorithm

The numerical solution of (3.100*), (3.101*) was provided by the method of splitting into * physical processes. Assume that $X = Y = 1$ (which can be done through scaling the variables x, y). Let

$$\Omega_{h_1} \equiv \left\{x_1 = i_1 h_1, \ \ x_2 = i_2 h_2, i_1 = \frac{1}{2}, \ldots, N_1 - \frac{1}{2}, i_2 = 0, \ldots, N_2\right\},$$

$$\Omega_{h_2} \equiv \left\{x_1 = i_1 h_1, \ \ x_2 = i_2 h_2, i_1 = 0, \ldots, N_1, i_2 = \frac{1}{2}, \ldots, N_2 - \frac{1}{2}\right\}.$$

Now we will consider the difference schemes using the method of splitting

$$\frac{\tilde{u}_{i+1/2,\,j}-u^{(n)}_{i+1/2,\,j}}{\Delta t}+L^1_h(\vec{u}^{(n)},u^{(n)})_{i+1/2,\,j}=\nu\Delta_h u^{(n)}_{i+1/2,\,j}\ in\ \Omega_{h_1},$$
$$\frac{\tilde{v}_{i,\,j+1/2}-v^{(n)}_{i,\,j+1/2}}{\Delta t}+L^2_h(\vec{v}^{(n)},v^{(n)})_{i,\,j+1/2}=\nu\Delta_h v^{(n)}_{i,\,j+1/2}\ in\ \Omega_{h_2},\tag{3.102}$$

$$\frac{u^{(n+1)}_{i+1/2,\,j}-\tilde{u}_{i+1/2,\,j}}{\Delta t}+\frac{p^{(n+1)}_{i+1,\,j}-p^{(n+1)}_{ij}}{h_2}+\frac{\xi(x)}{\varepsilon}v^{(n+1)}_{ij+1/2}=0,$$
$$\frac{v^{(n+1)}_{i,\,j+1/2}-\tilde{v}_{i,\,j+1/2}}{\Delta t}+\frac{p^{(n+1)}_{ij+1}-p^{(n+1)}_{ij}}{h_2}+\frac{\xi(x)}{\varepsilon}v^{(n+1)}_{ij+1/2}=0,\tag{3.103}$$

$$\frac{u^{(n+1)}_{i+1/2,\,j}-u^{(n+1)}_{i-1/2,\,j}}{h_1}+\frac{v^{(n+1)}_{i,\,j+1/2}-v^{(n+1)}_{i,\,j-1/2}}{h_2}=0,\tag{3.104}$$

$$v^{(n+1)}\Big|_{\gamma_h}=u^{(n+1)}\Big|_{\gamma_h}=0,\ \int_0^1 u^{(n+1)}dy\le Q,\tag{3.105}$$

with conditions of periodicity in x_1:L^1_h, L^2_h are the approximation of non-linear terms. To apply the difference schemes we will use the methods proposed in 3.3 of the present Chapter. Let us introduce a difference analogue of the flow function:

$$u^{(n+1)}_{i+1/2,j}=-\frac{\psi^{(n+1)}_{i+1/2,j+1/2}-\psi_{i+1/2,j-1/2}}{h_2};v^{(n+1)}_{i,j+1/2}=-\frac{\psi^{(n+1)}_{i+1/2,j+1/2}-\psi^{(n+1)}_{i-1/2,j+1/2}}{h_1}.\tag{3.106}$$

We can re-write (3.103) as follows:

$$\left(u^{(n+1)}_{i+1/2,\,j}+\frac{\Delta t}{\varepsilon}\xi(x)u^{(n+1)}_{i+1/2,\,j}\right)+\Delta t\frac{p^{(n+1)}_{i+1,\,j}-p^{(n+1)}_{ij}}{h_1}=\tilde{u}_{i+1/2,\,j},\tag{3.107}$$

$$\left(v_{i,j+1/2}^{(n+1)} + \frac{\Delta t}{\varepsilon}\xi(x)v_{i,j+1/2}^{(n+1)}\right) + \Delta t\frac{p_{i,j+1}^{(n+1)} - p_{ij}^{(n+1)}}{h_2} = \tilde{v}_{i,j+1/2}. \quad (3.108)$$

Differentiating (3.107) by difference in x_2, (3.108) – in x_1 and subtracting the result from each other we obtain

$$\left(u_{i+1/2,j}^{(n+1)} + \frac{\Delta t}{\varepsilon}\xi(x)u_{i+1/2,j}^{(n+1)}\right)_{x_2} - \left(v_{i,j+1/2}^{(n+1)} + \frac{\Delta t}{\varepsilon}\xi(x)v_{i,j+1/2}^{(n+1)}\right)_{x_1} = \tilde{u}_{i+1/2,jx_2} - \tilde{v}_{i,j+1/2x_1}. \quad (3.109)$$

We will note that after substitution of (3.106) equation (3.104) is identically fulfilled. In (3.109), using formula (3.106), we obtain the relationship for $\psi_{i+1/2,j+1/2}^{(n+1)}$

$$\left(-\psi_{i+1/2,j+1/2\bar{x}_2}^{(n+1)} - \frac{\Delta t}{\varepsilon}\xi(x)\psi_{i+1/2,j+1/2\bar{x}_2}^{(n+1)}\right)_{x_2} - \left(\psi_{i+1/2,j+1/2\bar{x}_1}^{(n+1)} - \frac{\Delta t}{\varepsilon}\xi(x)\psi_{i+1/2,j+1/2\bar{x}_1}^{(n+1)}\right)_{x_1} = \tilde{u}_{i+1/2,jx_2} - \tilde{v}_{ij+1/2x_1}. \quad (3.110)$$

Let us re-write (3.110) as follows:

$$\left(\left(1+\frac{\Delta t}{\varepsilon}\xi(x)\right)\psi_{i+1/2,j+1/2\bar{x}_2}^{(n+1)}\right)_{x_2} + \left(\left(1+\frac{\Delta t}{\varepsilon}\xi(x)\right)\psi_{i+1/2,j+1/2\bar{x}_1}^{(n+1)}\right)_{x_1} = \tilde{u}_{i+1/2,jx_2} - \tilde{v}_{ij+1/2x_1}. \quad (3.111)$$

For $\psi^{(n+1)}$ the following boundary conditions are valid:

$$\psi_{i+\frac{1}{2,3}\over 2}^{(n+1)} = 0, \psi_{i+1/2,N-1/2}^{(n+1)} = const,$$
$$\psi_{3/2,j+1/2}^{(n+1)} = \psi_{N-1/2,j+1/2}^{(n+1)}, \psi_{5/2,j+1/2}^{(n+1)} = \psi_{N+1/2,j+1/2}^{(n+1)}. \quad (3.112)$$

The stability of the difference schemes (3.111), (3.112) is obvious when the right hand side is known. We will assume that

$f \equiv (v_{i,\, j+1/2x_1} - \tilde{u}_{i+1/2,\, jx_2}) \in L_2(\Omega_1)$. Multiplying (3.111) by $2\Delta t h_1 h_2 \psi_{i+}$ $1/2,\, j+1/2^{(n+1)}$ and summing in i, j using the formula of summation by parts we obtain

$$\left\| \sqrt{1+\frac{\Delta t}{\varepsilon}\xi(x)\psi_{x_2}^{(n+1)}} \right\|^2 + \left\| \sqrt{1+\frac{\Delta t}{\varepsilon}\xi(x)\psi_{x_1}^{(n+1)}} \right\|^2 \leq C\|f\|^2_{L_2(\Omega_h)}. \quad (3.113*)$$

3.4.4 Sequence of Calculations

From equation (3.102) we can find $\tilde{u}_{i+1/2,\, j}, \tilde{v}_{i,\, j+1/2}$. Then implementing the iteration method from (3.111), (3.112) we will calculate $\psi^{(n+1)}_{i+1/2,\, j+1/2}$. Due to formula (3.106) we obtain $u^{(n+1)}_{i+1/2,\, j}, v^{(n+1)}_{i,\, j+1/2}$. We will dwell on solving the problems (3.111), (3.112) for sufficiently small ε.

As $\xi(x)$ is a discontinuous function the equation refers to a class of equations with fast alternating coefficients. The operator (3.111) and (3.112) is positively determined and self- conjugated. The ratio of the minimum to the maximum eigenvalue depends on ε. Consequently, equations (3.111), (3.112) are ill-posed. It is not reasonable to apply classical iteration techniques to solve problems (3.111), (3.112) as the rate of convergence of the iteration process depends on a small parameter. Consequently, to solve problems (3.111), (3.112) effectively, one has to develop special iteration methods where the convergence rate is a weak function of ε. The identical approach to solve the Dirichlet problem for an elliptical equation has been considered in papers [21, 28, 55, 97, 116, 118, 119].

In paper [97] the Richardson method of extrapolation was used to improve the accuracy of approximation using the method of virtual domains. In [21, 28, 136] the method of variable directions was proposed:

$$D^* \frac{\psi^{k+1/2} - \psi^k}{\omega} + A_1^{(\varepsilon)}\psi^{k+1/2} + A_2^{(\varepsilon)}\psi^k = f,$$

$$D^*_\varepsilon \frac{\psi^{k+1} - \psi^{k+1/2}}{\omega} + A_1^{(\varepsilon)}\psi^{k+1/2} + A_2^{(\varepsilon)}\psi^{k+1} = f,$$

where $A_\alpha^{(\varepsilon)}\psi = \left(\left(1+\frac{\Delta t}{\varepsilon}\xi(x)\right)t_{\overline{x}_\alpha}\right)_{x_\alpha}, \alpha = 1, 2$; which, as was mentioned in [21], converges to a steady-state solution irrespective of D^*_ε. For the right choice of the diagonal operator D^*_ε. The modified alternate-triangular iteration method (MATIM) was developed in papers [116; 118]. MATIM can be applied to solve a periodic problem at the boundary of the calculation domain in the following way [30]. We will write down the solution of the problem (3.111), (3.112) in the form

$$\psi^{n+1}_{i+\frac{1}{2},j+1/2} = \psi^{(0)}_{i+\frac{1}{2},j+1/2} + \alpha_{j+1/2}\psi^{(1)}_{i+\frac{1}{2},j+1/2} + \beta_{j+1/2}\psi^{(2)}_{i+\frac{1}{2},j+1/2}, \tag{3.113}$$

where $\psi^{(0)}_{i+\frac{1}{2},j+1/2}, \psi^{(1)}_{i+\frac{1}{2},j+1/2}, \psi^{(2)}_{i+\frac{1}{2},j+1/2}$ are the solutions of the following auxiliary problems:

$$\left(\left(1+\Delta t\frac{\xi(x)}{\varepsilon}\right)\left(\psi^{(0)}_{i+\frac{1}{2},j+1/2}\right)_{\overline{x}}\right)_x + \left(\left(1+\frac{\Delta t}{\varepsilon}\xi(x)\right)\left(\psi^{(0)}_{i+\frac{1}{2},j+1/2}\right)_{\overline{y}}\right)_y$$

$$= \tilde{u}_{i+1/2,jy} - \tilde{v}_{i,j+1/2,x}, \psi^{(0)}_{i+1/2,3/2} = 0, \psi^{(0)}_{i+\frac{1}{2},N-1/2} = const, \psi^{(0)}_{\frac{3}{2},j+1/2} \tag{3.114}$$

$$= \psi^{n}_{\frac{3}{2},j+1/2}, \psi^{(0)}_{N+\frac{1}{2},j+1/2} = \psi^{n}_{N+\frac{1}{2},j+1/2},$$

$$div_h\left(\left(1+\frac{\Delta t}{\varepsilon}\xi(x)\right)\nabla_h\psi^{(1)}\right) = 0, \psi^{(1)}_{i+\frac{1}{2},\frac{3}{2}} = 0, \psi^{(1)}_{i+\frac{1}{2},M-\frac{1}{2}} = 0,$$

$$\psi^{(1)}_{\frac{3}{2},j+1/2} = 1, \psi^{(1)}_{N+\frac{1}{2},j+1/2} = 0, div_h\left(\left(1+\Delta t\frac{\xi(x)}{\varepsilon}\right)\nabla_h\psi^{(2)}\right) = 0, \tag{3.115}$$

$$\psi^{(2)}_{i+1/2,3/2} = 0, \psi^{(2)}_{i+\frac{1}{2},M-1/2} = 0, \psi^{(2)}_{\frac{3}{2},j+1/2} = 0, \psi^{(2)}_{N+\frac{1}{2},j+1/2} = 1. \tag{3.116}$$

The coefficients $\alpha_{j+1/2}, \beta_{j+1/2}$ are found from the boundary condition (3.112). We obtain the following system of equations:

$$\begin{cases} \left(1-\psi^{(1)}_{N-1/2,\,j+1/2}\right)\alpha_{j+1/2} - \psi^{(2)}_{N-1/2,\,j+1/2}\beta_{j+1/2} = \psi^{(0)}_{N-1/2,\,j+1/2}, \\ -\psi^{(1)}_{5/2,\,j+1/2}\alpha_{j+1/2} + \left(1-\psi^{(2)}_{5/2,\,j+1/2}\right)\beta_{j+1/2} = \psi^{(0)}_{5/2,\,j+1/2}. \end{cases} \tag{3.117}$$

As is shown in (3.113)–(3.117) the problems (3.115), (3.116) are solved once, that is they do not complete a full iteration step in time. For each full iteration step it is necessary to solve problem (3.114), in which the values of the flow function in the lower time layer are taken as boundary conditions for $\psi^{(0)}_{\frac{3}{2},j+1/2}$ and $\psi^{(0)}_{N+\frac{1}{2},j+1/2}$. One can easily notice that the problems (3.114)–(3.116) have simpler boundary conditions than the problem (3.111), (3.112). To solve the problem (3.114)–(3.116) one can make use of the modified alternate-triangular method [116]. Its convergence rate does not depend on the value of coefficients $\left(1+\frac{\Delta t}{\varepsilon}\xi(x)\right)$. We will assume

$$a_{1,ij}=\left(1+\frac{\Delta t}{\varepsilon}(\xi(x)_{i,j+1/2})\right),\ a_{2,ij}=\left(1+\frac{\Delta t}{\varepsilon}(\xi(x)_{i+1/2,j})\right).$$

Let's write the difference scheme (3.114) in new designations

$$\Lambda\psi=\left(\tilde{u}_{i+\frac{1}{2},j,x_2}-\tilde{v}_{ij+1/2,x_1}\right),\psi_{i+1/2,3/2}=0,\psi_{i+\frac{1}{2},M-1/2}=const,$$

$$\psi_{\frac{3}{2},j+1/2}=\psi^{n}_{\frac{3}{2},j+1/2},\psi_{N+\frac{1}{2},j+1/2}=\psi^{n}_{N+\frac{1}{2},j+1/2},\tag{3.118}$$

Where $\Lambda=\tilde{\Lambda}_1+\tilde{\Lambda}_2$. First of all we can represent A in the form $\Lambda=\Lambda_1+\Lambda_2$:

$$\begin{aligned}\Lambda_1\psi&=-\sum_{\alpha=1}^{2}\left[\frac{a_\alpha}{h_\alpha}\psi_{\bar{x}_\alpha}+\frac{1}{2h_\alpha}\left(\frac{\overset{+}{a}_\alpha}{h_\alpha}-\frac{a_\alpha}{h_\alpha}\right)\psi_{i+\frac{1}{2},j+1/2}\right],\\ \Lambda_2\psi&=-\sum_{\alpha=1}^{2}\left[\frac{\overset{+}{a}_\alpha}{h_\alpha}\psi_{x_\alpha}+\frac{1}{2h_\alpha}\left(\frac{\overset{-}{a}_\alpha}{h_\alpha}-\frac{a_\alpha}{h_\alpha}\right)\psi_{i+\frac{1}{2},j+1/2}\right],\end{aligned}\tag{3.119}$$

where $\overset{+}{a}_1=a_{1,i+1,j},\overset{+}{a}_2=a_{2,i,j+1}$.

We assume that $A_\alpha\psi=-\Lambda_2\psi,\alpha=1,2$ for any $\psi\in\omega_h$. Then $A=A_1+A_2=-\Lambda$. To determine operator D we can use the following formula

$$d(x_1,x_2)=\sum_{\alpha=1}^{2}\left(\frac{\overset{+}{a}_\alpha}{h^2_\alpha\sqrt{\varphi_\alpha}}+\frac{1}{2h_\alpha}\left|\frac{\overset{+}{a}_\alpha-a_\alpha}{h_\alpha}\right|\right)\frac{1}{\sqrt{\varphi_\alpha}+\theta_\alpha},\tag{3.120}$$

$$\varphi_\alpha(x_\beta) = \max_{x_\alpha} v_1^{(\alpha)}(x), \theta_\alpha(x_\beta) = \max_{x_\alpha} v_2^{(\alpha)}(x).$$

The functions $v_1^{(\alpha)}(x)$ and $v_2^{(\alpha)}(x)$ are determined as solutions to problems

$$(a_1 v_{1\bar{x}_1}^{(1)})_{x_1} = -\rho_1^{(1)}, x_1 \in \omega_1(x_2), v_1^{(1)}\big|_{\gamma_h} = 0, \rho_1^{(1)} = \frac{\overset{+}{a}_{1,ij}}{h_1^2}; \tag{3.121}$$

$$(a_2 v_{1\bar{x}_2}^{(2)})_{x_1} = -\rho_1^{(2)}, x_2 \in \omega_2(x_1), v_1^{(2)}\big|_{\gamma_h} = 0, \rho_1^{(2)} = \frac{\overset{+}{a}_{2,ij}}{h_2^2}; \tag{3.122}$$

$$(a_1 v_{2\bar{x}_1}^{(1)})_{x_1} = -\rho_2^{(1)}, x_1 \in \omega_1(x_2), v_2^{(1)}\big|_{\gamma_h} = 0, \rho_2^{(1)} = \frac{1}{2h_1^2}\left|\overset{+}{a}_{1,ij} - a_{1,ij}\right|; \tag{3.123}$$

$$(a_2 v_{2\bar{x}_2}^{(2)})_{x_2} = -\rho_2^{(2)}, x_2 \in \omega_2(x_1), v_2^{(2)}\big|_{\gamma_h} = 0, \rho_2^{(2)} = \frac{1}{2h_2^2}\left|\overset{+}{a}_{2,ij} - a_{2,ij}\right|. \tag{3.124}$$

We have to determine only four functions: $v_1^{(1)}, v_1^{(2)}, v_2^{(1)}, v_2^{(2)}$. This can be done using the fitting technique. Knowing $v_1^{(\alpha)}(x)$ and $v_2^{(\alpha)}(x)$, we construct the grid functions of one variable $\varphi_\alpha(x_\beta)$, $\psi_\alpha(x_\beta)$, to be followed by determining $d(x)$ using formula (3.117). For the chosen $d(x)$ we have [116]

$$\delta = 1, \Delta = 4\max_{\alpha=1,2}\left(\max_{x_\beta \in \omega_\beta}\left(\sqrt{\varphi_\alpha(x_\beta)} + \psi_\alpha(x_\beta)\right)^2\right). \tag{3.125}$$

Now we can determine parameters ω_0 and the optimum set of Chebyshev parameters $\{\tau_k\}$. We will use the iteration scheme [116]

$$(D + \omega A_1)D^{-1}(D + \omega A_2)\frac{\psi^{k+1} - \psi^k}{\tau_{k+1}} + A\psi^k$$
$$= -\left(u_{i+\frac{1}{2},jx_2} - \tilde{v}_{ij+\frac{1}{2x_1}}\right), k = 0, 1, 2, \ldots, n.$$

3.4.5 Calculation Results

For numerical calculations of the symmetrical flow of viscous incompressible fluid streamlining a cylinder the parameter ε in (3.102) was chosen as equal to $\varepsilon = 10^{-6}$. The number of internal iterations to solve the problem (3.118) was calculated through the formula [116]:

$$n = \frac{\ln(2/\varepsilon_1)\sqrt[4]{\Delta}}{2\sqrt{2}}, \varepsilon \text{ is the required accuracy.}$$

The dimensions of the calculation domain are $x = 2,5; y = 1$. The centre of a cylinder having radius $r = 0,2$ was set at the point $x = 1, y = 0,5$. The calculation domain was split into 50×20 cells. In our calculations for $\tau = 10^{-4}$ and $\varepsilon_1 = 10^{-9}$ the number of iterations equalled $n = 367$. Solutions of the problem (3.115), (3.116) converge with accuracy of $\varepsilon_1 = 10^{-9}$ over 362 iterations.

In Fig. 3.8 the isolines of flow functions are shown for the viscous fluid flow streamlining a cylinder for $Re = 10$ and $Re = 50$.

It is seen from the calculation results that there are no reverse flow zones. Further, in Fig. 3.9, *a* we see that at $Re = 100$ the reverse flow zones develop behind the cylinder.

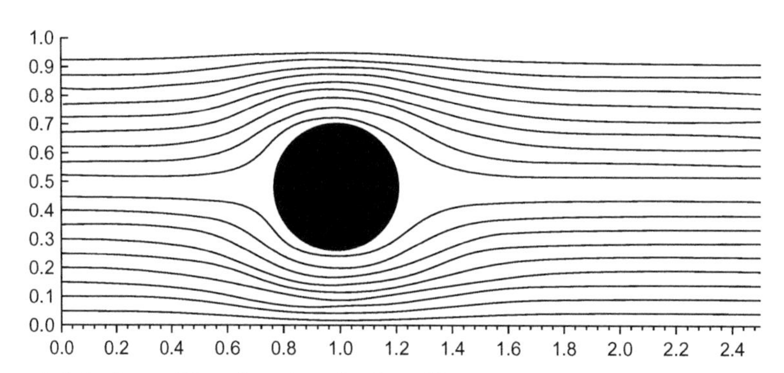

Figure 3.8 Isolines of flow functions for $Re = 10$.

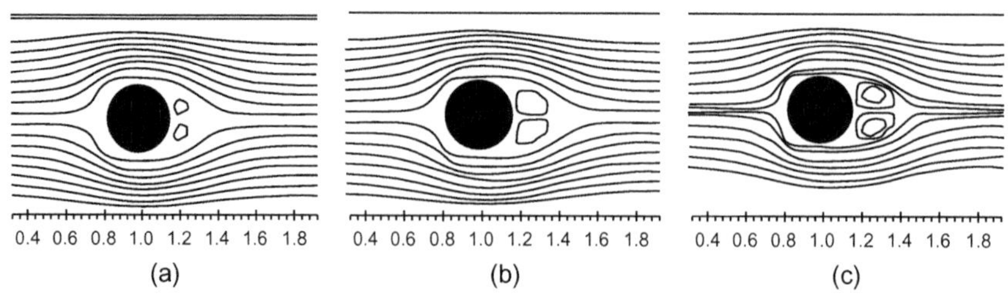

Figure 3.9 Isolines of flow functions.

When the Re number increases the volume of reverse flow zones becomes bigger (see Fig. 3.9 *b, c*).

In the case of incompressible fluid and for low or moderate Reynolds numbers the motion behind a symmetrical finite body may develop in two regimes: 1) symmetrical steady-state motion; 2) unsymmetrical periodic motion. At low Reynolds numbers experiments and calculations show [10, 11, 35] that we are dealing with the steady-state regime. It remains stable irrespective of the value of the instantaneous initial perturbation. In our calculations we assumed the perturbation to be as follows: the orthogonal velocity component in front of the cylinder was set equal to zero body on the symmetrical side. However, due to the small perturbation the finite number of iterations resulted in attaining the symmetrical steady-state regime.

The numerical solution of the non-stationary problem of the flow of viscous fluid streamlining a cylindrical body is interesting from the theoretical and practical point of view. To obtain a non-stationary flow scheme it is necessary to introduce a perturbation and to trace its evolution. We made an attempt to model a structure of a periodic flow. As the initial approximation we took the following function of the flow distribution:

$$\psi^0_{i+\frac{1}{2},j+1/2} = \sin\left(\frac{\pi}{2}y_j + 1/2\right), i = \overline{1, N}, j = \overline{2, M-1}.$$

In numerical simulation the solution of the problem (3.111), (3.112) was found using the following scheme of stabilizing correction:

$$(E - L_1)(E - \tau L_2)\frac{\psi^{(n+1,k+1)} - \psi^{(n+1,k)}}{\tau_0} + L\psi^{(n,k)} = \tilde{u}_{i+1/2jy} - \tilde{v}_{ij+1/2x}. \tag{3.127}$$

Solving the above problem for the vertical direction was made through the scalar fitting. The method of cyclic fitting was used to solve the problem in a horizontal direction. In numerical calculations we chose the following parameters: $Re = 250, \tau = 0,0005; \tau_0 = 0,01$.

The calculation domain was split into 50×20 cells. To obtain the approximate solution of the problem (3.106), (3.107) with accuracy $\varepsilon = 10^{-4}$ it took an average number of 20 iterations. The criterion to attain the steady state was as follows:

$$\left\| \frac{\psi^{(n+1,k+1)} - \psi^{(n+1,k)}}{\tau_0} \right\|_C \leq \varepsilon_1.$$

In practice these are asymmetric or non-stationary types of motion in the wake that are absolutely stable.

Figs. 3.10–3.12 show the isolines of the flow functions ($\psi = const$) for streamlining a cylinder by a viscous fluid at $Re = 250$ for different points in time.

A non-stationary flow is observed. A stable area is formed behind the cylinder in the bottom half-plane. When the "turbulence" is separated and fluid is discharged from this region the stable area is formed again to be followed by separation in the upper half- plane.

Thus, the periodic (self-oscillating) regime develops for the flow of fluid behind the cylinder. The period is $T \approx 1$.

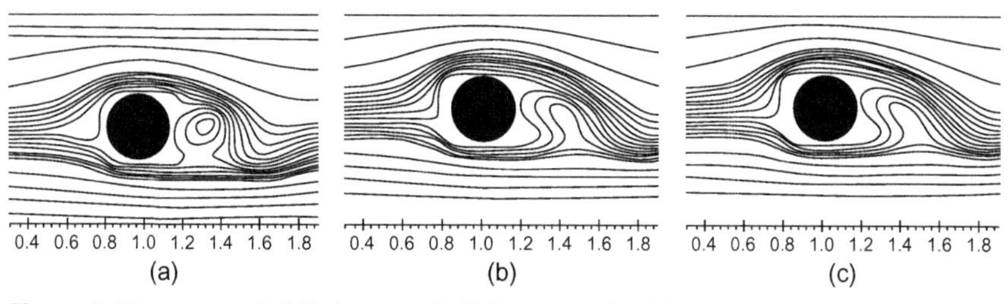

Figure 3.10 $a - t_1 = 0,443; b - t_2 = 0,618; c - t_3 = 0,623.$

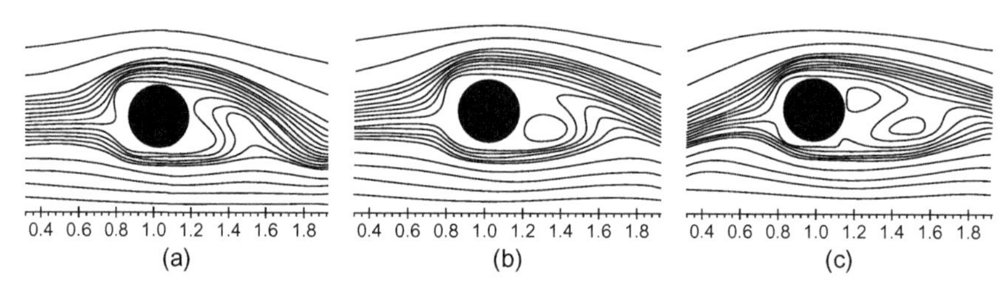

Figure 3.11 $a - t_4 = 0,623; b - t_5 = 0,893; c - t_6 = 1,143.$

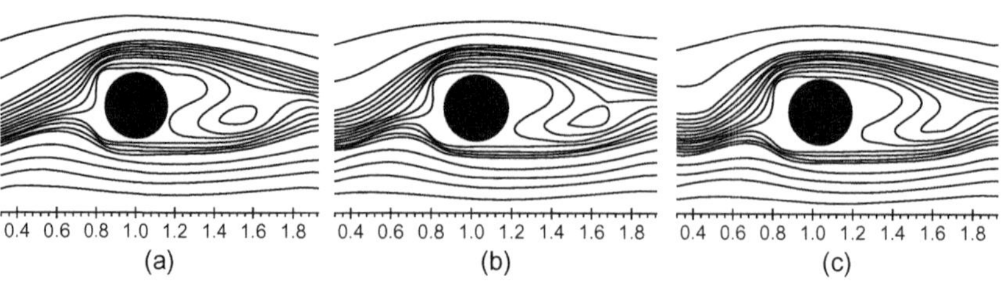

Figure 3.12 $a - t_7 = 1,193; b - t_8 = 1,243; c - t_9 = 1,1293.$

3.5 NUMERICAL SIMULATION OF THE OIL DISPLACEMENT PROCESS BASED ON THE N.E. ZHUKOVSKY MODEL

3.5.1 Introduction

The method was developed to determine the velocity and pressure distribution in formations near the bottom of a well using the N.E. Zhukovsky model [106]. The method is based on solving the problem through assessing the distribution of pressure in a formation having known permeability for the known pressure inside the well and an unknown fluid flow rate. The effect of boundary conditions of solid impermeable walls on the pressure was taken into account. The solution of the boundary problem may be used to evaluate practical well operating situations [9, 23, 26, 46, 47, 86, 126].

3.5.2 Calculation of the Pressure in Bottom-hole Zones

Formulation of the problem and algorithm for numerical solution. The model problem of a water displacing oil ("water flooding") having horizontal layered flow is considered. The problem is analysed in two-dimensions. The flow region corresponds to the vertical bed section [50]. Horizontal boundaries Γ_1, Γ_4 and vertical boundary Γ_3 of the formation are assumed to be impermeable; boundaries Γ_5 and Γ_6 are assumed to be isobars with known injection pressure (on Γ_5) and outlet pressure. The shaded area is occupied by the porous medium.

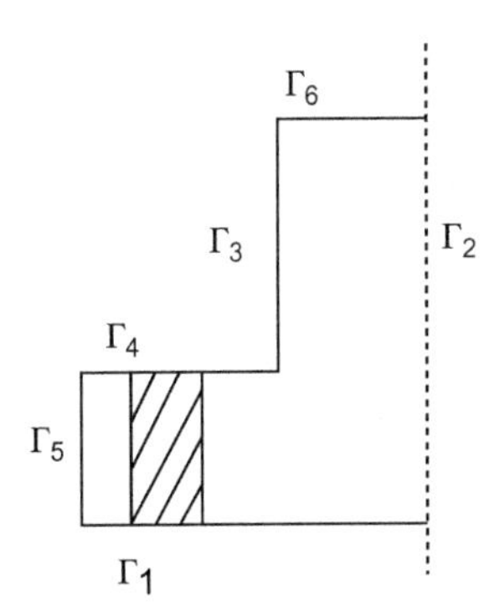

Figure 3.13 The problem of water displacing oil ("water flooding") with horizontal layered flow.

Corresponding equations have the form

$$\frac{\partial \vec{u}}{\partial t} + (\vec{u}\nabla)\vec{u} + \nabla p = \frac{1}{Re}\Delta\vec{u} - \lambda(x, y)\vec{u},\ div\vec{u} = 0. \tag{3.128}$$

Here $\vec{u}(u, v)$ is the velocity vector; p is the pressure; $\lambda(x, y)$ is the coefficient of resistance of a porous medium which is determined in the following way:

$$\lambda(x, y) = \begin{cases} \mu m/k & -in\ porous\ medium, \\ 0 & -in\ nonporous\ medium, \end{cases}$$

Where m is the porosity coefficient; Re is the Reynolds number; k is the permeability coefficient of porous material. The boundary conditions are:

$$\begin{gathered} \vec{u} = 0, (x, y) \in \Gamma_k, k = 1, 3, 4; p = p_0, v = 0, (x, y) \in \Gamma_5; \\ p = p_1, u = 0, (x, y) \in \Gamma_6, \frac{\partial v}{\partial x} = 0, u = 0, (x, y) \in \Gamma_2. \end{gathered} \tag{3.129}$$

Problem (3.128), (3.129) is solved by a finite-difference method on a hybrid grid. The known method of splitting into the underlying physical processes [12] was used to determine the velocity and pressure.

3.5.3 Approximation of the Boundary Conditions

At the lower boundary Γ_1 the conditions of laminar flow and no flow are set, such that $u^{n+1}_{i+1/2,1/2} = 0, v^{n+1}_{i,1/2} = 0$. We expand into the Taylor series the tangent velocity component $u(x, y)$ normal to y relative to Γ_1 in the wall-adjacent points $\left(x_{i+1/2}; \frac{h}{2}\right)$:

$$u^{n+1}_{i+1/2,1/2} = u^{n+1}_{i+1/2,1} - \frac{h_2}{2}\left(\frac{\partial u}{\partial y}\right)^{n+1}_{i+1/2,1} + \frac{h_2^2}{8}\left(\frac{\partial^2 u}{\partial y^2}\right)^{n+1}_{i+1/2,1} + O(h_2^3).$$

Let us consider the projection of equation of motion on Ox in the same point:

$$\left(\frac{\partial^2 u}{\partial y^2}\right)^{n+1}_{i+1/2,1} = Re\left(\frac{\partial p}{\partial x}\right)^{n+1}_{i+1/2,1} + O(Reh_2).$$

We will substitute $\left(\frac{\partial u}{\partial y}\right)_{i+1/2,1}^{n+1}$ by $\left(\frac{\partial u}{\partial y}\right)_{i+1/2,1}^{n}$ with accuracy $O(\tau)$:

$$u_{i+1/2,1}^{n+1} = \frac{h_2}{2}\left(\frac{\partial u}{\partial y}\right)_{i+1/2,1}^{n+1} - \frac{h_2^2}{8}Re\left(\frac{\partial p}{\partial x}\right)_{i+1/2,1}^{n+1}$$

Let's write down the following equality in the form

$$u_{i+1/2,1}^{n+1} = \tilde{u}_{i+1/2,1} - \tau_0\left(\frac{\partial p}{\partial x}\right)_{i+1/2,1}^{n+1}, \tau_0 = \frac{h_2^2}{8}Re,$$

where $\tilde{u}_{i+1/2,1} = \frac{h_2}{2}\left(\frac{\partial u}{\partial y}\right)_{i+1/2,1}^{n}$.

We approximate $\left(\frac{\partial u}{\partial y}\right)_{i+1/2,1}^{n}$ through values $u_{i+1/2j}^{n}$ in nodes $i+\frac{1}{2},\frac{1}{2}$; $i+\frac{1}{2},1; i+\frac{1}{2},2$ and obtain

$$\left(\frac{\partial u}{\partial y}\right)_{i+1/2,1}^{n} = \frac{-4u_{i+1/2,1/2}^{n} + 3u_{i+1/2,1}^{n} + u_{i+1/2,2}^{n}}{3h_2}.$$

Taking into account the boundary condition we find

$$\tilde{u}_{i+1/2,1} = \frac{3u_{i+1/2,1}^{n} + u_{i+1/2,2}^{n}}{6}, i = 2, 3, \ldots, N_1.$$

Equation for pressure for $j = 1$ is found from equation

$$\frac{u_{i+1/2,1}^{n} - u_{i+1/2,1}^{n+1}}{h_1} + \frac{v_{i,3/2}^{n+1}}{h_2} = 0, i = 2, 3, \ldots, N_1.$$

Equation for pressure has the form

$$\tau_0\frac{p_{i+1,1}^{n+1} - 2p_{i,1}^{n+1} + 2p_{i-1,1}^{n+1}}{h_1^2} + \tau\frac{p_{i,2}^{n+1} - p_{i,1}^{n+1}}{h_2^2} = \frac{\tilde{u}_{i+1/2,1}^{n+1} - \tilde{u}_{i-1/2,1}^{n+1}}{h_1} + \frac{\tilde{v}_{i,3/2}^{n+1}}{h_2}.$$

At the boundary Γ_4 when $1 \le i \le k, j = m + 1/2$ the boundary conditions are specified as $u_{i+\frac{1}{2},m+1/2}^{n+1} = v_{i,m+1/2}^{n+1} = 0$.

Reasoning in the same way we obtain

$$u^{n+1}_{i+1/2,m} = \tilde{u}_{i+1/2,m} - \tau_0 \frac{p^{n+1}_{i+1,m} - p^{n+1}_{i,m}}{h_1}, i = 2, 3, \ldots, k,$$

where $\tilde{u}_{i+1/2,m} = \frac{3u^n_{i+1/2,m} + u^n_{i+\frac{1}{2},m-1}}{6}$.

The respective continuity equations and relationships to determine pressure at $j = m$ have the form:

$$\tau_0 \frac{p^{n+1}_{i+1,m} - 2p^{n+1}_{i,m} + p^{n+1}_{i-1,m}}{h_1^2} + \tau \frac{p^{n+1}_{i,m} - p^{n+1}_{i,m-1}}{h_2^2} = \frac{\tilde{u}_{i+1/2,m} - \tilde{u}_{i-1/2,m}}{h_1} - \frac{\tilde{v}_{i,m-1/2}}{2};$$

$$\frac{u^{n+1}_{i+1/2,m} - u^{n+1}_{i-1/2,m}}{h_1} - \frac{v^{n+1}_{i,m-1/2}}{h_2} = 0.$$

At the boundary Γ_3:

$$i = k + \frac{1}{2}, m + \frac{1}{2} < j < N_2, u^{n+1}_{k+1/2,j} = v^{n+1}_{k+\frac{1}{2},j+1/2} = 0.$$

For the nodes near the boundary the numerical values of velocity can be determined as follows:

$$v^{n+1}_{k+1,j+1/2} = \tilde{v}_{k+1,j+1/2} - \tau_1 \frac{p^{n+1}_{k,j+1} - p^{n+1}_{k,j}}{h_2}, j = m + 1, m + 2, \ldots, N_2,$$

where $\tilde{v}_{k+1,j+1/2} = \frac{3v^n_{k+1,j+1/2} + v^n_{k+2,j+1/2}}{6}$, $\tau_1 = \frac{h_1^2 Re}{8}$.

The equation of continuity at $i = k + 1$ takes the form

$$\frac{u^{n+1}_{k+3/2,j}}{h_1} + \frac{v^{n+1}_{k+1,j+1/2} - v^{n+1}_{k+1,j-1/2}}{h_2} = 0, j = m + 2, \ldots, N_2.$$

Equations for pressure are as follows:

$$\frac{\tau}{h_1^2}(p^{n+1}_{k+2,j} - p^{n+1}_{k+1,j}) + \frac{\tau}{h_2^2}(p^{n+1}_{k+1,j+1} - 2p^{n+1}_{k+1,j} + p^{n+1}_{k+1,j-1})$$

$$= \frac{\tilde{u}_{k+3/2,j}}{h_1} + \frac{\tilde{v}_{k+1,j+1/2} - \tilde{v}_{k+2,j-1/2}}{h_2}.$$

The boundary Γ_2 is the symmetry axis, $i = N_1 + \frac{1}{2}, 1/2 \leq j \leq N$. The difference analogue of the boundary conditions on symmetry axis can be written as:

$$v^{n+1}_{N_1-1,j+1/2} = v^{n+1}_{N_1+1,j+1/2}, u^{n+1}_{N_1+3/2j} = -u^{n+1}_{N_1-1/2j}.$$

The equation of continuity in the vicinity of the symmetry axis has the form

$$-\frac{u^{n+1}_{N_1-1/2,j}}{h_1} + \frac{v^{n+1}_{N_1,j+1/2} - v^{n+1}_{N_1,j-1/2}}{h_2} = 0, j = 2, 3, \ldots, N_2 - 1.$$

Pressure value in the nodes next to symmetry axis is given by formula

$$p^{n+1}_{N_1,j} = \left[\frac{p^{n+1}_{N_1-1,j}}{h_1^2} + \frac{p^{n+1}_{N_1,j+1} + p^{n+1}_{N_1,j-1}}{h_2^2} - f\right]\frac{1}{\frac{1}{h_1^2} + \frac{1}{h_2^2}}; j = 2, 3, \ldots, N_2 - 1,$$

where $f = \frac{1}{\tau}\left[-\frac{\tilde{u}_{N_1-1/2,j}}{h_1} + \frac{\tilde{v}^{n+1}_{N_1,j+1/2} - \tilde{v}_{N_1,j-1/2}}{h_2}\right].$

Pressure in the corner (point $i = k + 1, j = m$) is calculated from the relationship

$$\frac{1}{h_1^2}(\tau(p^{n+1}_{k+2,m} - p^{n+1}_{k+1,m}) - \tau_0(p^{n+1}_{k+1,m} - p^{n+1}_{k,m})) + \frac{1}{h_2^2}(\tau_1(p^{n+1}_{k+1,m+1} - p^{n+1}_{k+1,m})$$

$$-\tau_0(p^{n+1}_{k+1,m} - p^{n+1}_{k+1,m-1})) = \frac{\tilde{u}_{k+3/2,m} - \tilde{u}_{k+1/2,m}}{h_1} + \frac{\tilde{v}_{k+1,m+1/2} - \tilde{v}_{k+\frac{1}{2},m-1/2}}{h_2}.$$

3.5.4 Calculation Results

Computer simulation was carried out for various *Re* numbers, permeability coefficients, and geometries.

The domain of numerical integration of the system of the difference equations (3)−(5) was covered by a grid having a size 21×41 and 31×41. All calculations were made for the same initial approximation until the following inequality was satisfied $\left\| u^{n+1} - u^n \right\| \geq \varepsilon, \varepsilon = 2 \times 10^{-4}$.

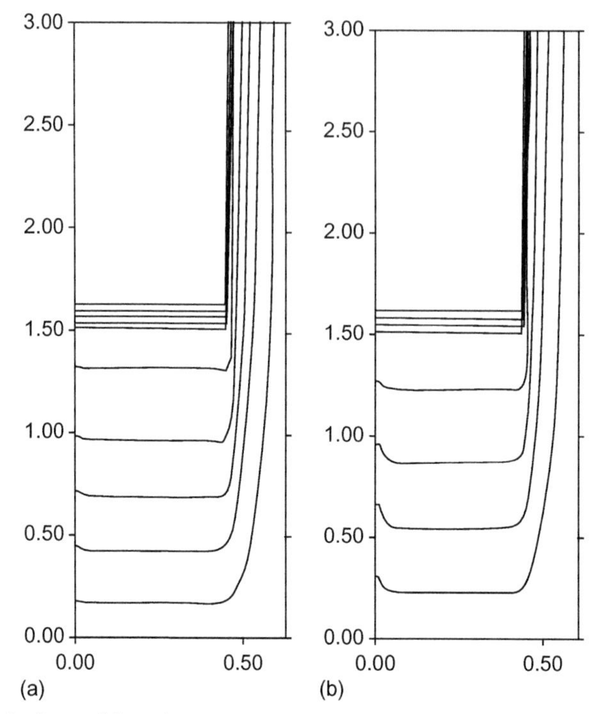

Figure 3.14 Isolines of flow functions at $Re = 100$.
a – fluid flow in a formation without porous medium; b – a calculated region partially filled by porous medium; the size of the grid is 21×41; iteration parameter $t = 0,001$; steps in spatial variables: $h_1 = 0,015, h_2 = 0,0025$.

Fig. 3.14 shows isolines of the flow functions for $Re = 100$. As Fig. 3.14, *a* shows the fluid flow region is not occupied by the porous medium.

The numerical simulation shows that the pressure values decrease in direction of the fluid flow. In front of the porous barrier the pressure increases and velocities equilibrate. A dramatic change in velocity profile takes place at the outlet of the layer. The maximum velocity is observed next to the corner whereas the conditions of complete laminar flow are met on the wall.

For $0 \leq x \leq l_1, 0 \leq y \leq l_2$ the simulation domain is filled with porous material – (a) and with the medium – (b); the grid size is 31×41; iteration parameter $t = 0,001$.

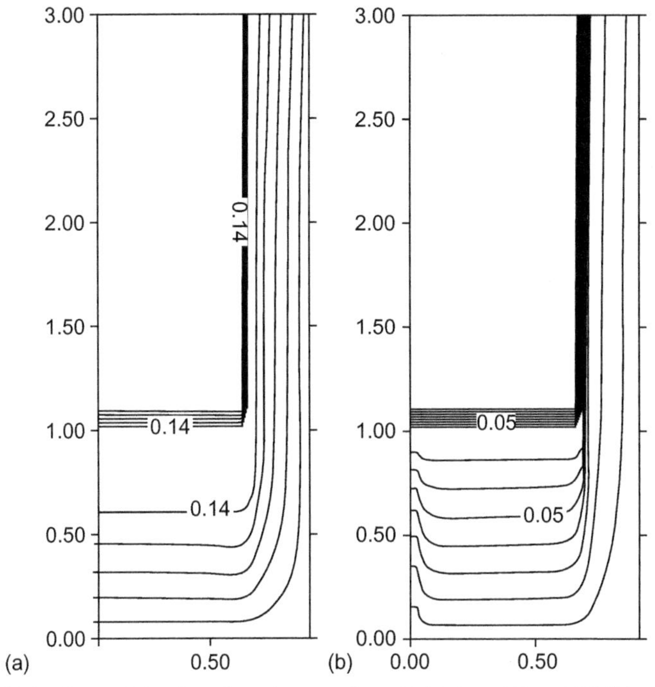

Figure 3.15 Isolines of the flow functions at $Re = 100$ (a) and $Re = 500$ (b).

Isolines of flow functions for $Re = 100$ and $Re = 500$ are given in Fig. 3.15. With increase of Re number and growing resistance of the porous layer the distribution of the pressure becomes complicated. In the middle section of the bed under the porous layer there are areas of apparent increase in pressure. Inside the porous part the pressure is almost constant across the channel section, whereas it significantly reduces downstream in the filtered fluid.

Using the numerical results we can reach the following conclusions:

- stability of the fluid occurs in front of the porous layer for any parameters controlling the flow regime;
- the flow through the porous layer is close to one-dimensional;
- the velocity increases near the wall behind the porous layer.

3.6 NUMERICAL MODEL TO FIND FORMATION PRESSURE

In this paragraph we examine the problem of viscous fluid flowing through an L-shaped domain having set the pressure $p = p_{out}$ at the outlet and zero tangent component of the velocity vector at the inlet and outlet (uniform flow). This problem models finding formation pressure $p = p_{form}$ using its measured value $p = p_{outlet}$ at the well head of a producing oil well which is important from a practical point of view.

We prove the convergence of the difference approximation of this problem thereby validating its general solvability in time. A computer model was proposed to simulate more complicated problem when the total pressure $q_{outlet} = p + \frac{1}{2}\left|\vec{u}\right|$ is used instead of pressure p_{outlet}. In this case, instead of using the Navier-Stokes the Zhukovsky model is used.

3.6.1 Substantiation of Numerical Algorithm to Determine Bottom Hole Pressure

Let the oil flow in through the section AB and out through section DM (Fig. 3.16). Flow is caused by the pressure drop at the boundaries AB and DM.

The problem is described by the system of differential equations:

$$\frac{\partial \vec{u}}{\partial t} + (\vec{u}\nabla)\vec{u} = \mu\Delta\vec{u} - \nabla p, \, div\vec{u} = 0 \tag{3.130}$$

with initial-boundary conditions

$$\vec{u}\Big|_{t=0} = \vec{u}_0(x), \tag{3.131}$$

$$\text{AB: } p = p_0 = const \text{ (unknown)}, \;\; v = 0; \text{BCD:} \vec{u} = 0,$$

$$\text{DM: } p = p_1 = const \text{ (known)}, u = 0; \int_M^D v dx = Q/2, \tag{3.132}$$

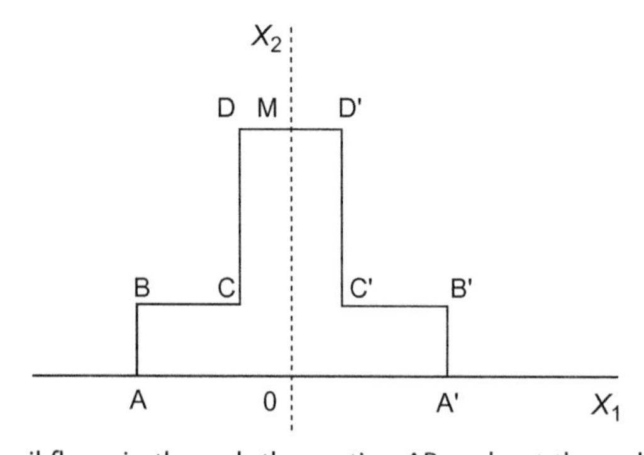

Figure 3.16 The oil flows in through the section AB and out through section DM.

where Q is the flow rate;

AO: $\vec{u} = 0$; OM: $u = 0, v_x = 0$ is the symmetry condition.

Let us introduce flow function assuming $u = \psi_{x_2}, v = -\psi_{x_1}$.

We will re-write the system of equations (3.130) in variables ψ, ω:

$$\frac{\partial \omega}{\partial t} + J(\psi, \omega) = \mu \Delta \omega, \Delta \psi = \omega,$$
$$J(\psi, \omega) = \psi_{x_2} \frac{\partial \omega}{\partial x_1} - \psi_{x_1} \frac{\partial \omega}{\partial x_2}, \omega = u_{x_2} - v_{x_1}. \tag{3.133}$$

Conditions (3.131), (3.132) take the form:

$$\omega\big|_{t=0} = \omega_0(x); \text{AB:} \mu \frac{\partial \omega}{\partial x_1} = \frac{\partial}{\partial x_1}(\psi_{x_1}, \psi_{x_2}), \frac{\partial \psi}{\partial x_1} = 0;$$

$$\text{AO:} \psi = 0, \frac{\partial \psi}{\partial x_2} = 0; \text{OM:} \psi = 0, \omega = 0; \tag{3.134}$$

$$\text{MD:} \frac{\partial \psi}{\partial x_2} = 0; \mu \frac{\partial \omega}{\partial x_2} = -\frac{\partial}{\partial x_2}(\psi_{x_1}, \psi_{x_2}); \text{DBC:} \psi = \frac{Q}{2} \cdot |DM|, \frac{\partial \psi}{\partial n} = 0.$$

The pressure on AB is calculated using the equations (3.130):

$$dp = \nu(\omega_{x_2} dx_1 - \omega_{x_1} dx_2) - \frac{\partial}{\partial t}\psi_{x_2} dx_1 + \frac{\partial}{\partial t}\psi_{x_1} dx_2$$
$$-\left(\frac{\partial}{\partial x_1}(\psi_{x_2})^2 + \frac{\partial}{\partial x_2}(\psi_{x_1} \cdot \psi_{x_2})\right) dx_1 + \left(\frac{\partial}{\partial x_1}(\psi_{x_1} \cdot \psi_{x_2}) + \frac{\partial}{\partial x_2}(\psi_{x_1})^2\right) dx_2. \tag{3.135}$$

We will integrate (3.135) along an arbitrary curve connecting boundaries DM and AB:

$$p_{\text{AB}} = P_{\text{DM}} + \oint_\gamma \left(\mu \omega_{x_2} - \frac{\partial}{\partial t}\psi_{x_2} - \frac{\partial}{\partial x_1}(\psi_{x_2})^2 + \frac{\partial}{\partial x_2}(\psi_{x_1} \cdot \psi_{x_2})\right) dx_1$$
$$+ \oint_\gamma \left(-\mu \omega_{x_1} + \frac{\partial}{\partial t}\psi_{x_1} + \frac{\partial}{\partial x_1}(\psi_{x_1} \cdot \psi_{x_2}) + \frac{\partial}{\partial x_2}(\psi_{x_1})^2\right) dx_2, \tag{3.136}$$

In this manner we can find the pressure at bottom-hole. Sometimes instead of the boundary condition (3.132) one can use:

AB: $p + \frac{u^2 + v^2}{2} = p_0 = const$ (unknown), $v = 0$;

BC, CD, AO – $\vec{u} = 0$ – laminar flow condition,

DM: $p + \frac{u^2 + v^2}{2} = p_1 = const$ (known), $u = 0$, $\int_M^D v dx = Q/2$,

OM: $v = 0$; $v_{x_1} = 0$.

Then (3.134) take the form

$$\text{AB:}\mu\frac{\partial\omega}{\partial x_1} = \psi_{x_2}\cdot\omega - \frac{1}{2}(\psi_{x_2})^2_{x_2}, \frac{\partial\psi}{\partial x_1} = 0; \text{AO:}\psi = 0, \frac{\partial\psi}{\partial x_2} = 0; \quad (3.137)$$

$$\text{DM:}\mu\frac{\partial\omega}{\partial x_2} = -\psi_{x_1}\omega + \frac{1}{2}(\psi_{x_1})^2_{x_1}, \frac{\partial\psi}{\partial x_2} = 0; \text{BCD:}\psi = \frac{Q}{2}\cdot|\text{DM}|, \frac{\partial\psi}{\partial n} = 0.$$

For simplicity let us consider a linear stationary problem in a rectangular domain. The fluid is flowing in through section AB, and leaving the domain through section DM.

In this case DM and AO are parallel.

The functions $\omega(x), \psi(x), x = (x_1, x_2)$ satisfy the boundary problem:

$$\Delta\omega = f, \Delta\psi = \omega, \quad (3.138)$$

$$\text{AB:}\frac{\partial\omega}{\partial x_1} = 0, \frac{\partial\psi}{\partial x_1} = 0, \text{AO:}\psi = 0, \frac{\partial\psi}{\partial x_2} = 0, \quad (3.139)$$

$$\text{DM:}\frac{\partial\omega}{\partial x_2} = 0, \frac{\partial\psi}{\partial x_2} = 0, \text{DB:}\psi = 0, \frac{\partial\psi}{\partial x_2} = 0,$$

where f is the known smooth function.

Multiplying the second equation of the system (3.138) by ψ and integrating, we obtain

$$\int_\Omega |\Delta\psi|^2 dx = \int_\Omega f\cdot\psi dx. \quad (3.140)$$

We will introduce the class $m(\Omega) = \{\psi \in C^4(\Omega), \frac{\partial\psi}{\partial n}|_{\partial\Omega} = 0, \psi = 0$ on AO and DB, $\frac{\partial\psi}{\partial n} = 0$ on AB and DM$\}$.

The closure $m(\Omega)$ in $W_2^2(\Omega)$ is designated as $\hat{W}_2^2(\Omega)$.

Consider $\hat{W}_2^{-2}(\Omega)$ having norm $\|f\|_{\hat{W}_2^{-2}(\Omega)} = \sup_{\|\varphi\|_{\hat{W}_2^{-2}(\Omega)}=1} |(f \cdot \varphi)|$.

We will construct the grid in a way that boundaries AB and AO are located between nodes $(0,j)$ and $(1,j)$ and $(N_0 - 1, j), (N_1, j)$:

$$\Delta_h \omega_{ij} = f_{ij}, \Delta_h \psi_{ij} = \omega_{ij}, \tag{3.141}$$

$$\text{AB:} \omega_{0j} = \omega_{1j}, \psi_{0j} = \psi_{1j}, j = \overline{1, N-1},$$

$$\text{DM:} \omega_{Nj} = \omega_{N-1j}, \psi_{Nj} = \psi_{N-1j}, j = \overline{1, N-1}, \tag{3.142}$$

$$\text{AO:} \psi_{i0} = 0; \omega_{i0} = \frac{2}{h^2}\psi_{i1}, i = \overline{1, N-1}, \text{DB:} \psi_{i0} = 0;$$
$$\omega_{iN} = \frac{2}{h^2}\psi_{N-1}, i = \overline{1, N-1}.$$

Lemma 1. *The problem(3.141),(3.142) has a unique solution complying with the following estimate*

$$\|\Delta_h \psi\|^2 \le C\|f\|_{h_0}^2. \tag{3.143}$$

Proof. Inequality (3.143) is obtained by multiplying (3.141) by ψ_{ij} and using the Helder inequality. Using the known methods [25] the problem (3.141), (3.142) can be reduced to the difference system of equations:

$$\Delta_h \omega_{ij} + C_{ij}(x)\psi_{ij} = f_{ij}, \Delta_h \psi_{ij} = \omega_{ij}, \tag{3.144}$$

where $C_{ij}(x) = \begin{cases} -\frac{2}{h^4}; i = \overline{1, N-1}, j = \overline{1, N-1}, \\ 0. \end{cases}$

At boundaries AB, DM the conditions for ω_{ij}, ψ_{ij} remain the same for AO, DM. Thus,

$$\text{AO:} \psi_{i0} = 0, \omega_{i0} = 0; \text{DB:} \psi_{i0} = 0, \omega_{iN} = 0. \tag{3.145}$$

The difference equations (3.144) with conditions (3.145) have quickly alternating coefficients, therefore we should make use of special iteration techniques:

$$\frac{\omega^{n+1/2}-\omega^n}{\Delta t}=\Delta_h\omega^n+C_{ij}(x)\psi^{n+\frac{1}{2}}+f_{ij};\ \Delta_h\psi^{n+\frac{1}{2}}=\omega^{n+1/2};$$
$$\frac{\omega^{n+1}-\omega^{n+1/2}}{\Delta t}=\Delta_h(\omega^{n+1}-\omega^n) \tag{3.146}$$

with relevant boundary conditions.

The solution (3.146) is provided by the modified alternate-triangular iteration method.

To solve the nonlinear problem we can use the following scheme:

$$\frac{\omega^{n+1/2}-\omega^n}{\Delta t}+(\omega_{ij}^n\psi_{ijx_2}^{n+1/2})_{x_1}-(\omega_{ij}^n\psi_{ijx_1}^{n+1/2})_{x_2}=\nu\Delta_h\omega_{ij}^n+C_{ij}(x)\psi_{ij}^{n+\frac{1}{2}}+f_{ij}; \tag{3.147}$$

$$\Delta_h\psi^{n+1/2}=\omega^{n+1/2};$$

$$\frac{\omega^{n+1}-\omega^n}{\Delta t}=\nu\Delta_h(\omega^{n+1}-\omega^n)$$

with conditions (3.142) and (3.145). The scheme (3.147), (3.142), and (3.145) is stable in space $L_2(\Omega_h)$ [25].

3.6.2 The Case of Zhukovsky Model

Consider the following system of nonlinear equations describing the fluid flow through a porous media:

$$\frac{\partial\vec{u}}{\partial t}+(\vec{u}\nabla)\vec{u}+\nabla p=\mu\Delta\vec{u}-\lambda(x,y)\vec{u},\ div\vec{u}=0. \tag{3.148}$$

The initial conditions are:

$$\vec{u}\big|_{t=0}=\vec{u}_0(x,y). \tag{3.149}$$

The boundary conditions are given as:

$$\text{AB}: p=p_0=const\ \text{(unknown)},\ v=0;\ \text{BD}: u=v=0; \tag{3.150}$$

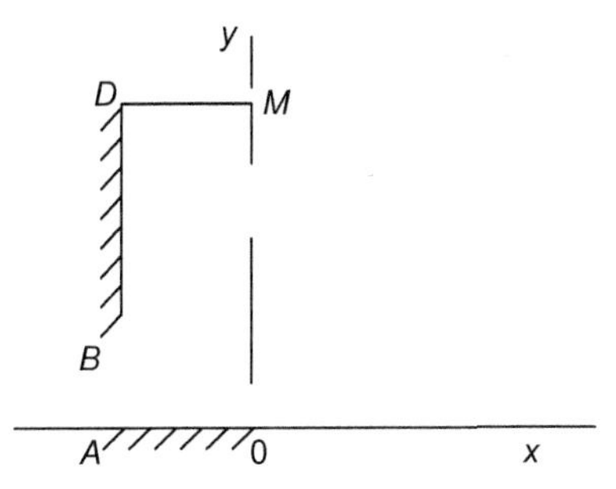

Figure 3.17 Fluid flow through a porous media.

DM: $p = p_1 = const$ (known), $u = 0$; $\int_M^D v dx = Q/2$, where Q is the fluid flow rate;

$$\text{OM}: u = 0; \frac{\partial v}{\partial x} = 0 \text{ (symmetry condition)}; \text{AO}: u = v = 0.$$

Here $\lambda(x, y) = m\mu k^{-1}$; m – porosity; k- permeability of porous medium; μ – viscosity coefficient.

We will re-write the system of equations (3.148) using variables: flow function and turbulence

$$\frac{\partial \omega}{\partial t} + \frac{\partial u\omega}{\partial x} + \frac{\partial v\omega}{\partial y} = \mu\left(\frac{\partial^2 \omega}{\partial x^2} + \frac{\partial^2 \omega}{\partial y^2}\right) - div(\lambda(x, y) grad\psi); \Delta\psi = \omega, \quad (3.151)$$

where $u = \frac{\partial \psi}{\partial y}, v = -\frac{\partial \psi}{\partial x}, \omega = \frac{\partial u}{\partial y} - \frac{\partial v}{\partial x}$.

At the boundary AB we will obtain the boundary values ψ, ω, using condition (3.150) for AB and the principal equation for AB:

$$\left(u\frac{\partial v}{\partial x} + v\frac{\partial v}{\partial y} + \frac{\partial p}{\partial y}\right)_{AB} = \mu\left(\frac{\partial^2 v}{\partial x^2} + \frac{\partial^2 v}{\partial y^2}\right)_{AB} - (\lambda(x, y) \cdot v)_{AB};$$

$$\left(u\frac{\partial v}{\partial x}\right)_{AB} = -\left(\mu\frac{\partial \omega}{\partial x}\right)_{AB}; \left(\mu\frac{\partial \omega}{\partial x} - \omega\frac{\partial \psi}{\partial y}\right)_{AB} = -\frac{1}{2}\frac{\partial}{\partial y}\left(\frac{\partial \psi}{\partial y}\right)^2_{AB}; \quad (3.152)$$

$$\left(\frac{\partial \psi}{\partial x}\right)_{AB} = 0. \quad (3.153)$$

Reasoning in the same manner we obtain the boundary conditions at the outlet boundary DM:

$$\left(\mu\frac{\partial\omega}{\partial y}+\omega\frac{\partial\psi}{\partial x}\right)_{DM}=\frac{1}{2}\frac{\partial}{\partial x}\left(\frac{\partial\psi}{\partial x}\right)^2_{DM}; \tag{3.154}$$

$$\left(\frac{\partial\psi}{\partial y}\right)_{DM}=0. \tag{3.155}$$

For the rest of the boundary we obtain the boundary conditions stemming from (3.150):

$$\text{BD}:\psi=\frac{Q}{2}\cdot\left|DM\right|,\frac{\partial\psi}{\partial x}=0, \tag{3.156}$$

$$\text{AO}:\psi=0,\frac{\partial\psi}{\partial y}=0,\text{OM}:\psi=0,\omega=0.$$

Cover the region ABDMO with the uniform grid:

$$\Omega_h=\left\{\begin{matrix} x_i=(i-1)h_1, \quad y_j=(j-1)h_2, i=1,\ldots,N_1, j=1,2,\ldots,N_2 \\ h_1=\left|AO\right|(N_1-1), h_2=\left|MO\right|(N_2-1)\end{matrix}\right\}$$

We will substitute the system of equations (3.151) by the difference relationships [102]:

$$\begin{aligned}&\omega_t^{n+1}=L_{1h}\omega^{n+1}+L_{2h}\omega^{n+1}-div_h(\lambda grad_h\psi^n),\\&\Delta_h\psi^{n+1}=\omega_{ij}^{n+1}\end{aligned} \tag{3.157}$$

Here the difference operators are determined in the following way:

$$\begin{aligned}&L_{1h}\omega=\mu(A_{k+1/2}\omega_x)_{\bar{x},ij}-0{,}5(u_{k+1/2,j}\omega_{x,ij}+u_{k-1/2,j}\omega_{\bar{x},ij}),\\&L_{2h}\omega=\mu(B_{k+1/2}\omega_y)_{\bar{y},ij}-0{,}5(v_{i,k+1/2,}\omega_{y,ij}+v_{i,k-1/2,j}\omega_{\bar{y},ij}),\end{aligned} \tag{3.158}$$

$$A_{k+1/2}=1+\frac{1}{\mu}\left|\frac{u_{k+1/2,j}h_1}{2}\right|, B_{k+1/2}=1+\frac{1}{\mu}\left|\frac{v_{ik+1/2}h_2}{2}\right|$$

$$div_h(\lambda grad_h\psi)=(\lambda_{ij}\psi_x)_{\bar{x},ij}+(\lambda_{ij}\psi_y)_{\bar{y},ij}.$$

The scheme to represent the implicit difference scheme (3.157) is similar to that discussed in paper [25]:

$$\frac{\omega_{ij}^{n+1/3}-\omega_{ij}^{n}}{\tau}=L_{1h}\omega^{n+1/3}+L_{2h}\omega^{n}-div_h(\lambda grad_h\psi^n);$$

$$\frac{\omega_{ij}^{n+2/3}-\omega_{ij}^{n+1/3}}{\tau}=L_{1h}\omega^{n+1/3}+L_{2h}\omega^{n+2/3}-div_h(\lambda grad_h\psi^n); \tag{3.159}$$

$$\frac{\omega_{ij}^{n+1}-\omega_{ij}^{n+2/3}}{\tau}=-\,div_h(\lambda grad_h(\psi^{n+1}-\psi^n));$$

$$\Delta_h\psi^{n+1}=\omega_{ij}^{n+1},i=2,3,\ldots,N_1-1,j=2,3,\ldots,N_2-1.$$

We obtain the boundary conditions for the turbulence at the inlet and outlet approximating (3.152), (3.154) though integral interpolation method:

$$\mu\int_{x_1}^{x_2}\frac{\partial\omega}{\partial x}dx-\int_{x_1}^{x_2}\frac{\partial\psi}{\partial y}dx=-\frac{1}{2}\int_{x_1}^{x_2}\frac{\partial}{\partial y}\left(\frac{\partial\psi}{\partial y}\right)^2dx;$$

$$\mu\frac{\omega_{2j}^{n+1/3}-\omega_{1j}^{n+1/3}}{h_1}-\frac{\psi_{1,j+1/2}-\psi_{1,j-1/2}}{h_2}\cdot\omega_{1j}^{n+1/3}=-\frac{1}{2}\frac{(\psi_{1j+1}^n-\psi_{1j}^n)^2-(\psi_{1j}^n-\psi_{1j-1}^n)^2}{h_2^3}; \tag{3.160}$$

$$\omega_{2j}^{n+1/3}-\left(1+\frac{h_1}{2\mu}\cdot\frac{(\psi_{1j+1}-\psi_{1j-1})}{h_2}\right)\omega_{1j}^{n+1/3}=-\frac{h_1}{2\mu}\cdot\frac{(\psi_{1j+1}-\psi_{1j})^2-(\psi_{1j}-\psi_{1j-1})^2}{h_2^3}.$$

Integrating (3.154), we find the boundary condition at the outlet:

$$\int_{y_{N_2-1}}^{y_{N_2}}\left(\mu\frac{\partial\omega}{\partial y}+\omega\frac{\partial\psi}{\partial y}\right)dy=\int_{y_{N_2-1}}^{y_{N_2}}\frac{\partial\left(\frac{\partial\psi}{\partial x}\right)^2}{\partial x}dy;$$

$$\mu\frac{\omega_{i,N_2}^{n+2/3}-\omega_{i,N_2-1}^{n+2/3}}{h_2}+\omega_{i,N_2-1}^{n+2/3}\cdot\frac{\psi_{i+1/2,N_2}^n-\psi_{i-1/2,N_2}^n}{h_2}=\frac{1}{2}\frac{(\psi_x^n)_{i+1/2,N_2}^2-(\psi_x^n)_{i-1/2,N_2}^2}{h_1}; \tag{3.161}$$

$$\omega_{i,N_2}^{n+2/3} - \left(1 - \frac{h_2}{2\mu} \cdot \frac{(\psi_{i+1,N_2}^n - \psi_{i-1,N_2}^n)}{h_1}\right)\omega_{i,N_2-1}^{n+2/3}$$

$$= \frac{h_2}{2\mu h_1^3}\left[(\psi_{i+1,N_2}^n - \psi_{i,N_2}^n)^2 - (\psi_{i,N_2}^n - \psi_{i-1,N_2}^n)^2\right], i = 2, 3, \ldots, N_1 - 1.$$

On the solid walls BD and AO we can make use of the Tom formula [96] with relaxation. The difference scheme (3.159) is developed as follows: using the scalar fitting with boundary conditions (3.160), (3.161) and the known conditions on the solid wall and on the symmetry axis from the first two relationships we calculate: $\omega_{ij}^{n+1/3}, \omega_{ij}^{n+2/3}$, $i = 2, 3, \ldots, N_1 - 1; j = 2, 3, \ldots, N_2 - 1$.Then from the last two difference equations we derive the difference equation to determine the flow function distribution on the five-point template:

$$\frac{1}{h_1^2}\left[(1+\tau\lambda)_{i+1/2j}(\psi_{i+1,j}^{n+1} - \psi_{i,j}^{n+1}) - (1+\tau\lambda)_{i-1/2j}(\psi_{i,j}^{n+1} - \psi_{i-1,j}^{n+1})\right] +$$

$$\frac{1}{h_2^2}\left[(1+\tau\lambda)_{i,j+1/2}(\psi_{i,j+1}^{n+1} - \psi_{i,j}^{n+1}) - (1+\tau\lambda)_{i,j-1/2}(\psi_{i,j}^{n+1} - \psi_{i,j-1}^{n+1})\right] = \quad (3.162)$$

$$\omega_{ij}^{n+2/3} + \tau \cdot div_h(grad_h \psi_{ij}^n), i = 2, 3, \ldots, N_1 - 1; j = 2, 3, \ldots, N_2 - 1$$

with the boundary conditions:

$$\text{AB}: \psi_{x,1j}^{n+1} = 0, j = 1, 2, \ldots, m, \text{BD}: \psi_{1j}^{n+1} = const, j = m+1, m+2, \ldots, N_2,$$

$$\text{DM}: \psi_{y,iN_2}^{n+1} = 0, i = 1, 2, \ldots, N_1, \text{OM}: \psi_{N_1,j}^{n+1} = 0, j = 1, 2, \ldots, N_2, \quad (3.163)$$

$$\text{AO}: \psi_{i,1}^{n+1} = 0, i = 1, 2, \ldots, N_1.$$

The problem (3.162), (3.163) is solved using the iteration method. The value ω_{ij}^{n+1} is determined from the difference relationship:

$$\omega_{ij}^{n+1} = \Delta_h \psi_{ij}^{n+1}, i = 2, 3, \ldots, N_1 - 1; j = 2, 3, \ldots, N_2 - 1.$$

Iteration process (3.159) stops when the following condition is met

$$\left\| \omega_j^{n+1} - \omega^n \right\|_C \le \varepsilon, \varepsilon > 0. \tag{3.164}$$

Having found the distribution of flow function and turbulence in the calculation domain the pressure values on AB can be calculated by formula (3.136). From the system of equations (3.148) we have

$$\begin{cases} \dfrac{\partial p}{\partial x} = \mu\omega_y - (\psi_y)_x^2 + (\psi_x \cdot \psi_y)_y - \lambda(x, y) \cdot \psi_y - (\psi_y)_t, \\ \\ \dfrac{\partial p}{\partial y} = -\mu\omega_x + (\psi_x \cdot \psi_y)_x - (\psi_x)_y^2 + \lambda(x, y) \cdot \psi_x - (\psi_x)_t. \end{cases} \tag{3.165}$$

Accounting for our assumption we obtain from (3.150)

$$P_B = p_{AB} = p_0(\text{unknown}); p_D = p_{DM} = p_1(\text{known}). \tag{3.166}$$

Further integrating along the boundary line BD, we find

$$p_D - p_B = \int_{\mathrm{BD}} \left(\frac{\partial p}{\partial x} dx + \frac{\partial p}{\partial y} dx \right), p_B = p_D - \int_{\mathrm{BD}} \left(\frac{\partial p}{\partial x} dx + \frac{\partial p}{\partial y} dx \right).$$

Using parametric equation BD: $(x = 0, y = t, y_B \le t \le y_D)$, we will write

$$p_B = p_D - \int_{y_B}^{y_D} \frac{\partial p}{\partial y} \bigg|_{x=0} dy,$$

and accounting for (3.165) we can write down this relationship in the form

$$p_B = p_D - \int_{y_B}^{y_D} (-\mu\omega_x + (\psi_x \cdot \psi_y)_x - (\psi_x)_y^2 + \lambda(x, y)\psi_x + (\psi_x)_t)\big|_{x=0} dy. \tag{3.167}$$

We will determine the integral on the right hand side of (3.167) applying the known formulas of numerical integration. This method yields the pressure value only at the inlet boundary AB.

In order to determine the pressure distribution within the whole calculation domain using calculated values ψ, ω and the known pressure p_{DM}, consider the finite-difference analogue of the second equation (3.165):

$$\frac{p_{ij+1}^{n+1} - p_{ij}^{n+1}}{h_2} = -\mu \cdot \frac{\omega_{i+1/2j}^{n+1} - \omega_{i-1/2j}^{n+1}}{h_1} +$$
$$\frac{\left(\psi_{i+1j}^{n+1} - \psi_{ij}^{n+1}\right)\left(\psi_{i+\frac{1}{2}j+1/2} - \psi_{i+\frac{1}{2}j-1/2}\right) - \left(\psi_{ij}^{n+1} - \psi_{i-1j}^{n+1}\right)\left(\psi_{i-\frac{1}{2}j+1/2}^{n+1} - \psi_{i-\frac{1}{2}j-1/2}^{n+1}\right)}{h_1^2 h_2}$$
$$-\lambda_{ij} \cdot \frac{\psi_{i+1/2j}^{n+1} - \psi_{i-1/2j}^{n+1}}{h_1} + \left(\frac{\psi_{i+1/2j}^{n+1} - \psi_{i-1/2j}^{n+1}}{h_1}\right)_{\bar{t}} \quad i = 2, 3, \ldots, N_1 - 1, j = 2, 3, \ldots, N_2 - 1. \tag{3.168}$$

We can find the pressure distribution from the relationship (3.168) describing the difference scheme. The calculations were conducted in the domain $x = 0,5 y = 2$ and using above finite-difference method the number of uniform grid nodes in the two- dimensional region was 21×41. The iteration parameter was taken as $\tau = 0,0001$. The procedure for optimising the iteration parameter using the upper relaxation method was used for numerical solution of equation (3.162). This equation was approximately solved with an accuracy of $\varepsilon = 10^{-6}$. The calculation results are represented as tables and graphs. Figs. 3.18, 3.19 show the profiles of velocity component U for different sections, Fig. 3.20 depicts the profiles of velocity component V, Fig. 3.21 shows the pressure values. In Fig. 3.22 the isolines of the flow functions are displayed for $Re = 100$ and $Re = 250$.

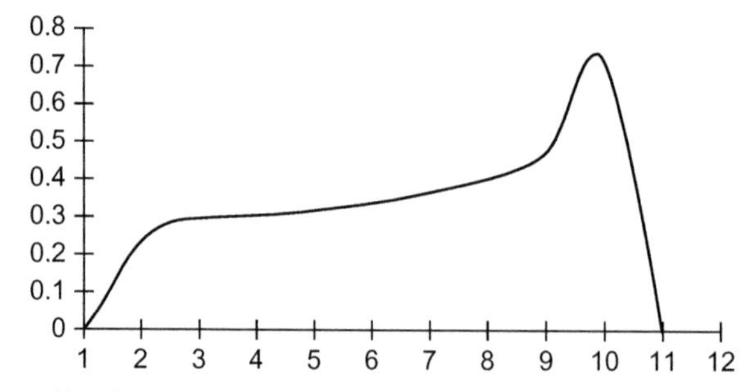

Figure 3.18 Profile of velocity U at the inlet.

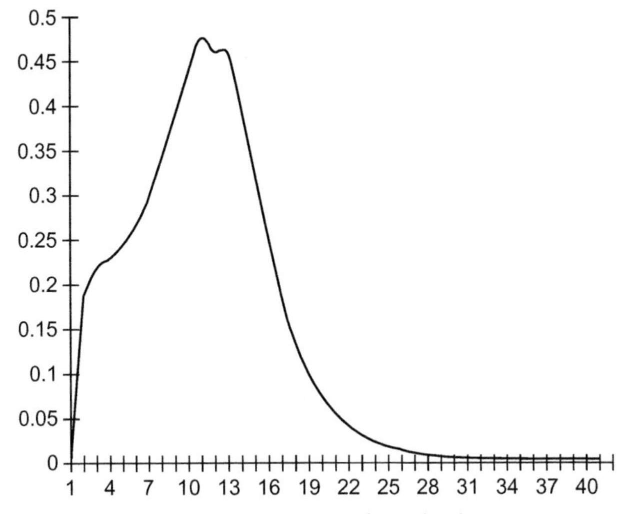

Figure 3.19 Profile of velocity component U in layer $(11,j)$.

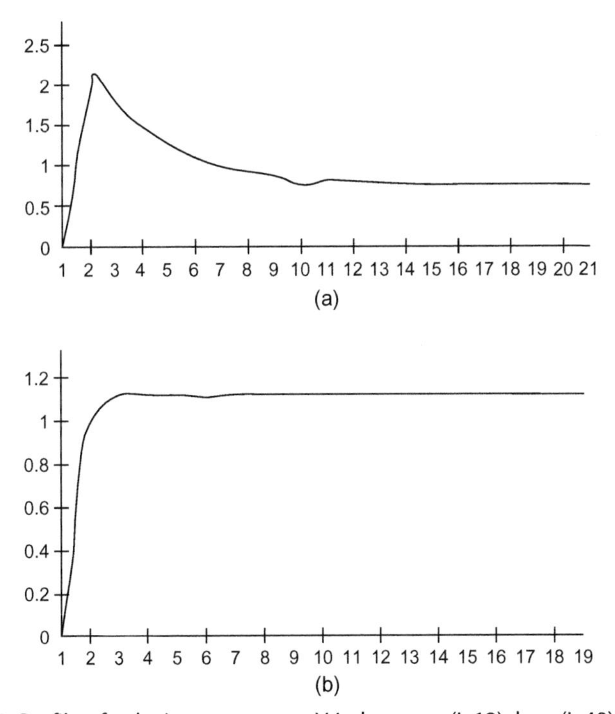

Figure 3.20 Profile of velocity component V in layer: a - (i, 12), b – (i, 40).

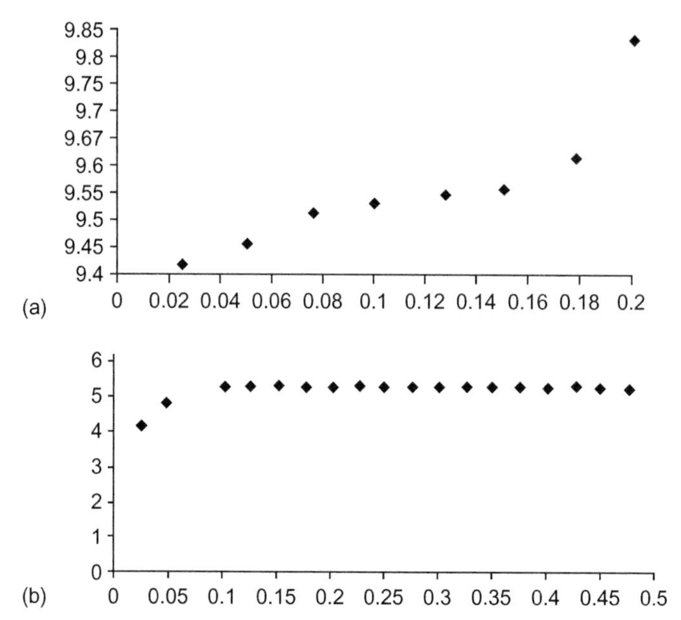

Figure 3.21 Pressure value in nodes of grid in calculation domain: a – (1, j), b – (i, 34).

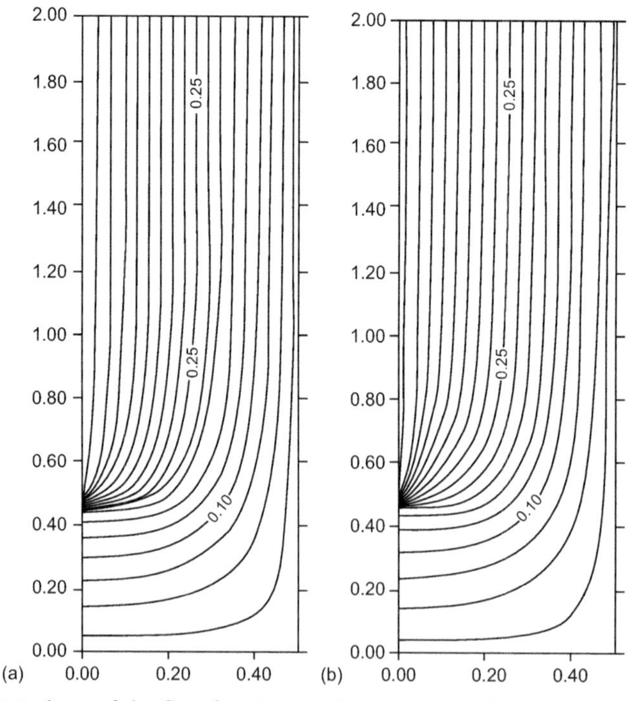

Figure 3.22 Isolines of the flow functions at $Re = 100$ (a) and $Re = 250$ (b).

3.7 CONVECTIVE WARM-UP OF AN INHOMOGENEOUS POROUS MEDIUM

The numerical method of a convection problem solution in a double connected region with nonlocal boundary conditions for temperature is proposed. The problem is solved for the variables flow function – turbulence by the method of virtual domains.

We will consider a vertical oil formation section $\Omega = \Omega_1 \cup \Omega_0$with an impermeable inclusion Ω_0.

Formation warm-up takes place through the well γ_3 (Fig. 3.23). In particular, thermal treatment of a well is carried out with the assistance of hot water or steam injection.

3.7.1 Formulation of the Problem

In a double connected region Ω_1 (see Fig. 3.23) there arises heat-conducting fluid flow which is described by the following equations of thermal convection in dimensionless form [104, 127, 128]:

$$\vec{u}_t + (\vec{u}\nabla)\vec{u} = \frac{1}{Re}\Delta\vec{u} - \nabla p - \vec{\Gamma}\theta, div\vec{u} = 0, \tag{3.169}$$

$$\theta_t + (\vec{u}\nabla)\theta = \chi\Delta\theta,$$

where $\vec{\Gamma} = (0, \Gamma)$- Grashoff number; $\vec{u} = (u, v)$ – velocity vector; θ – temperature; χ- quantity inverse to Prandtl number.

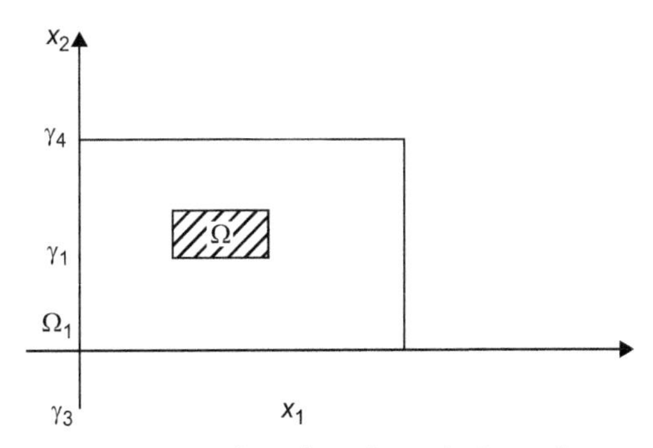

Figure 3.23 Formation warm-up takes place through the well γ_s.

At the boundary $\partial\Omega_0$, the temperature $\theta\big|_{\partial\Omega_0} = \theta_1$ is considered the unknown constant, possibly dependent on time and satisfying the condition

$$\int_{\partial\Omega_0} \frac{\partial\theta}{\partial n} dl = 0. \tag{3.170}$$

For the initial-boundary problem:

$$\theta\big|_{\gamma_3} = \varphi(x_1), \theta\big|_{\gamma_1 \cup \gamma_2 \cup \gamma_4} = 0, \vec{u}\big|_{\partial\Omega} = 0, \tag{3.171}$$

$$\vec{u}\big|_{t=0} = \vec{u}(x), \theta\big|_{t=0} = \theta_0(x). \tag{3.172}$$

For conditions (3.170)–(3.172) the boundaries of the investigated regions are designated as follows: γ_1, γ_2 – left and right vertical boundaries of region Ω_1; γ_3, γ_4 – lower and upper horizontal boundaries of region Ω_1; $\partial\Omega_0$- boundary of region Ω_0; $\Omega = \Omega_0 \cup \Omega_1$; $\partial\Omega = \partial\Omega_0 \cup \gamma_1 \cup \gamma_2 \cup \gamma_3 \cup \gamma_4$.

3.7.2 Problem in Variables Turbulence – Flow Function

Introducing the flow function $u = \psi_{x_2}, v = -\psi_{x_1}$ and turbulence $\omega = u_{x_2} - u_{x_1}$ and writing equations (3.169) in these variables:

$$\begin{cases} \dfrac{\partial\omega}{\partial t} + \psi_y \dfrac{\partial\omega}{\partial x_1} + \psi_x \dfrac{\partial\omega}{\partial x_2} = \Delta\omega + \Gamma\theta_x, \Delta\psi = \omega, \\ \dfrac{\partial\theta}{\partial t} + \psi_y \dfrac{\partial\theta}{\partial x_1} + \psi_x \dfrac{\partial\theta}{\partial x_2} = \chi\Delta\theta \end{cases} \tag{3.173}$$

From the boundary conditions (3.171) we obtain the boundary conditions for the flow function. Let us assume $\psi(0,0) = 0$, then from the condition $v\big|_{\gamma_3} = \psi_{x_1}\big|_{\gamma_3} = 0$ it follows that $\psi\big|_{\gamma_3} = 0$ from condition $u\big|_{\gamma_1} = \psi_{x_2}\big|_{\gamma_1} = 0$ follows that $\psi\big|_{\gamma_1} = 0$. Analogously we obtain $\psi\big|_{\gamma_4} = \psi\big|_{\gamma_2} = 0$.

From condition $u\big|_{\gamma_3 \cup \gamma_4} = \psi_y\big|_{\gamma_3} = \psi_y\big|_{\gamma_4} = 0$ we have $\dfrac{\partial\psi}{\partial\vec{n}}\Big|_{\gamma_3 \cup \gamma_4} = 0$. In the same manner from $u\big|_{\gamma_1 \cup \gamma_2} = \psi_x\big|_{\gamma_1} = \psi_x\big|_{\gamma_2} = 0$ we obtain $\dfrac{\partial\psi}{\partial\vec{n}}\Big|_{\gamma_1 \cup \gamma_2} = 0$. Thus, we have

$$\begin{gathered} \psi\big|_\gamma = 0, \frac{\partial\psi}{\partial n}\bigg|_\gamma = 0; \gamma = \cup_{k=1}^4 \gamma_k; \\ \omega\big|_{t=0} = \omega_0(x), \theta\big|_{t=0} = \theta_0(x), \end{gathered} \tag{3.174}$$

Where $\vec{n}$- is the normal vector to the boundary of domain Ω_1.

From equalities $u|_{\partial\Omega_0} = \left.\frac{\partial\psi}{\partial y}\right|_{\partial\Omega_0} = 0, v|_{\partial\Omega_0} = -\left.\frac{\partial\psi}{\partial x}\right|_{\partial\Omega_0} = 0$ we find that $\left.\frac{\partial\psi}{\partial\vec{n}}\right|_{\partial\Omega_0} = 0, \left.\frac{\partial\psi}{\partial\tau}\right|_{\partial\Omega_0} = 0.$

Let $\vec{\tau}(\tau_{x_1}, \tau_{x_2})$ – is the tangent vector to boundary $\partial\Omega_0$. Multiplying the first equation (3.169) by τ_{x_1}, the second – by τ_{x_2}, adding them and calculating the boundary value at $\partial\Omega_0$:

$$\Delta u\tau_{x_1} + \Delta v\tau_{x_2} - \Gamma\theta\tau_{x_2}|_{\partial\Omega_0} = \left.\frac{\partial p}{\partial\tau}\right|_{\partial\Omega_0}. \tag{3.175}$$

Under the assumption that the pressure p – is a single-valued function, the left part of (3.175) becomes zero. Then from the condition of pressure being unique it follows that:

$$\int_{\partial\Omega_0} \frac{\partial\omega}{\partial n} dl = 0. \tag{3.176}$$

Let $\vec{n}$ – be the normal to $\partial\Omega_0$. Providing here a more detailed proof of condition (3.176) and making the following transformations:

$$\Delta u = \frac{\partial}{\partial x_2}(u_{x_2} - v_{x_1}) + \frac{\partial}{\partial x_1}\left(\frac{\partial u}{\partial x_1} + \frac{\partial v}{\partial x_2}\right),$$

$$\Delta v = \frac{\partial}{\partial x_2}(u_{x_2} - v_{x_1}) + \frac{\partial}{\partial x_2}\left(\frac{\partial u}{\partial x_1} + \frac{\partial v}{\partial x_2}\right), div\vec{u} = 0,$$

$$\Delta u\tau_{x_1} + \Delta v\tau_{x_2} = \frac{\partial}{\partial x_2}(u_{x_2} - v_{x_1})\tau_{x_1} - \frac{\partial}{\partial x_2}(u_{x_2} - v_{x_1})\tau_{x_2} = \left|\begin{array}{c}\tau_{x_1} = n_{x_2}\\ \tau_{x_2} = -n_{x_1}\end{array}\right|$$

$$= \frac{\partial}{\partial x_2}(u_{x_2} - v_{x_1})n_{x_2} + \frac{\partial}{\partial x_2}(u_{x_2} - v_{x_1})n_{x_1} = \frac{\partial}{\partial n}(u_{x_2} - v_{x_1}) = \frac{\partial\omega}{\partial n}.$$

From (3.175) we have $\frac{\partial\omega}{\partial n} - \Gamma\theta\tau_{x_2}|_{\partial\Omega_0} = \left.\frac{\partial p}{\partial\tau}\right|_{\partial\Omega_0}.$

Using the condition of pressure uniqueness we find

$$\int_{\partial\Omega_0} \frac{\partial p}{\partial\tau} d\tau = \int_{\partial\Omega_0} \frac{\partial\omega}{\partial n} d\tau - \Gamma\int_{\partial\Omega_0} \theta_1\tau_{x_2} d\tau = 0. \tag{3.177}$$

We will introduce the function φ, assuming $\theta_1 \tau_{x_2} = \frac{\partial \varphi}{\partial x_1} x_1 - \frac{\partial \varphi}{\partial x_2} = \frac{\partial \varphi}{\partial \tau}$. This brings us to the equalities $\frac{\partial \varphi}{\partial x_1} = 0, \frac{\partial \varphi}{\partial x_2} = \theta_1$, which we integrate and obtain

$$\varphi = \theta_1 x_2 \big|_{\partial \Omega_0}, \theta \big|_{\partial \Omega_0} = \theta_1 = const, \frac{\partial \varphi}{\partial x_1} = 0.$$

Consequently, $\theta_1 \tau_{x_2} = \frac{\partial \varphi}{\partial \tau}$, where φ – is the single-valued function. As a result, from (3.177) we obtain the relationship.

$$\int_{\partial \Omega_0} \frac{\partial p}{\partial \tau} d\tau = -\Gamma \int_{\partial \Omega_0} \frac{\partial \varphi}{\partial \tau} d\tau + \int_{\partial \Omega_0} \frac{\partial \omega}{\partial n} d\tau, \tag{3.178}$$

which is equivalent to (3.176).

Thus, we have obtained the following problem:

$$\begin{cases} \dfrac{\partial \omega}{\partial t} + \psi_{x_2} \dfrac{\partial \omega}{\partial x_1} - \psi \dfrac{\partial \omega}{\partial x_2} = \dfrac{1}{Re} \Delta \omega + \Gamma \theta_{x_1}, \Delta \psi = \omega, \\ \dfrac{\partial \theta}{\partial t} + \psi_{x_2} \dfrac{\partial \theta}{\partial x_1} - \psi \dfrac{\partial \theta}{\partial x_2} = \chi \Delta \theta, \end{cases} \tag{3.179}$$

$$\begin{cases} \omega|_{t=0} = \omega_0(x), \theta|_{t=0} = \theta_0(x); \psi|_\gamma = 0, \left. \dfrac{\partial \psi}{\partial n} \right|_\gamma = 0, \gamma = \cup_{k=1}^4 \gamma_k, \\ \left. \dfrac{\partial \psi}{\partial \tau} \right|_{\partial \Omega_0} = 0, \left. \dfrac{\partial \psi}{\partial n} \right|_{\partial \Omega_0} = 0, \displaystyle\int_{\partial \Omega_0} \dfrac{\partial \omega}{\partial n} dl = 0; \int_{\partial \Omega_0} \dfrac{\partial \theta}{\partial n} dl = 0, \theta|_{\gamma_3} = \varphi(x), \theta|_{\gamma_1 \cup \gamma_2 \cup \gamma_4} = 0. \end{cases} \tag{3.180}$$

3.7.3 Difference Problem

Let us apply the method of virtual domains to (3.179) and (3.180). The concept of the method is as follows: an auxiliary problem with a small parameter is solved in the domain Ω_0, a certain conformity condition is set in $\partial \Omega_0$.

Let us consider an auxiliary problem with a small parameter in Ω [122, 123]:

$$\begin{aligned} &\frac{\partial \omega^\varepsilon}{\partial t} + \psi^\varepsilon_{x_2} \frac{\partial \omega^\varepsilon}{\partial x_1} - \psi^\varepsilon_{x_1} \frac{\partial \omega^\varepsilon}{\partial x_2} = \frac{1}{Re} \Delta \omega^\varepsilon - div \left(\frac{\xi(x)}{\varepsilon} \nabla \psi^\varepsilon \right) + \Gamma \theta^\varepsilon_{x_1}, \\ &\frac{\partial \theta^\varepsilon}{\partial t} + \psi^\varepsilon_{x_2} \frac{\partial \theta^\varepsilon}{\partial x_1} - \psi^\varepsilon_{x_1} \frac{\partial \theta^\varepsilon}{\partial x_2} = div(\chi^\varepsilon \nabla \theta^\varepsilon), \Delta \psi^\varepsilon = \omega^\varepsilon; \end{aligned} \tag{3.181}$$

$$\omega^{\varepsilon}\big|_{t=0} = \omega_0(x), \theta^{\varepsilon}\big|_{t=0} = \theta_0(x), \theta^{\varepsilon}\big|_{\gamma_1 \cup \gamma_4 \cup \gamma_2} = 0,$$
$$\frac{\partial \psi^{\varepsilon}}{\partial n}\bigg|_{\gamma} = \psi^{\varepsilon}\big|_{\gamma} = 0, \theta^{\varepsilon}\big|_{\gamma_3} = \varphi(x), \gamma = \cup_{k=1}^{4}\gamma_k, \tag{3.182}$$

$$\xi(x) = \begin{cases} 1 \text{ in } \Omega_0, \\ 0 \text{ in } \Omega_1, \end{cases} \quad \chi^{\varepsilon} = \begin{cases} 1 \text{ in } \Omega_1, \\ \varepsilon \text{ in } \Omega_0. \end{cases} \tag{3.183}$$

We will cover domain Ω with the uniform grid:

$$\Omega_h = \{(x_{1i}, x_{2j}); i = 0, 1, \ldots, N; j = 0, 1, \ldots, M; x_{1i} = (i-1)\cdot h_1; x_{2j} = (j-1)\cdot h_2; Nh_1 = 1; Mh_2 = 1\}.$$

For problem (3.181)–(3.183) let us construct the difference scheme

$$\omega_t^n + L_{1h}\omega^{n+1} + L_{2h}\omega^{n+1} = -\,div_h\left(\frac{\xi_{i,j}}{\varepsilon}\nabla_h\psi^n\right) + \Gamma\cdot\theta_{x_1}^n, \Delta_h\psi = \omega,$$
$$\theta_t^n + \frac{1}{2}\left[(\psi_{x_2}^n + |\psi_{x_2}^n|)\cdot\theta_{x_1}^{n+1} + (\psi_{x_2}^n - |\psi_{x_2}^n|)\cdot\theta_{\bar{x}_1}^{n+1}\right] - \tag{3.184}$$
$$\frac{1}{2}\left[(\psi_{x_1}^n + |\psi_{x_1}^n|)\cdot\theta_{x_2}^{n+1} + (\psi_{x_1}^n - |\psi_{x_1}^n|)\cdot\theta_{\bar{x}_2}^{n+1}\right] = div_h(\chi\nabla_h\theta^{n+1}),$$

Where

$$L_1\omega = \frac{1}{Re}(A_{k+1/2}\omega_{x_1})_{\bar{x}_1} - \frac{1}{2}(u_{k+1/2j}\omega_{x_1} + u_{k-1/2j}\omega_{\bar{x}_1}),$$

$$L_2\omega = \frac{1}{Re}(B_{k+1/2}\omega_{x_2})_{\bar{x}_2} - \frac{1}{2}(v_{ik+1/2}\omega_{x_2} + v_{ik-1/2}\omega_{\bar{x}_2}),$$

$$A_{k+1/2} = 1 + \frac{Reh_1}{2}\left|u_{k+1/2j}\right|, B_{k+1/2} = 1 + \frac{Reh_2}{2}\left|v_{ik+1/2}\right|.$$

The boundary conditions can be approximated. For the turbulence we can take the Thom condition. For example, the condition at the boundary γ_1 may be written as

$$\frac{\partial\psi}{\partial x_1}\bigg|_{\gamma_1} = 0, \psi\big|_{\gamma_1} = 0.$$

We expand $\psi_{1,j}$ into the Taylor series in the vicinity of $(0, j)$:

$$\psi_{1,j} = \psi_{0,j} + h_1\left(\frac{\partial\psi}{\partial x_1}\right)_{0,j} + \frac{h_1^2}{2}\left(\frac{\partial^2\psi}{\partial x_1^2}\right)_{0,j} + O(h_1^3).$$

As $\omega_{0,j} = \left(\frac{\partial^2\psi}{\partial x_1^2}\right)_{0,j}$, then $\omega_{0,j} = \frac{2(\psi_{1,j} - \psi_{0,j})}{h_1^2}$. Due to the fact that $\psi_{0,j} = 0$, we obtain $\omega_{0,j} = \frac{2}{h_1^2}\psi_{1,j}$.

In the same manner we can calculate the approximation for the other boundaries:

$$\begin{aligned} \gamma_1{:}\omega_{0,j} &= \frac{2}{h_1^2}\psi_{1,j}, \gamma_2{:}\omega_{i,0} = \frac{2}{h_2^2}\psi_{i,1}, \\ \gamma_3{:}\omega_{N,j} &= \frac{2}{h_1^2}\psi_{N-1,j}, \gamma_4{:}\omega_{i,M} = \frac{2}{h_2^2}\psi_{i,M-1}. \end{aligned} \tag{3.185}$$

For the numerical modelling of scheme (3.184) a three-level difference scheme was used:

$$\frac{\omega_{i,j}^{n+1/3} - \omega_{i,j}^{n}}{\tau} + L_1\omega^{n+1/3} + L_2\omega^{n} = -div\left(\frac{\xi_{i,j}}{\varepsilon}\nabla_h\psi^n\right) + \Gamma\theta_{x_1}^{n},$$

$$\frac{\omega_{i,j}^{n+2/3} - \omega_{i,j}^{n+1/3}}{\tau} + L_1\omega^{n+1/3} + L_2\omega^{n+2/3} = -div\left(\frac{\xi_{i,j}}{\varepsilon}\nabla_h\psi^n\right) + \Gamma\theta_{x_1}^{n},$$

$$\frac{\omega_{i,j}^{n+1/3} - \omega_{i,j}^{n+2/3}}{\tau} = -div\left(\frac{\xi_{i,j}}{\varepsilon}\nabla_h(\psi^{n+1} - \psi^n), \nabla_h\psi^{n+1,k+1}\right) = \omega^{n+1},$$

$$\begin{aligned} &\frac{\theta_{i,j}^{n+1,k+1/2} - \theta_{i,j}^{n+1,k}}{\tau} + \frac{1}{2}((\psi_{x_2}^{n+1} + |\psi_{x_2}^{n+1}|)\cdot\theta_{x_2}^{n+1,k+1/2} \\ &+ (\psi_{x_2}^{n+1} - |\psi_{x_2}^{n+1}|)\cdot\theta_{\overline{x}_2}^{n+1,k+1/2}) - \frac{1}{2}((\psi_{x_1}^{n+1} + |\psi_{x_1}^{n+1}|)\cdot\theta_{x_2}^{n+1,k} \\ &+ (\psi_{x_1}^{n+1} - |\psi_{x_1}^{n+1}|)\cdot\theta_{\overline{x}_2}^{n+1,k}) = (\chi\theta_{x_1}^{n+1,k})_{\overline{x}_1} + (\chi\theta_{x_2}^{n+1,k})_{\overline{x}_2}, \end{aligned}$$

$$\frac{\theta_{i,j}^{n+1,k+1/2}-\theta_{i,j}^{n+1,k}}{\tau}+\frac{1}{2}((\psi_{x_2}^{n+1}+|\psi_{x_2}^{n+1}|)\cdot\theta_{x_2}^{n+1,k+1/2}$$
$$+(\psi_{x_2}^{n+1}-|\psi_{x_2}^{n+1}|)\cdot\theta_{\bar{x}_2}^{n+1,k+1/2})-\frac{1}{2}((\psi_{x_1}^{n+1}+|\psi_{x_1}^{n+1}|)\cdot\theta_{x_2}^{n+1,k+1}$$
$$+(\psi_{x_1}^{n+1}-|\psi_{x_1}^{n+1}|)\cdot\theta_{\bar{x}_2}^{n+1,k+1})=(\chi\theta_{x_1}^{n+1,k})_{\bar{x}_1}+(\chi\theta_{x_2}^{n+1,k+1})_{\bar{x}_2},$$

3.7.4 Results

The proposed method was used to solve the problem of natural convection in the double connected domain under nonlocal temperature boundary conditions. The calculation domain (Fig. 3.23) was covered with a grid of 31×31. Initially the program was tested for the convection problem in a simply connected region. In Figs. 3.24 and 3.25 the isolines and isotherms are plotted for $\Gamma = 2500$ and $Pr = 0,2$.

This was followed by obtaining the results of numerical calculations of the problem in the double connected region at $\Gamma = 300 - 3500$, $Pr = 0,2 - 1$. Figs. 3.26 and 3.27 show the isolines of flow functions and isotherms plotted for $\Gamma = 3500$, $Pr = 1$.

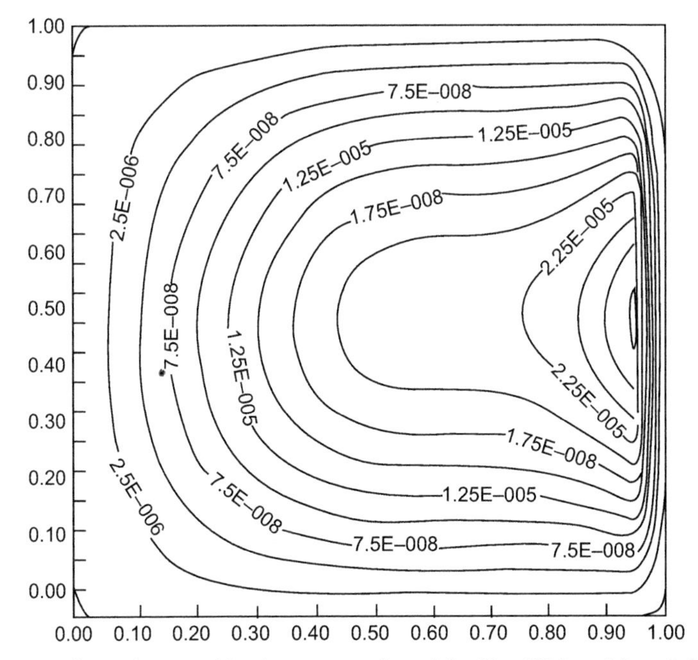

Figure 3.24 The isolines and isotherms are plotted for $\Gamma = 2500$ and $Pr = 0,2$.

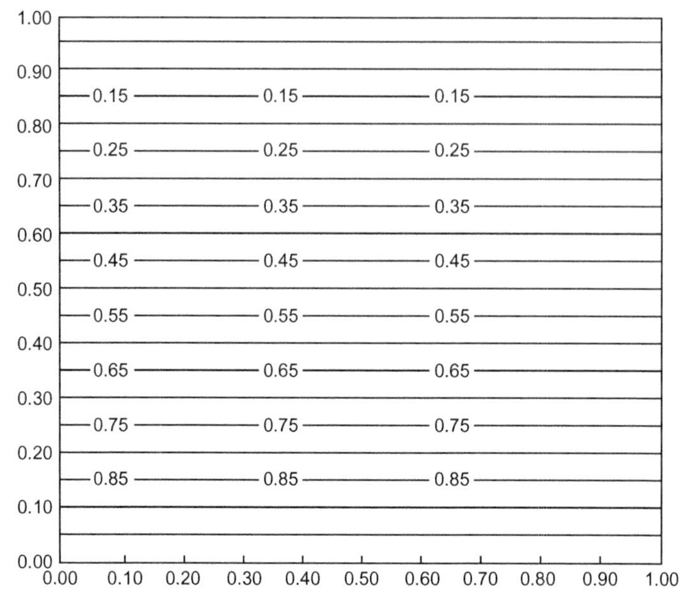

Figure 3.25 The isolines and isotherms are plotted for $\Gamma = 2500$ and $Pr = 0, 2$.

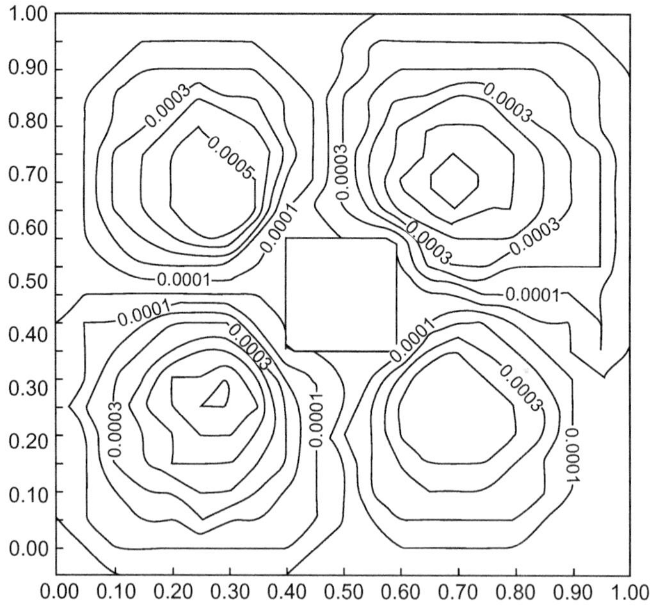

Figure 3.26 The isolines of flow functions and isotherms plotted for $\Gamma = 3500$, $Pr = 1$.

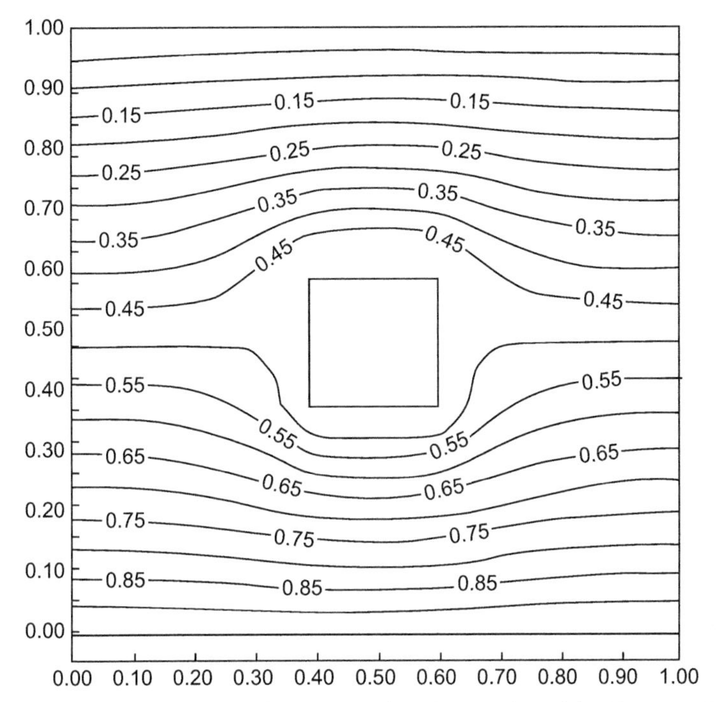

Figure 3.27 The isolines of flow functions and isotherms plotted for $\Gamma = 3500, Pr = 1$.

The simulation in a simply connected region is in agreement with previous results published in [127].

3.8 THE SPLITTING OF PHYSICAL PROCESSES UNDER NON-ISOTHERMAL TWO-PHASE FLUID FLOW IN A POROUS MEDIUM

In this section we discuss the MLT-model, which was initially described in [15, 16]. It accounts for effect of the thermal processes on the flow of two-phase fluid as a function of the temperature, viscosity coefficients and the capillary properties of the two-phase fluid.

The peculiarity of the MLT-model is that all equations involved (but the Darcy and Laplace laws) are derived from the laws of conservation formulated for the mechanics of solids. In particular, the motion of the interface between the two-phase fluid and fixed boundaries (Stefan type problem) can be described in the framework of this model.

The purpose of this section is to determine the relationship between the degree of smoothness of the solution of the initial-boundary problem for the MLT-model and the coefficients in the equations as well as at the boundary conditions. Similar to [3, 6] the results obtained are then used to prove the convergence of the iteration method for solving the MLT-problem and finding the convergence rate.

3.8.1 Formulation of the MLT-problem

The mathematical model of nonisothermal flow includes the equation for the equilibrium temperature θ and the ML-model which uses the Darcy laws linking the velocities of the phases $\vec{v}_k$ and ∇p_k, the Laplace law $p_0 - p_2 = p_c(x, \theta, s)$ (here p_c is the capillary pressure, p_0, p_2 are the phase pressures) with the phase continuity laws.

Introducing the average pressure

$$p = p_0 - \int_0^s b_2(\theta, s) p_{cs} ds$$

and making relevant transformations in the Muskat-Leverett equations we obtain the following MLT-model [3, 15, 16]:

$$m_k u_{kt} + div\vec{v}_k = 0, k = 1, 2, 3, \ -\vec{v}_k = A_k \nabla u_k + B_k \nabla \theta - b_k \vec{v} + \vec{f}_k. \quad (3.186)$$

Here $u_1 = \theta(x, t)$ is the equilibrium temperature of the fluids and porous medium; $u_2(x, t) = (s_2 - s_2^0)(1 - s_0^0 - s_2^0)^{-1}$ and $s_2(x, t)$ designate the dynamic and total saturation of the aqueous phase; $s_k^0 = const \in (0, 1), s_0^0 + s_2^0 < 1$ are the average residual phase saturations; $u_3(x, t) = p$ is the average pressure of the mixture; $\vec{v}_2$ and $\vec{v}_3 = \vec{v}$ are rates of phase flow and the mixture velocity; $m_2 = m_2(x)$ is the effective porosity; $m_1 = 1, m_3 = b_3 = B_1 = \vec{f}_1 = 0$, $b_1 = u_1 = \theta, b_2 = b_2(u_1, u_2)$. The properties of tensors A_k and B_k, those of vectors $\vec{f}_k, k = 1, 2, 3$, and functions $m_2(x)$ and $b_2(u_1, u_2)$ will be described below.

Occasionally the function $u_2^* = m_2 u_2$ is used instead of u_2. For this function the coefficients $m_2 = 1, A_2^* = m_2 A_2, B_2, -b_2$ and $\vec{f}_2^{**} = \vec{f}_2 + A_2 u_2 \nabla(m_2^{-1})$ in equation (3.186)($k = 2$) have the properties of the initial equation coefficients.

Let us assume that $Q = \Omega \times [0, T], \Omega \subset R^3$ is a confined domain, $\partial\Omega = \cup_1^3 \Gamma^k$, $\Sigma^k = \Gamma^k \times [0, T], \Gamma^1$ and Γ^2 model the locations of injection, production and contact with a homogeneous immobile fluid. Here Γ^3 is an impermeable boundary.

The initial-boundary problem for $\vec{u}=(u_1,u_2,u_3)$ may be represented in the form

$$\begin{cases} m_i u_i = m_i u_i^0, (x,t)\in\Sigma^0\cup\Sigma^1; \vec{v}_i\vec{n}=\beta_i\psi^0, \\ A_t\nabla\theta\vec{n}=\beta_0(\theta^0-\theta), \vec{v}_k\vec{n}=0, (x,t)\in\Sigma^3; \\ (x,t)\in\Sigma^2; \end{cases} \tag{3.187}$$

Where
$i=1,2,3;\ k=1,2;\ \beta_1=-b_1,\ \beta_2=-b_2,\ \beta_3=0;\ \Sigma^0=\{t=0, x\in\Omega\}$. If $\Gamma^1=\varnothing$, then the law of mass conservation in Ω leads to conditions:

$$\int_\Omega u_3(x,t)dx=\int_\Omega \psi^0(x,t)dx=0, t\in[0,T]. \tag{3.188}$$

We will call the problem (3.186), (3.187) a MLT-problem.

3.8.2 The Regular MLT – Problem

Similar to paper [3] we will make use of the following designations for norms

$$|u|_{q,\Omega}=|u|_{L_q(\Omega)}, |u|_{q,r,Q}=|u|_{L_{q,r}(Q)}, |u|_{V^2(Q)}=|u|_{2,\infty,Q}+|\nabla_x u|_{2,Q}$$

and Banach spaces $L_q(\Omega), L_{q,r}(Q), V_2(Q), W_q^1(\Omega), W_q^{1,0}(Q)$. We will assume

$$|u|_\Omega^{(1)}=|u|_{C^l(\Omega)}, \alpha=(l-[l])\in[0,1), |u|_Q^{(l,l/2)}=|u|_{C^{l,l/2}(Q)}.$$

Introducing the vector of coefficients $\Phi(x,u_1,u_2)=(m_2,\Phi_1,\Phi_2,\Phi_3)$ in the system (3.186); $\Phi_k(x,u_1,u_2)=(A_k,B_k,-b_k,\vec{f}_k), k=1,2$ is the vector of coefficients of the elliptical operator $(-div\vec{v}_k)$, a $\varphi^0(x,t,u_1,u_2)=(\vec{u}^0,\psi^0),(x,t)\in\overline{Q}$ composes the vector of coefficients that form part of the initial and boundary data satisfying the conditions [3] and [15]:

i. $|\Phi|_G^{(0)}\le M_0, M_0^{-1}\le(m_2,\beta_0,(\overline{A}_k\xi,\xi))\le M_0^{-1}, |\xi|=1,$

$$G=\Omega\times(\theta_*,\theta^*)\times(0,1); \overline{A}_k=A_k, k=1,3, \overline{A}_2=\alpha^{-1}(u_2)A,$$

$$0<\alpha(u_2)\le M_0, u_2\in(0,1); (\alpha,A_2,B_2,\vec{f}_2)\big|_{u_2=0,1}=0, b_2(0)=0;$$

ii. $\varphi^0 = (\vec{u}^0, \psi^0)$ for $(x,t) \in \overline{Q}$ has the following properties

$$0 < \delta_0 \le \vec{u}_2^0(x,t) \le 1 - \delta_0; 0 < \theta_* \le u_1^0(x,t) \le \theta^* < \infty;$$

$$(\|u_{1t}^0, u_{2t}^0\|_{1,Q}, \|\varphi^0, \varphi_x^0\|_{2,Q}, \|\psi^0\|_{2,\Sigma^2}) \le M_1.$$

In order to draw a valid analogy with the classical principle to find the maximum saturation $u_2(x,t)$ let us assume that

$$0 < \delta \le u_2(x,t) \le 1 - \delta, \delta \in (0,1), (x,t) \in \overline{Q}. \tag{3.189}$$

The latter provides a uniform parabolic feature of equation (3.186) when $k = 2$

$$\left|\ln(A_2\xi, \xi)\right| \le M_2, (x, u_1, u_2) \in \overline{G}, \tag{3.190}$$

Similar to [3] let the following conditions be met

$$\vec{f}_2\vec{n}\big|_\Sigma = 0, \frac{\partial}{\partial\theta}(b_2, \vec{F}_2) = div_x\vec{F}_2 = 0, \forall u_2 \notin (\delta, 1-\delta), \tag{3.190*}$$

where $\Sigma = \Sigma^1 \cup \Sigma^2, \vec{F}_2 = B_2\nabla\theta + \vec{f}_2, \delta > 0$ is a small constant. In the physical sense (3.190*) implies that when a single fluid flows $(u_2 \le \delta, (1-u_2) \le \delta)$ the equilibrium temperature $u_1 = \theta$ should be close to that of the flowing fluid, and the porous medium saturated with this fluid should be homogeneous.

Definition The vector $\vec{u} = (u_1, u_2, u_3)$ is called a regular solution of the MLT-problem when for all points $\overline{Q}$ we have $u_1(x,t) \in [\theta_*, \theta^*], u_2(x,t) \in [\delta, 1-\delta]$, conditions (3.190) are valid; $(u_1, u_2) \in V_2(Q), u_3 \in W_2^{1,0}(Q)$ and (u_1, u_2, u_3) satisfy the standard integral identities, and Σ^1, Σ^2 in (3.187) satisfy the assumptions [3, 15].

The authors of [15] prove the existence of a weak solution of the MLT-problem (without condition (3.190)).

3.8.3 Proof of the Regular Solution Smoothness

Theorem 1 *Let us assume that assumptions i), ii) are fulfilled,(3.190), provided*

$$\left|\Phi\right|_G^{(l)} \le M_3, [l] \le 2, \alpha = l - [l] > 0; \left|B_3\right|_G^{(0)} \le \varepsilon, \tag{3.191}$$

where $\varepsilon > 0$ *is a small number. Then there is such* α_0 *that if* $\alpha_0 > 0$*, then*

$$\left|\vec{u}\right|^{(l_0,l_0/2)} \leq N(Q'), l_0 = [l] + \alpha_0, \overline{Q}' \subset Q. \tag{3.192}$$

In this case $Q' = Q$*, if in addition we have*

iii. $\partial\Omega \in H_*^{l_1}[3], \quad \left|\varphi^0\right|^{(l_1,l_1/2)} \leq M_4, l_1 = l + 1$

and if the conformity conditions are fulfilled up to the order of $[l]$.

Proof Under conditions *iii)* we can use a standard continuation into the wider domain where $\Omega^* \supset \Omega$ is determined. Then the estimation (3.192) is valid for $Q' = Q$. Consequently we can confine our case to $\overline{Q}' \subset Q$.

Inequation (3.186) for $u_3(x,t)$ we substitute $B_3\nabla\theta_h$ instead of the term $B_3\nabla\theta$ [θ_h is the Steklov averaging $\theta(x,t) = u_1$]. Select $h > 0, N > 0$ and $\varepsilon > 0$ so that

$$\varepsilon\left(\left\|\theta_h\right\|_Q^{(2,0)} + \left\|\theta_{hxxx}\right\|_{2,Q} \leq N\right). \tag{3.193}$$

As before the solution found to the problem can be designated through $\vec{u} = (u_1, u_2, u_3)$. Due to the restriction $\vec{f}_0 = B_3\nabla\theta$ we obtain from equation (3.186)($k = 3$) we obtain $u_3 \in C^{\alpha_0}(\Omega') \cap W_{q_0}^1(\Omega'), \alpha_0 > 0, q_0 \geq 2$. Using the equation (3.186), ($k = 1$), we obtain $u_1 \in C^{\alpha_0}$ (theorem 8.1 [3]).

Thus, we have

$$\left|u_k\right|_{Q'}^{(\alpha_0)} \leq N(Q'), \alpha_0 > 0, k = 1, 3. \tag{3.194}$$

We can write (3.186) in the form

$$-m_k u_{kt} + div(A_k \nabla u_k) = h_k, k = 1, 2, 3, \tag{3.195}$$

where $h_k = \sum_{i,j} q_{ij}^k u_{ix} u_{jx} - \gamma_k \Delta\theta (\gamma_1 = 0)$ is a smooth function with carrier $\Omega_\rho, \left|\Omega_\rho\right| = \rho$. When (3.194) is valid and if we consider the integral identities corresponding to (3.195) using the multiple inequalities taken from [3] for (u_{ix}, u_{jx}) we find

$$J \leq C \sum_{k=1}^{3} (\varepsilon_k \delta^{-1} \rho^{2\alpha} + \delta) \left\| u_{kxx} \xi^2 \right\|_{2,Q}^2 + c_0(\rho), \tag{3.196}$$

where $J = \sum_1^3 J_k(u_k), J_k(u_k) = \left\| u_{kxx} \xi^2 \right\|_{2,Q}^2 + m_k \left\| u_k \xi^2 \right\|_{V(Q)}^2,$

$$\left\| f \right\|_{V(Q)}^2 = \left\| f_x \right\|_{2,Q}^2 + \left\| f_t \right\|_{2,Q}^2, \text{and } \varepsilon_2 = 0, \varepsilon_1 = \varepsilon_3 = 1.$$

Choosing $\rho > 0$ and $\delta > 0$ and splitting Q' into a finite number of domains Q_ρ, we obtain from (3.196)

$$\left\| \vec{u}_{xx} \right\|_{2,Q}^{2} + \sum_{j=1}^{2} \left\| u_j \right\|_{V(Q')}^{2} \leq N(Q'). \tag{3.197}$$

Expression (3.197) yields that (3.194) is valid for $k = 2$. Now examining the equation (3.186), $k = 3$, in which $A_3 \in C^{\alpha}(Q')$ and $|\vec{f}_0| \leq M, (\vec{f}_0 = B_3 \nabla \theta_h + \vec{f}_3)$, we obtain $u_{3x} \in L_{q,\infty}(Q')$ $\forall q \in (1, \infty)$ (theorem 4.2 [3]). Then, using (3.186), we find $(u_1, u_2) \in L_q(Q')$ for $k = 1, k = 2$ (theorem 5.3 [3]) that is

$$\left\| \vec{u}_x \right\|_{q,Q'} \leq N(q, Q') \forall q \in (1, \infty). \tag{3.198}$$

Further we differentiate the equations (3.195) for functions $(u_1, u_2^*, u_3) = \vec{u}, u_2^* = m_2 u_2$ having variable $x_i, i = 1, 2, 3$, assuming that $f^i = f_{x_i}$ (omitting asterisk *). Then u_k^i satisfies the same equations (3.195) $(m_1 = m_2 = 1)$ on the right hand side:

$$h_{ix} = div \vec{G}_{ik} + F_i(\gamma_k), \vec{G}_{i1} = (\theta \vec{f}_0)^i + \lambda^i \nabla \theta, \vec{G}_{i3} = \vec{f}_0^{\,i},$$

$$\vec{G}_{i2} = (b_2 \vec{f}_0 + B_2 \nabla \theta + \vec{f}_2)^i, F_i(\gamma_k) = [div(\gamma_k B_3 \nabla u_3)]^i; \gamma_1 = \theta, \gamma_2 = b_2, \gamma_3 = 0.$$

For the functional $J_k(u_{kx}^i)$ in (3.196) we obtain

$$J_k(u_{kx}) \leq c_k \left\| h_k \xi^2 \right\|_{2,Q}^{2} + c_k^0(Q'),$$

$$\left\| h_k \xi^2 \right\| = \sup_i \left\| h_{ik} \xi^2 \right\|. \tag{3.199}$$

Introducing h_k let us consider the function:

$$R_i(\gamma) = [div \gamma \vec{f}_0]^i, \gamma = u_1, b_2, 1.$$

We will assume that $\vec{f} \equiv f_x, g_{x_i} \equiv g_x$, then

$$R_i(\gamma) = (\gamma f_2)_{xx} + \gamma(B_{3xx}\theta_{hx} + 2B_{3x}\theta_{hxx} + B_3\theta_{hxxx}).$$

In accordance with (3.193) we obtain $(\gamma f_2)_{xx} \in L_2(Q')$ and $(B_3 \theta_{hxxx}) \in L_\infty(Q')$ on the basis of estimations (3.197), (3.198). Reasoning along these lines we find for $R_i(\gamma)$ and hence for $(h_k \xi^2)$

$$\left\| h_k \xi^2 \right\|_{2,Q}^{2} \leq \delta c_1 \left\| \vec{u}_{xxx} \xi^2 \right\|_{2,q}^{2} + c_0(Q', \delta).$$

Summing up estimations obtained in terms of k for chosen small $\delta > 0$, we have

$$\left\| \vec{u}_{xxx} \right\|_{2,Q}^{2} + \sum_{j=1}^{2} \left\| u_j \right\|_{V(Q')}^{2} \leq N(Q'). \qquad (3.200)$$

We obtain from (3.186) that $\vec{u}_x \in C^{\alpha_0}(\Omega'), \alpha_0 > 0$. Returning to equations (3.186) in u_1, u_3, and later in u_2, we obtain (3.192), when $l_0 = 2 + \alpha_0, \alpha_0 > 0$. This allows for a limiting transition of the averaging parameter h. Further improvement in the degree of smoothness $\vec{u}([l] > 2)$ is provided by the standard method of differentiating equations (3.186). Thus, the theorem is proved.

3.8.4 Approximation of the MLT-problem (AMLT-problem)

$$\begin{cases} m_k u_{kt}^{i+1} = div(\Phi_k^i \nabla_k u^{i+1}), k = 1, 2 \\ -\nabla \cdot \vec{v}^{i+1} = div(A_3 \nabla u_3^{i+1} + B_3^i \nabla \theta^i + \vec{f}_3^{\,i}) = 0 \end{cases} \qquad (3.201)$$

where functions $\vec{u}^{i+1}$ satisfy the conditions (3.187)($\Sigma^3 = \varnothing$). Here $\Phi_k^i = \Phi_k(x, u_1^i, u_2^i), \Phi_k = (A_k, B_k, -b_k, \vec{f}_k)$, are the coefficients of the elliptical operator ($-div\vec{v}_k$). For u_k we introduce the designation $\nabla_k u = (\nabla u_k, \nabla\theta, \vec{v}, 1)$. As a result we have $\vec{v}_k = -\Phi_k \nabla_k u$. In the AMLT-problem the linear equations are solved in sequence: first for u_3, then for u_1 and finally, for u_2. A similar iteration technique is used to prove theorem 1. Consequently, together with the proof of Theorem 1 the validity of the following assertion is confirmed.

Theorem 2 *Provisions of theorem 1 are valid for solutions of $\vec{u}^{i+1}$ of the AMLT-problem (3.201).*

3.8.5 Estimation of the Approximate Convergence Rate

Theorem 3 *Let $B_3 = 0 and [l] = 2 under$ conditions of theorem 1 provided that $\delta\Omega = \Gamma^1 or$ $\delta\Omega = \Gamma^2$. Then when $i \to \infty functions$ $\vec{u}^{i+1} converge$ to the classical solution of $\vec{u}^{i+1}, \alpha_0 > 0$ of the non-linear MLT-problem (3.187) In this case*

$$\left\| u_{3i+1}, \nabla u_{3i+1} \right\|_{2,\infty,Q} + \left\| u_{1i+1}, u_{2i+1} \right\|_{V_2(Q)} \leq \varepsilon, \qquad (3.202)$$

$$\left\| \vec{u}_{i+1} \right\|_{\infty,Q} \leq \varepsilon^{\beta}, \beta \in (0, 1), \qquad (3.203)$$

where $\vec{u}_{i+1} = \vec{u}^{i+1} - \vec{u}, \varepsilon = c\{(cT)^i/i!\}^{1/2}$, *and constant* c *depends only on the given data of the problem.*

Proof Similar to (3.201) we have

$$m_k \frac{\partial}{\partial t} u_{ki+1} = div\{\Phi_k \nabla_k u_{i+1} + \vec{F}_k^i\}, k = 1, 2, 3, \tag{3.204}$$

where $\vec{F}_k^i = \Phi_{ki} \nabla_k u^{i+1}, \Phi_{ki} = \Phi_k(x, u_1^i, u_2^i) - \Phi_k(x, u_1, u_2)$.

According to (3.191) we find that

$$|\Phi_{ki}| \le M_3(|u_{1i}| + |u_{2i}|),$$

and further we obtain for $\vec{F}_3^i$:

$$\|F_3^i\|_{2,\Omega} \le c_0 (\|u_{1i}\|_{2,\Omega} + \|u_{2i}\|_{2,\Omega}) \equiv c_0 y_i(t).$$

Multiplying (3.204), $k = 3$ by u_{3i+1} and integrating within Ω, we finally obtain

$$\|u_{3i+1}, \nabla u_{3i+1}, \vec{v}_{3i+1}\|_{2,\Omega} \le c y_i(t). \tag{3.205}$$

Now consider $\vec{F}_1^i$ and $(B_2 \nabla \theta_{i+1} + \vec{F}_2^i)$ as given vectors. Using (3.205), we have

$$\|u_{ki+1}\|_*^2 \le c \int_0^t (\|u_{ki+1}\|_{2,\Omega}^2 + y_i(\tau) + l_k \|\nabla u_{1i+1}\|_{2,\Omega}^2) d\tau,$$

$$\|f\|_*^2 = \|f\|_{2,\Omega}^2 + \int_0^t \|f\|_{2,\Omega}^2 d\tau; k = 1, 2; l_1 = 0, l_2 = 1.$$

Substituting expression $\int_0^t \|\nabla u_{1i+1}\|_{2,\Omega}^2 d\tau$ in the inequality obtained and making $k = 1$ for $\forall i$ making $k = 2$ we obtain

$$\|u_{ki+1}\|_*^2 \le c \int_0^t (y_{i+1}(\tau) + y_i(\tau)) d\tau, k = 1, 2. \tag{3.206}$$

Summing (3.206) for$k = 1$ and (3.206) when $k = 2$ leads to inequality

$$y_{i+1}(t) \le c_0 \int_0^t (y_{i+1}(\tau) + y_i(\tau))d\tau, y_{i+1}(0) = 0,$$

from which according to the Gronuoll inequality [3] we find

$$y_{i+1}(t) \le c \int_0^t y_i(\tau)d\tau, \quad (c = c_0 e^{c_0 T}). \tag{3.207}$$

Applying the method of induction from (3.207) for

$$z_i = \sup_{0 \le t \le T} (\tau) = \|u_{1i}\|^2_{2,\infty,Q} + \|u_{2i}\|^2_{2,\infty,Q}$$

we obtain the following estimate:

$$z_i \le \varepsilon(i) z_0, \varepsilon(i) = \frac{(cT)^i}{i!},$$

Substituting it in (3.205), (3.206) results in (3.202). Using the interpolation inequality

$$\sup_Q |u| \le c(\beta)(\|u\|_{V_2(Q)})^{\beta}(\|u\|_{c^{\alpha}(\Omega)})^{1-\beta}, \beta \in (0,1),$$

we obtain the estimate (3.203). This proves the theorem.

Let's consider now (3.201) as a system of linear equations in which vector Φ^i composed of coefficients depending on the parameter i and, in accordance with theorem 3, values Φ^i converge when $i \to \infty$. Let us split the domain $\Omega, \Omega = \cup_1^N \Omega_i, |\Omega_i| \le M$ and interval $[0, T]$ into smaller parts having dimensions h_0. Designate the relevant normalized space as $X = X(\Omega)$ describing its finite dimension analogue as $X_h = X(\Omega_h)$, $h = (h_0, h_1)$.

Then

$$\|u\|_{X_h} \le \|u\|_X \le |u|_{\Omega}^{(l)}, [l] \ge 2.$$

Using the following assumption, similar to that used in [6].

Theorem 4 For solutions $\vec{u}_h$ of the algebraic system of equations approximating to the linear problems (3.201), (3.187) for $\vec{u}^{i+1}$, the following estimation is valid

$$\|\vec{u}^{i+1} - \vec{u}_h\|_{X_h} \le M(h); M(h) \to 0 \text{ at } |h| \to 0,$$

where $M(h)$ does not depend on i. Then there is such l that $l \geq 2$ in (3.191), for which $\vec{u}_h$ converges to solution $\vec{u}$ of the initial nonlinear problem (3.186), (3.187) having the following convergence rate:

$$\left\| \vec{u} - \vec{u}_h \right\|_X \leq c\varepsilon^{\gamma} + M(h); \gamma \in (0, 1] \tag{3.208}$$

3.9 FLOW OF TWO IMMISCIBLE INHOMOGENEOUS FLUIDS IN POROUS MEDIA

Now we prove the theorem of existence of general solutions in time for initial-boundary problems using the mathematical model describing the process of flow of two immiscible inhomogeneous fluids (for example, water to steam or oil to gas). This approach was proposed by the authors in [15, 17]. In these papers the conventional conditions of the density taken from the theory of flow of a two-phase fluid [15] were replaced by the immiscibility condition of these fluids that has proved reliable in the models used in oceanological and hydrological studies [145].

The mathematical model corresponding to the above assumptions is developed using the system of composite differential equations including the uniformly parabolic equation for the temperature, degenerating elliptical-parabolic system to describe for saturation in one of the fluids and the average pressure as well as the hyperbolic system of equations for the density transfer.

For the first time the model of flow of a single inhomogeneous fluid was proposed in [91], to be followed by analysis of the validity of the initial-boundary problems [91, 124]. Similar problems were examined in [19] for the flow model of two inhomogeneous fluids. The validity of the initial-boundary problems set for the models of nonisothermal flow of two homogeneous fluids having constant density was investigated in papers [15, 16]. In this section we focus on similar problems formulated for the general field development model [15, 19, 91]. The results obtained are reported in paper [18].

3.9.1 Equation of the Model

Let $s_i, i = 1, 2$ be the phase saturations of the porous space, $s_1 + s_2 = 1, m$ is the porosity, $\alpha_i = ms_i, i = 1, 2, \alpha_3 = 1 - m$ are the volume concentrations of fluids and that of the solid phase (or the porous space frame), ρ_i, p_i, u_i are the density, pressure and velocity of fluid flow respectively, $v_i = \alpha_i u_i = ms_i u_i, i = 1, 2$ are the rates of phase flowand $v = v_1 + v_2$ is the rate of flow of the mixture. It is assumed that rocks are nondeformable and the thermal equilibrium is observed in each point of the porous medium, i.e. the temperatures θ_i coincide in phases – $\theta_i = \theta, i = 1, 2, 3$.

It is assumed that the phase motion obeys the laws of flow set forth in the Muskat-Leverett model [2]:

$$\frac{\partial m s_i \rho_i}{\partial t} + div \rho_i v_i = 0, \tag{3.209}$$

$$v_i = -K_0(x) \frac{\sigma_i(s)}{\mu_i(\theta)} (\nabla p_i - \rho_i g), \tag{3.210}$$

$$p_2 - p_1 = \gamma(\theta) \cos \alpha(\theta) \left(\frac{m(x)}{|K_0(x)|} \right)^{\frac{1}{2}} J(s) \equiv p_c(x, \theta, s), s \equiv s_1. \tag{3.211}$$

Here $K_0(x)$ is the permeability tensor in the absolute medium; $\sigma_i(s)$ are the relative phase permeabilities; $\sigma_1(s_*) = \sigma_2(s^*) = 0, s_*, 1 - s^*$ are the residual phase saturations; μ_i are the phase viscosities; g is the gravity constant; γ is the surface tension coefficient; α is the wetting angle; $J(s)$ is the Leverett function for capillary pressure. Further, we designate

$$k_i(s, \theta) = \sigma_i(s) \mu_i^{-1}(\theta), k = k_1 + k_2.$$

Instead of the fluid state equations written at the bottom of (3.209), (3.210), we will make use of the conditions of fluid immiscibility:

$$\frac{\partial \rho_i}{\partial t} + u_i \nabla \rho_i = 0, i = 1, 2, \tag{3.212}$$

They imply fixed fluid densities ρ_i along the trajectory of motion. Taking into account only the convective heat transfer and the heat conductivity the energy balance equation for the mixture can be written in the form

$$\frac{\partial \theta}{\partial t} + div(v\theta - \lambda(x, \theta, s) \nabla \theta) = 0. \tag{3.213}$$

Here λ is the temperature conductivity of the mixture (two fluids and porous frame). In [15] equation (3.213) was obtained from the general equations of the balance of the energy of components when the mixture is kept under thermal equilibrium [95] provided the density was the same for the phases and their specific heat was identical. It was also assumed that

$$\lambda = \sum_{i=1}^{3} \alpha_i \lambda_{\cdot i} \tag{3.214}$$

The equations (3.213), (3.214) derived in paper [15] remain valid for a more general case of an inhomogeneous incompressible fluid under the condition that the product of the phase densities multiplied by their specific heat remains the same.

3.9.2 Transformation of Equations. Formulation of the Problem

Differentiating (3.209) and using (3.212), we obtain

$$\rho_i(ms_{i_t} + div v_i) + ms_i(\rho_{i_t} + u_i \nabla \rho_i) = 0,$$

which leads to the equations in s_i, having the same form as equations derived for $\rho_i = const$ [2]:

$$ms_{i_t} + div v_i = 0, i = 1, 2. \tag{3.215}$$

Since $s_2 = 1 - s_i$, then the above system is equivalent to the system compiled for s, v_1 and v:

$$ms_t + div v_i = 0, div v = 0. \qquad .(3.216)$$

By analogy with [2] we will introduce the average pressure

$$p = p_2 + \int_S^{S^*} b_1 \frac{\partial}{\partial \xi} p_c(x, \theta, \xi) d\xi, b_i = \frac{k_i}{k}, i = 1, 2.$$

After making relevant transformations in equations (3.210), (3.211), (3.216) we obtain the following system of equations for $s, p, \theta, \rho_1, \rho_2$:

$$ms_t = div[K_0(a_1 \nabla s - a_2 \nabla \theta + f_1) - b_1 v] \equiv - div v_1(s, p, \rho_i, \theta),$$

$$0 = div K_0 k(\nabla p + f_2 + a_3 \nabla \theta) \equiv - div v(s, p, \rho_i, \theta),$$

$$\frac{\partial \theta}{\partial t} + div(v\theta - \lambda(x, \theta, s)\nabla \theta) = 0, \tag{3.217}$$

$$\frac{\partial m s_i \rho_i}{\partial t} + div \rho_i v_i = 0, i = 1, 2,$$

where

$$a_1 = \left|p_{cs}\right| a_0,\, a_2 = p_{c\theta} a_0,\, a_0 = b_1 k_2,\, a_3 = -k_1 p_{c\theta} - \int_s^{s^*} \frac{\partial}{\partial \theta}(b_1 p_{cs}) ds,$$

$$f_1 = a_0[(\rho_1 - \rho_2)g - \nabla_x p_c], f_2 = \int_s^{s^*} b_2 \nabla_x p_{cs} ds - g(b_1 \rho_1 + b_2 \rho_2).$$

Let $\Omega \subset R_3$ – be the confined domain $\Omega_T = \Omega \times [0, T], \partial\Omega = S$, $\Gamma = S \times [0, T], S = S_1 \cup S_2, \Gamma_i = S_i \times [0, T]$. To determine the functions we are looking for, let us consider the following initial-boundary problem:

$$(s, \theta, \rho_1, \rho_2)\big|_{i=0} = (s, \theta, \rho_1, \rho_2)_0(x), x \in \Omega, \tag{3.218}$$

$$v_i n\big|_\Gamma = 0, \theta\big|_{\Gamma_1} = \theta_0(x, t), \lambda \frac{\partial \theta}{\partial n}\bigg|_{\Gamma_2} = \beta(\theta_0 - \theta). \tag{3.219}$$

Here n is the unit vector of the outer normal to Γ; $\beta(s)$ is the heat-transfer coefficient for a three-component mixture which for the previously discussed cases of deriving equation (3.213) can be written as follows

$$\beta = \sum\nolimits_{i=1}^{3} \alpha_i \beta_i (\rho_i c_i)^{-1},$$

β_i-is the heat-transfer coefficient of the i-th phase.

3.9.3 Assumptions. Determination of the Generalized Solution

Let the following be met for the coefficients of the equation and the initial-boundary conditions are as follows:

i. $(\nabla_{\theta,s}(k_i, p_c), p_c \theta_s) \in C(G)$, G is a closed domain in space of the variables (x, s, θ), and

$$M^{-1} \leq (m, p_{cs}, (K_0 \xi, \xi), \beta, \lambda) \leq M,$$

$$|\bar{\xi}| = 1, \left\|\nabla_x p_c\right\|_{\infty,\Omega} \leq M.$$

ii. $0 < (k_1, k_2) < M, s \in (0, 1), a_0\big|_{s=s_*, s^*} = k_1(s_*) = k_2(s^*) = 0$,and

$$\left|\ln(a_1 a^{-1}), \ln(a_2 a^{-1}), b_1 k_1^{-1}, f_1 a^{-1}, f_2 k_2^{-1}, a_3\right| \leq M_0(M), a = k_1 k_2.$$

iii. the functions in (3.218), (3.219) have the properties

$$s_* \le s_0(x,t) \le s^*, \theta_* \le \theta_0(x,t) \le \theta^*, 0 < m_i \le \rho_{i0}(x) \le M_i < \infty,$$

$$(\|\theta_{0t}, s_{0t}\|_{1,\Omega_T}, \|\nabla\theta_0, \nabla s_0\|_{2,\Omega_T}) \le M..$$

Designations of functional spaces and norms are taken from paper [83].

Definition *We shall call a set of functions(3.213)* the generalized solution of the problem (3.217)–(3.219), if:

i. $0 < s_* \le s \le s^* < 1, 0 < m_i \le \rho_i \le M_i < \infty, i = 1,2$ in Ω_T
ii. $u(s) = \int_{s_*}^{s} \sqrt{a_1(x,\xi)}d\xi \in W_2^{1,0}(\Omega_T)$
iii. $(p, \nabla p) \in L_2(\Omega), \theta \in V_2(\Omega_T) \cap L_\infty(\Omega_T)$
iv. the boundary conditions in Γ_1 are satisfied almost everywhere;
v. the following integral identities are fulfilled:

$$(ms, \varnothing_t) - (a_1\nabla s - a_2\nabla\theta + f_1, K_0\nabla\varnothing) + (b_1 v, \nabla\varnothing) = -(ms_0, \varnothing)_{\Omega_0}, \tag{3.220}$$

$$(\theta, \psi_t) - (\lambda\nabla\theta - v\theta, \nabla\psi) = -(\beta(\theta_0 - \theta), \psi)_{\Gamma_2} - (\theta_0, \psi)_{\Omega_0}, \tag{3.221}$$

$$(\nabla p + a_3\nabla\theta + f_2, K_0\nabla\eta) = 0, \tag{3.222}$$

$$(\rho_i, ms_i\zeta_l + v_i\nabla\zeta) = -(ms_{i0}\rho_{i0}, \zeta)_{\Omega_0}. \tag{3.223}$$

Here $(\cdot,\cdot)$-is the scalar product in $L_2(\Omega_T)$, and the test functions in $\varnothing, \psi, \eta, \zeta$ satisfy the conditions $\varnothing, \psi, \zeta \in W_2^1(\Omega_T), \psi = 0$ in $\Gamma_1, \varnothing = \psi = \zeta = 0$ for $t = T, \eta \in (0, T; W_2^1(\Omega))$.

3.9.4 Regularization of Problem (3.217)–(3.219)

We will introduce functions $S(s), \Theta(\theta), R_i(\rho_i)$ which extrapolate values s, θ, ρ_i outside the intervals $[s_*, s^*], [\theta_*, \theta^*], [m_i, M_i]$ using the boundary values. Further, instead of direct dependence of the coefficients in equations (3.217) on s, θ, ρ_i we consider complex functions depending on S, Θ, R_i. Getting rid of the degeneration in s we substitute a_1 by $a_1 + \delta = a_\delta$ in (3.217). In order to ensure that the coefficients remain smooth we will apply the averaging in terms of x and t with steps of h

and τ. Let $\varepsilon=(\delta,h,\tau)$ be the vector of the regularization parameters. Then, from (3.222) having $a_{3h\tau}\nabla\Theta_{h\tau}, f_{2h\tau}, K_{0h}$ and according to results obtained in [83] we have $p^{\varepsilon}\in C^{2+\alpha,1}_{x,t}, vv\in C^{1+\alpha,1}_{x,t}$. From (3.221) having $\lambda_{ht}, \beta_{ht}, \theta_{0ht}$ we obtain $\theta_{\varepsilon}\in C^{2+\alpha,1+\alpha/2}\equiv C^{2+\alpha}_{*}, \nabla\theta\in C^{1+\alpha}_{*}(\Omega_T)$. And finally, from (3.220) having $a_{\delta h\tau}, a_{2h\tau}\nabla\theta, b_{1h\tau}v, f_{1h\tau}, K_{0h}, s_{0h}$ we will have $s^{\varepsilon}\in C^{2+\alpha}_{*}$. It is worth mentioning that after truncating in ρ_i and f_i, the first three equations in the system (3.217) fully comply with the properties of coefficients in paper [15]. Therefore, we obtain the following estimates for uniformity of the regularization parameters ε:

$$\|\nabla\theta\|_{2,Q_T}+\|\nabla p\|_{2,Q_T}\leq C[\|\theta_{0t}\|_{1,Q_T}+\|\nabla p_0\|_{2,Q_T}+\|a_3\|_{2,Q_T}]\equiv N_1,$$
$$\|\sqrt{a}\nabla s\|_{2,Q_T}\leq C\left[N_1+\|\nabla s_0\|_{2,Q_T}+\|f_1 a_{\delta}^{-1/2}\|_{2,Q_T}\right], \tag{3.224}$$

$$s_*\leq s\leq s^*, \theta_*\leq\theta\leq\theta^*.$$

The last inequalities imply that (3.220)–(3.222) are valid for real coefficients rather than for truncated ones because of $S(s)$ and $\Theta(\theta)$. Due to the fact that coefficients are smooth we have inclusions $v_1^{\varepsilon}\in C^{1+\alpha,1}, v^{\varepsilon}\in C^{1+\alpha,1}, v_2^{\varepsilon}=v^{\varepsilon}-v_1^{\varepsilon}\in C^{1+\alpha,1}$. As s does not become zero, then $u_i^{\varepsilon}=v_i^{\varepsilon}/(ms_i^{\varepsilon})\in C^{1+\alpha,1}$. It is not difficult to see that the functions $\rho_i^{\varepsilon}(x,t)=\rho_{i0h}(y_i^{\varepsilon}(x,t,0))$, where

$$\frac{dy_i^{\varepsilon}}{d\tau}=u_i^{\varepsilon}(y,\tau), y\big|_{\tau=t}=x,$$

satisfy the equations (3.212) for u_i^{ε} [equations (4^{ε})] and initial conditions (3.218). Identities (3.220), (3.222) are equivalent to equations (3.215) with s_i^{ε} and v_i^{ε} [equations (7^{ε})] for the smooth solutions of the regularized problem. Multiplying (4^{ε}) by ms_i^{ε} and (7^{ε}) by ρ_i^{ε} and adding, we obtain the equations (1^{ε}) [(1) $s_i^{\varepsilon}, \rho_i^{\varepsilon}, v_i^{\varepsilon}$], from which the identities (3.223) are easily obtained for the regularized problem. Representing solution as ρ_i^{ε} we readily obtain that $\rho_i^{\varepsilon}\in[m_i, M_i]$ and consequently, truncation in ρ_i does not perturb the coefficients of the problem (3.217)–(3.219).

3.9.5 Limiting Transition

Making use of (3.224) in the same way as was discussed in papers [2, 15] we can obtain strong compactness $s^{\varepsilon}, p^{\varepsilon}, \theta^{\varepsilon}$ in $L_p(\Omega_T), \forall p\geq 1, v^{\varepsilon}$ in $L_2(\Omega_T)$ from (3.220)–(3.222) and weak compactness v_i^{ε} in $L_p(\Omega_T)$.

Multiply (4^{ε}) by $ms_i^{\varepsilon} q^{-1} \rho_{i\varepsilon}^{q-1}$ (for the time being we will omit the index ε), $q \geq 1$ being an integer and integrate the expression using Ω_t:

$$\begin{aligned} 0 &= (ms_i, (\rho_i^q)_t)_{\Omega_t} + (v_i, \nabla \rho_i^q)_{\Omega_t} \\ &= -(\rho_i^q, ms_i + div\, v_1)_{\Omega_t} + \int_{\Omega} m(s_i \rho_i^q\big|_{t=1} - s_i \rho_i^q\big|_{t=0})dx. \end{aligned}$$

Due to (7^{ε}) we obtain

$$\int_{\Omega} ms_i \rho_i^q dx = \int_{\Omega} ms_{i0} \rho_{i0}^q dx = const. \tag{3.225}$$

Let us consider $z_{i\varepsilon} = (ms_{i\varepsilon})^{1/q} \rho_{i\varepsilon}$. It is obvious that $z_{i\varepsilon} \to z_{i,*}$ – is weak in $L_{\infty}(\Omega_T)$when $\varepsilon \to 0$, and due to (3.225) $\|z_{i\varepsilon}\|_{q,\Omega} \equiv \|z_{i0}\|_{q,\Omega}$. In particular, for $q = 2$ we have $\|z_{i\varepsilon}\|_{2,\Omega} \to \|z_{i0}\|_{2,\Omega}$, consequently $z_{i\varepsilon} \to z_i$ is strong in $L_2(\Omega)$, which implies that $\forall p > 2$ in L_p as well. As in this case s_{ε} strongly converges in $L_p, p > 2$ and $0 < s_* \leq s \leq s^* < 1$, then $\rho_{i\varepsilon} = z_{i\varepsilon}(m, s_{i\varepsilon})^{-1}$ also converges strongly in $L_p, p > 2$. Strong convergence of $s_{i\varepsilon}, \rho_{i\varepsilon}$ provides a limiting transition in identity (3.223). Limiting transitions in (3.220)–(3.222) are conducted in the same manner [15].

Thus, the following assertion is proved.

Theorem Let the following assumptions be fulfilled i)–iii). Then the problem (3.217)–(3.219) has, at least, one general solution $(s, p, \theta, \rho_1, \rho_2)$ in $\Omega_T, \forall T > 0$.

3.10 NUMERICAL SOLUTION OF THE PROBLEM OF TWO-DIMENSIONAL TWO-PHASE FLOW

3.10.1 Introduction

Various fluid dynamic problems modelling water flooded oil production were considered taking into account the capillary forces [9, 41, 51, 72, 76, 80, 84, 110, 111, 139, 141, 142, etc.]. Nevertheless, water flood calculations remain tedious even using the approximation techniques. One of the reasons is that the system of non-linear differential equations in the second order partial derivatives in pressure and saturation is used to describe the two-phase flow of the oil–water mixture featuring capillary effects. Nonlinearity effects – both in pressure and saturation – are extremely complicated to describe. In many papers the one-dimensional problems were considered in more detail. The two-dimensional- and especially three-dimensional problems are especially difficult to solve.

In this section we consider the method of numerical solution applied to the two- dimensional problem of water oil drive taking into account capillary forces. This is carried out using the Muskat-Leverett mathematical model with "reduced" pressure (see Chapter 1, Section 1.2).

The nonlinear elliptical equation for pressure contains, on the right hand side, the terms describing the oil and water flow rates. A parabolic equation to determine water saturation is convenient to obtain the numerical solution using the variable orientation method. The initial water saturation values are taken as known for the initial domain, whereas the Newmann boundary conditions are fixed for the pressure.

To draw a comparison, the numerical calculations of the same problem were conducted using the Duglis, Pichman and Rechford schemes [51]. Of course, the results obtained were sufficiently close but the principal algorithm is closer to the scheme developed by Duglis, Pichman and Rechford. Solvability of this problem and the issues of convergence were studied in paper [2].

3.10.2 Formulation of the Problem

Let us consider the following initial-boundary problem for the system of equations inside a cylinder $Q=\{D\times(0<t<T)\}$ with boundary γ, describing the flow of the two-phase incompressible fluid:

$$\begin{cases} m\dfrac{\partial s}{\partial t} = div(K_0(x)a(x,s)\nabla s - b(s)\vec{v} + \vec{F}(x,s)) + q_2, \\ div(-K_0(x)k(s)\nabla p - b(s)\vec{v} + \vec{f}(x,s)) + q_1 + q_2 = 0, \\ \vec{v} = -K_0(x)k(s)\nabla p - b(s)\vec{v} + \vec{f}(x,s) \end{cases} \tag{3.226}$$

$$s(x_1,x_2,0) = s_0(x_1,x_2), \left.\frac{\partial p}{\partial n}\right|_\gamma = 0, \left.\frac{\partial s}{\partial n}\right|_\gamma = 0. \tag{3.227}$$

Numerical modelling is conducted using the assumption that the medium is isotropic and homogeneous, hence $K_0(x) = const$. The relative phase permeabilities $k_1(s), k_2(s)$ were determined using the following formulas:

$$k_1(s) \approx s^{\varepsilon}, k_2(s) = (1-s)^{\delta}, \varepsilon, \delta \in (0,1);$$

$$k(s) = k_1(s) + k_2(s) = s^{\varepsilon} + (1-s)^{\delta}, b(s) = \frac{k_1(s)}{k(s)} = \frac{s^{\varepsilon}}{s^{\varepsilon} + (1-s)^{\delta}}; \quad (3.228)$$

$$a(x,s) = -\overline{p}_c(x) \cdot J'(s). \frac{k_1 \cdot k_2}{k}. \quad (3.229)$$

Here $\overline{p}_c(x)$ is expressed through the Laplace formula:

$$\overline{p}_c(x) = \sigma \cdot cos\theta \sqrt{\frac{m}{K_0}}, \quad (3.230)$$

where σ – is the interphase tension coefficient; m – is the porosity coefficient; K_0 – is the flow coefficient; θ – is the boundary; $J(s)$ – is the Leverett function; decreasing from ∞ to 0 when s is changing from 0 to 1. Usually it is assumed that $J'(s) = c \cdot \ln 1/s$ or $J'(s) = const/s^{\alpha}$, where $0 < \alpha < 1$. Consequently,

$$a(x,s) = \overline{p}_c(x) \frac{const \cdot s^{\varepsilon-\alpha}(1-s)^{\delta}}{s^{\varepsilon} + (1-s)^{\delta}}; \quad (3.231)$$

$$\vec{F}(x,s) = -\frac{k_1(s) \cdot (\rho_2 - \rho_1)\vec{g}}{k_1(s) + k_2(s)} = \frac{s^{\varepsilon}(\rho_1 - \rho_2)\vec{g}}{s^{\varepsilon} + (1-s)^{\delta}}. \quad (3.232)$$

If $\overline{p}_c(x) \neq const$, then

$$\vec{f}(x,s) = K_0(x).\nabla\overline{p}_c(x) \cdot k(s) \cdot \int_s^1 J'(\xi)\frac{k_1(\xi)}{k(\xi)}d\xi + K_0(x)k_2(s)J(s)\nabla\overline{p}_c(x) + (\rho_2 - \rho_1)\vec{g},$$

$$\vec{F}(x,s) = -K_0(x) \cdot \frac{k_1(s) \cdot k_2(s)}{k(s)} J(s)\nabla\overline{p}_c(x) - \frac{k_1(s)}{k(s)}(\rho_2 - \rho_1)\vec{g}.$$

From here on s implies that $s = \frac{s_1 - s_1^0}{1 - s_1^0 - s_2^0}$ – the dynamic water saturation; s_1^0 is the residual water saturation; s_2^0 – is the residual oil saturation; s_1 is the true water saturation.

We consider a two dimensional (in orthogonal plane $\vec{g}$) problem of an incompressible two-phase fluid flowing at $\overline{p}_c = const$. In this case $\vec{F}(x, s) = (0, 0), \vec{f}(x, s) = (0, 0)$ and equations (3.226) take the form

$$\begin{cases} m\dfrac{\partial s}{\partial t} = div\ K_0(x)a(x, s)\nabla s - \vec{v}\dfrac{\partial b}{\partial s}\nabla s + q_2; \\ \\ 0 = div\ (K_0(x)K(s)\nabla p) + q_1 + q_2; \vec{u} = -K_0(x)K(s)\nabla p. \end{cases} \tag{3.233}$$

To solve the problem (3.227), (3.233) numerically we will construct the following difference scheme of variable directions to express water saturation:

$$\begin{aligned} m\frac{s_{ij}^{(n+1/2)} - s_{ij}^{(n)}}{\tau} &= (d_{i+1/2j}s_{\overline{x}_1,ij}^{(n+1/2)})_{x_1} + (d_{ij+1/2}s_{\overline{x}_2,ij}^{(n)})_{x_2} - (bu^{(n)})_{\overline{x}_1,i+1/2j} \\ &\quad - (bv^{(n)})_{\overline{x}_2,ij+1/2} + q_{2,ij}^{(n)}, \\ m\frac{s_{ij}^{(n+1)} - s_{ij}^{(n+1/2)}}{\tau} &= (d_{i+1/2j}s_{\overline{x}_1,ij}^{(n+1/2)})_{x_1} + (d_{ij+1/2}s_{\overline{x}_2,ij}^{(n+1)})_{x_2} - (bu^{(n)})_{\overline{x}_1,i+1/2j} \\ &\quad - (bv^{(n)})_{\overline{x}_2,ij+1/2} + q_{2,ij}^{(n)}, \end{aligned} \tag{3.234}$$

where $d_{ij} = K_0 \cdot a(x_{ij}, s_{ij}^{(n)}), x_{ij} = (x_{1i}, x_{2j})$.

The calculation of pressure is provided by solution of the elliptical equation through the following method:

$$\frac{p_{ij}^{(n+1/2)} - p_{ij}^{(n)}}{0,5\tau_1} = (\tilde{d}_{i+1/2j}p_{\overline{x}_1,ij}^{(n+1/2)})_{x_1} + (\tilde{d}_{ij+1/2}p_{\overline{x}_2,ij}^{(n)})_{x_2} + (q_1^{(n)} + q_2^{(n)})_{ij}; \tag{3.235}$$

$$\frac{p_{ij}^{(n+1)} - p_{ij}^{(n+1/2)}}{0,5\tau_1} = (\tilde{d}_{i+1/2j} p_{\overline{x}_1,ij}^{(n+1/2)})_{x_1} + (\tilde{d}_{ij+1/2} p_{\overline{x}_2,ij}^{(n+1)})_{x_2} + (q_1^{(n)} + q_2^{(n)})_{ij}.$$

The velocity distribution is determined using the Darcy law:

$$u_{i+1/2j}^{n+1} = -\tilde{d}_{ij} p_{x_1,ij}^{n+1},\ v_{ij+1/2}^{n+1} = -\tilde{d}_{ij} p_{x_2,ij}^{n+1}, \tag{3.236}$$

where $\tilde{d}_{ij} = K_0 \cdot k(s_{ij}^{(n)})$.

The calculation algorithm can be specified as follows: the method of scalar fitting is used to find $s^{(n+1/2)}, s^{(n+1)}$. The iteration technique is used to calculate the corresponding $p_{ij}^{(n+1)}$.

3.10.3 Analysis of Numerical Results

The algorithm developed was used to formulate a model capable of simulating the dynamics of oil displacement using the measured water saturation. Some examples, with results for various well locations, are given below.

For instance, the following nine-spot oil production pattern was considered (Fig. 3.28).

Eight injection wells having the same injection rates q_2 (squares) are plotted on the uniform grid $\{(x_i, y_j), x_i = ih, y_j = jh, i = 0.30, j = 0.20\}$. The production well (circle) having oil flow rate q_1 and water flow rate q_2, proportional to the mobilities of corresponding phases is placed in the centre. If q_0 is the quantity of water

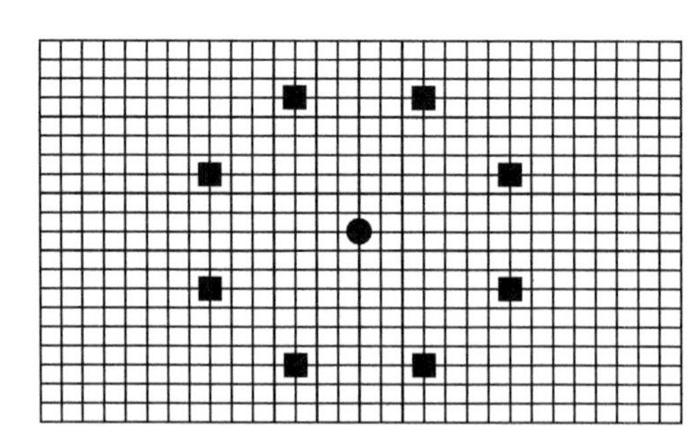

Figure 3.28 A nine-spot oil production pattern is considered.

injected into the formation then the following values are given for the production well:

$$q_1 = -\frac{k_1 q_0}{(k_1 + k_2)}, \quad q_2 = -\frac{k_2 q_0}{(k_1 + k_2)},$$

Correspondingly for injection wells the values

$$q_1 = 0, \quad q_2 = \frac{q_0}{8} \text{ are known.}$$

We will assume that at an initial point $s_0 = const$, the porosity is constant through over the calculation domain $m = 0,375$, the permeability $k = 1$, and dynamic viscosities are $\mu_1 = 0,00928, \mu_2 = 0,00115$.

During water injection the distribution of water saturation has the following profile (Fig. 3.29). As can be seen the maximum values correspond to the locations of the water injection wells. We stopped the calculation at $t = 0,2792$, when the time step was $\Delta t = 0,000358$.

Table 3.1 contains the values in nodes *(i,j)*, where, due to symmetry $i = 0, \ldots, Nx/2, j = 0, \ldots, Ny/2$.

The corresponding total pressure profile is as shown in Fig. 3.30.

In Table 3.2 pressures are reported as dimensionless values.

To study the symmetry in a selected domain using the same parameters as mentioned in the first example we took 8 producing wells having identical flow rates q_1 and q_2, placing an injection well having the injection rate of q_0 in the centre (Fig. 3.31).

Indeed, the water saturation pattern has a symmetry relative to the centre (Fig. 3.32) as the formation was taken to be homogeneous, and the water saturation at initial moment – constant ($t = 0,8234, \Delta t = 0,00358$).

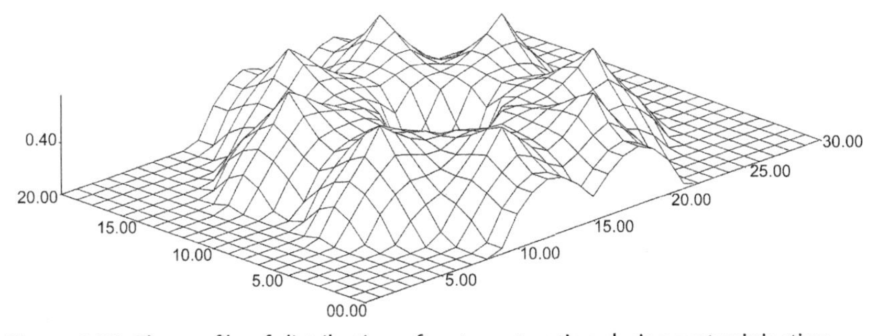

Figure 3.29 The profile of distribution of water saturation during water injection.

Table 3.1 Values in nodes (i,j), where, due to symmetry $i = 0, \ldots, \frac{Nx}{2}, j = 0, \ldots, Ny/2$

I\I	0	1	2	3	4	5	6	7	8	9	10	11	12	13	14	15
0	0.21	0.21	0.21	0.21	0.21	0.21	0.21	0.21	0.21	0.23	0.34	0.40	0.43	0.41	0.37	0.30
1	0.21	0.21	0.21	0.21	0.21	0.21	0.21	0.21	0.21	0.23	0.34	0.40	0.43	0.41	0.37	0.30
2	0.21	0.21	0.21	0.21	0.21	0.21	0.21	0.21	0.21	0.29	0.37	0.43	0.48	0.45	0.40	0.37
3	0.21	0.21	0.21	0.21	0.21	0.21	0.21	0.21	0.25	0.33	0.40	0.47	0.56	0.50	0.44	0.41
4	0.21	0.21	0.21	0.21	0.21	0.21	0.22	0.29	0.33	0.35	0.40	0.46	0.53	0.49	0.46	0.43
5	0.21	0.21	0.21	0.21	0.21	0.22	0.31	0.37	0.40	0.40	0.40	0.44	0.49	0.48	0.46	0.44
6	0.21	0.21	0.21	0.21	0.21	0.29	0.37	0.43	0.47	0.46	0.44	0.43	0.46	0.47	0.45	0.45
7	0.21	0.21	0.21	0.21	0.23	0.32	0.40	0.47	0.56	0.52	0.48	0.45	0.45	0.45	0.43	0.41
8	0.21	0.21	0.21	0.21	0.21	0.29	0.38	0.44	0.51	0.50	0.48	0.46	0.44	0.41	0.31	0.24
9	0.21	0.21	0.21	0.21	0.21	0.23	0.33	0.39	0.44	0.46	0.46	0.45	0.43	0.31	0.21	0.21
10	0.21	0.21	0.21	0.21	0.21	0.21	0.26	0.36	0.41	0.43	0.44	0.45	0.38	0.23	0.21	0.21

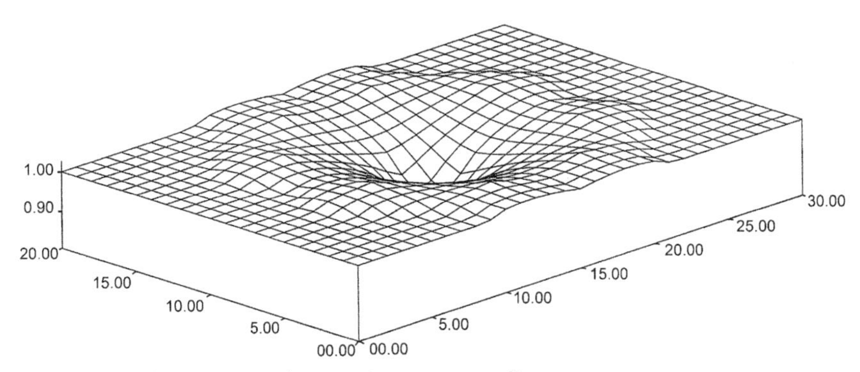

Figure 3.30 The corresponding total pressure profile.

In the next example the wells are located in four rows (Fig. 3.33). Such a pattern can be used to describe domains elongated along the *Ox* axis.

In this case the values of porosity *m*, permeability *k*, viscosities μ_1, μ_2 are the same as those given in the previous examples. The water flow rates in the injection wells are assumed to be equal and for producing wells the sum of flow rates of oil q_1 and water q_2 is proportional to the mobilities of the corresponding phases. For such location of wells the distribution of water saturation has the following profile (Fig. 3.34).

Table 3.3 lists the numerical values of water saturation plotted in Fig. 3.34 for $i = 1 \ldots Nx/2, j = 1 \ldots Ny/2$.

In this section we give a numerical solution of the two-dimensional problem of the flow of an incompressible two-phase fluid.

Various examples were used to show that the model proposed allows forecasting the formation of isolated immobile domains in a reservoir that occur due to the nonhomogeneity of the formation or due to water blocking access to the producing well.

By choosing the optimum location of injection and producing wells and varying their operating regimes it is possible to increase the volume of oil production.

Software called "Muskat" with a service interface was developed to solve the two-dimensional ML equations. It is capable of calculating the water saturation and pressure at any point in time for given locations of wells and their production rates.

Module "Gr.bas" is responsible for plotting one- and two-dimensional graphs (in colour) during the calculations.

Table 3.2 Pressures (reported as dimensionless values)

I\I	0	1	2	3	4	5	6	7	8	9	10	11	12	13	14	15
0	1,00	1,00	1,00	1,00	1,00	1,00	1,00	1,00	1,00	1,00	1,02	1,02	1,02	1,02	1,02	1,01
1	1,00	1,00	1,00	1,00	1,00	1,00	1,00	1,00	1,00	1,00	1,02	1,02	1,02	1,02	1,02	1,01
2	1,00	1,00	1,00	1,00	1,00	1,00	1,00	1,00	1,00	1,01	1,02	1,02	1,03	1,02	1,02	1,01
3	1,00	1,00	1,00	1,00	1,00	1,00	1,00	1,00	1,01	1,02	1,02	1,02	1,03	1,02	1,01	1,01
4	1,00	1,00	1,00	1,00	1,00	1,00	1,00	1,01	1,02	1,02	1,02	1,02	1,01	1,01	1,00	1,00
5	1,00	1,00	1,00	1,00	1,00	1,00	1,02	1,02	1,02	1,02	1,01	1,01	1,00	1,00	0,99	0,99
6	1,00	1,00	1,00	1,00	1,00	1,01	1,02	1,02	1,02	1,02	1,01	1,00	0,99	0,98	0,97	0,97
7	1,00	1,00	1,00	1,00	1,00	1,02	1,02	1,03	1,03	1,02	1,00	0,99	0,98	0,96	0,95	0,94
8	1,00	1,00	1,00	1,00	1,00	1,01	1,02	1,02	1,02	1,01	1,00	0,98	0,96	0,94	0,92	0,90
9	1,00	1,00	1,00	1,00	1,00	1,00	1,02	1,02	1,01	1,00	0,99	0,97	0,95	0,92	0,89	0,87
10	1,00	1,00	1,00	1,00	1,00	1,00	1,01	1,02	1,01	1,00	0,99	0,97	0,94	0,90	0,87	0,81

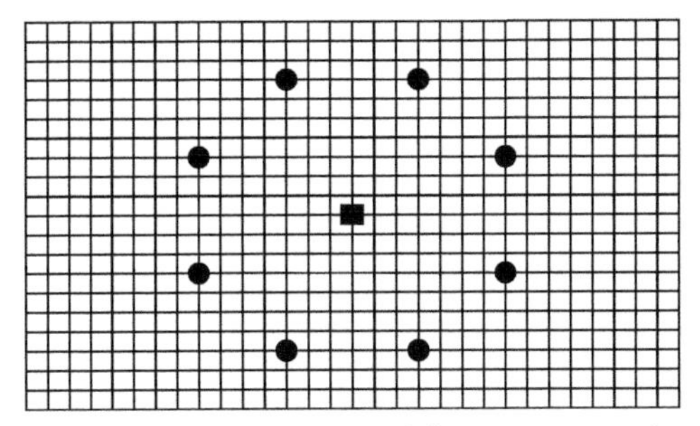

Figure 3.31 8 producing wells having identical flow rates q_1 and q_2, placing an injection well having the injection rate of q_0 in the centre.

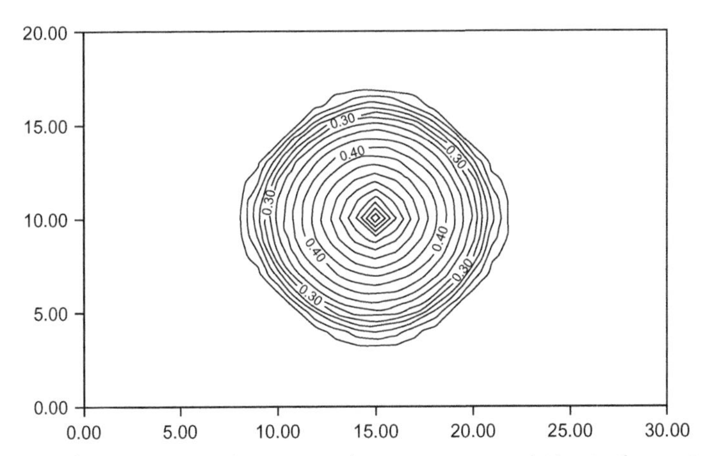

Figure 3.32 The water saturation pattern has a symmetry relative to the centre.

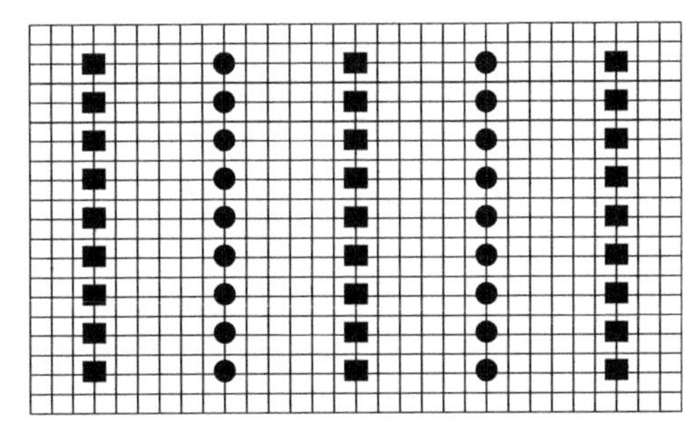

Figure 3.33 Wells are located in four rows.

Table 3.3 lists the numerical values of water saturation plotted in Fig. 3.34 for $i = 1, \ldots, \frac{Nx}{2}, i = 1, \ldots, Ny/2$

I\J	0	1	2	3	4	5	6	7	8	9	10	11	12	13	14	15
0	0.21	0.21	0.21	0.22	0.21	0.21	0.21	0.21	0.21	0.21	0.21	0.21	0.21	0.21	0.21	0.22
1	0.21	0.21	0.21	0.22	0.21	0.21	0.21	0.21	0.21	0.21	0.21	0.21	0.21	0.21	0.21	0.22
2	0.21	0.21	0.22	0.30	0.22	0.21	0.21	0.21	0.21	0.21	0.21	0.21	0.21	0.21	0.22	0.30
3	0.21	0.21	0.21	0.23	0.21	0.21	0.21	0.21	0.21	0.21	0.21	0.21	0.21	0.21	0.21	0.23
4	0.21	0.21	0.22	0.30	0.22	0.21	0.21	0.21	0.21	0.21	0.21	0.21	0.21	0.21	0.22	0.30
5	0.21	0.21	0.21	0.23	0.21	0.21	0.21	0.21	0.21	0.21	0.21	0.21	0.21	0.21	0.21	0.23
6	0.21	0.21	0.22	0.30	0.22	0.21	0.21	0.21	0.21	0.21	0.21	0.21	0.21	0.21	0.22	0.30
7	0.21	0.21	0.21	0.23	0.21	0.21	0.21	0.21	0.21	0.21	0.21	0.21	0.21	0.21	0.21	0.23
8	0.21	0.21	0.22	0.30	0.22	0.21	0.21	0.21	0.21	0.21	0.21	0.21	0.21	0.21	0.22	0.30
9	0.21	0.21	0.21	0.23	0.21	0.21	0.21	0.21	0.21	0.21	0.21	0.21	0.21	0.21	0.21	0.23
10	0.21	0.21	0.22	0.30	0.22	0.21	0.21	0.21	0.21	0.21	0.21	0.21	0.21	0.21	0.22	0.30

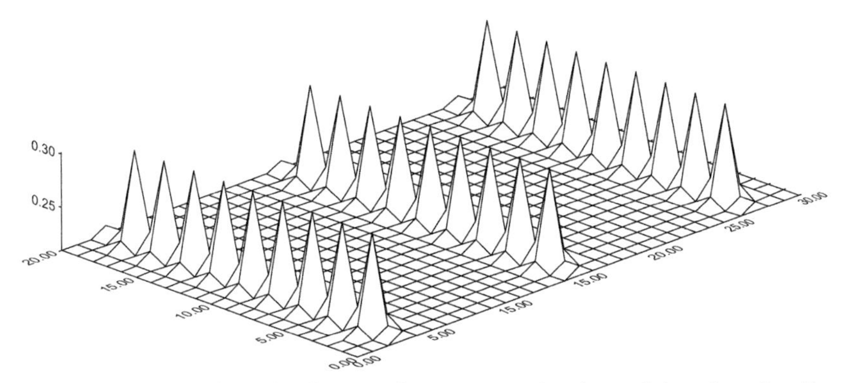

Figure 3.34 Profile of the distribution of water saturation for such location of wells.

3.11 GEOLOGICAL AND MATHEMATICAL MODELS OF THE RESERVOIR

3.11.1 Introduction

In this section we provide a comparative analysis of various mathematical models of reservoirs, give the limits of their application and conduct the fluid dynamic analysis of the results of numerical calculation. Also, we consider the issues of applying the mathematical models and calculating reservoir parameters.

Development of oil and gas fields is highly dependent on the development of methods for mathematically modelling real reservoirs and describing the complicated geological processes.

The characteristic parameter for the inflow of the heavy Kumkol oil is the paraffin pour point [61] θ_k. At this temperature precipitating wax sediments at the surface of the pores resulting in a dramatic decrease in the oil production rate. To maintain the formation temperature hot water is injected into the formation. The process of production of heavy oil using steam flooding can be described using the model of two-phase flow.

The book [61] outlines the principles of developing various mathematical models to describe reservoirs. It also explains how specific models developed by the authors and their colleagues are applied in computer simulation of oil field development.

The aim of this section is to solve the following problems of computer simulation which had not been discussed in [61] or were only mentioned briefly.

1. identification of oil formation parameters (developing a geological model);
2. adaptation of mathematical models through comparison of the field data (*in site experiments*);
3. comparative analysis of the various mathematical models and definition of their applicability range;
4. numerical approximation of mathematical models and fluid dynamic analysis of the results of numerical calculation.

3.11.2 The Main Mathematical Flow Models

3.11.2.1 Darcy, Navier-Stokes and Zhukovsky Models

Let $m(x)$ and $k(x)$ – are the coefficients of porosity and permeability of the porous medium ($x \in R^3$); $\mu = const > 0$ and $\rho = 1$ – are the homogeneous fluid viscosity and density; p_0 and $\vec{u}$ are the pressure and the velocity vector of the moving fluid particles; $p = p_0 + \rho g h$ and $\vec{v} = m\vec{u}$ are the fluid head and the flow rate. Then the main models describing the flow of the homogeneous fluid in porous medium may be reduced to the following two-parameter model:

$$\varepsilon\left(\frac{d\vec{u}}{dt} - \mu\Delta\vec{u}\right) + \nabla p + \delta\lambda m\vec{u} = 0, \nabla \cdot \vec{u} = 0. \tag{3.237}$$

For $\varepsilon = 1, \delta = 0$ the system of equations (3.237) becomes the Navier-Stokes model (NS), for $\varepsilon = 0, \delta = 1$ – we have the Darcy model (D), and for $\varepsilon = \delta = 1$ – the Zhukovsky model (Zh). Here $d/dt = \partial/\partial t + (\vec{u} \cdot \nabla)$; $\lambda = \mu/k$ is the Zhukovsky coefficient and correspondingly $m\lambda\vec{u}$ is the Zhukovsky force (see Chapter 1).

For some values of parameters of the porous medium and the flowing fluid the Irmey hypothesis is valid. According to the hypothesis $\Delta\vec{u} = -\gamma\vec{u}$ ($\gamma(x) > 0$), which, in this case, indicates that the Darcy and Navier-Stokes models are close (see Chapter 1).

3.11.2.2 Live Oil Model

Let p_H – be the saturation pressure of the live oil having reduced density $\rho = 1$ at $p > p_H$. The process of gas drive when $p < p_H$ will be accounted by a special equation of state for an oil and gas mixture (Chapter 1):

$$\rho(p) = \begin{cases} 1, p \geq p_H, \\ \{1 + \delta(p - p_H), p < p_H(\delta = const > 0). \end{cases}$$

Assuming that the fluid flow is one-dimensional and planar (orthogonal to the gravity vector) we obtain the following equations:

$$m\frac{\partial \rho(p)}{\partial t} - \frac{\partial}{\partial x}\left(\rho k\frac{\partial p}{\partial x}\right) = 0, v = -k\frac{\partial p}{\partial x}. \tag{3.238}$$

3.11.2.3 The Muskat-Leverett and the Buckley-Leverett Models

The equations comprising the main mathematical models of the flow of a two-phase fluid may be presented as a single parameter family of models:

$$\vec{v}_i = -k_i(\nabla p_i + \rho_i \vec{g}), ms_{it} + div\vec{v}_i = 0, p_2 - p_1 = \varepsilon p_c, \tag{3.239}$$

where $\vec{v}_i, p_i$, and ρ_i are the phase flow, pressure and density $(i = 1, 2)$; s_i, μ_i and $k_i(s)$ are the phase saturations, viscosities and permeabilities; $k_i = kk_i/\mu_i$, where k - is the average permeability.

For $\varepsilon = 0$ expression (3.239) becomes the simplest Buckley-Leverett model not taking into account capillary forces, for $\varepsilon = 1$ the Muskat-Leverett model (ML).

Model (3.239) is transformed into the following system of equations in the dynamic saturation $s(x, t)$ and the medium pressure $p(x, t)$:

$$\begin{cases} ms_t - div(\varepsilon a\nabla s - b\vec{v} + \vec{f}_0) = 0, \\ div\vec{v} \equiv -div(k_0\nabla p + \vec{f}_1) = 0, \end{cases} \tag{3.240}$$

where $a(x, s) > 0$ for $s \in (0, 1)$ and $a(x, s) = 0, s = 0, 1; k_0(x, s) \geq \delta_0 > 0$ when $s \in [0, 1], (|b_s|, |b_x| < \infty)$.

3.11.2.4 The Displacement Model

The previously described models were developed in such a manner that they cannot describe the flow of a two-phase fluid under conditions when a porous medium is undersaturated with a displacing phase (flow into a "dry" formation). In this case we have to use the displacement model (Chapter 1, Section 1.3), which provides the same system of equations of the type (3.240) with a potential for dynamic saturation $s(x, t)$ to take negative values.

3.11.2.5 The Boundary Layer Model

One of the disadvantages of the Muskat-Leverett model (the one that accounts for capillary forces) deals with the "edge effect" or poor accuracy in the boundary conditions set for the producing well. To make sure that the boundary conditions of the producing well are generated in the course of the problem solution (as is the case with the Buckley-Leverett model) simultaneously taking into account capillary forces, one can make use of the boundary layer model (Chapter 1, Section 1.3). For the planar steady flows of a two-phase fluid in a porous medium the boundary layer model simply becomes an integration of the following saturation equation $s(x, y)$:

$$|b'(s)|\frac{\partial s}{\partial x} = \frac{\partial}{\partial y}\left(a(s)\frac{\partial s}{\partial y} + b'Q(t)\frac{\partial s}{\partial y}\right), \tag{3.241}$$

where $Q(t)$ is a given flow rate in the mixture.

3.11.3 Geological Model (GM) of Oil Formation

Among the mathematical flow models (Chapter 1) there are more than 30 describing various processes in reservoirs.

To conduct numerical calculations using any of these models to forecast field development parameters it is necessary to know the underlying parameters: $m = m(x)$ is porosity; $k = k(x)$ is permeability (in general case k-tensor); $\bar{k}_i(s), \mu_i, \rho_i$ are permeabilities, viscosities and densities of the phases; $p_c(x, s)$ is capillary pressure. Besides, to estimate the parameters, i.e. the flow rates, pressures and saturations of the phases $\vec{v}_i, p_i, s_i (i = 1, 2)$ it is necessary to collect well data at reservoir boundaries and under initial conditions.

The basis for developing a model is provided by inputting geological and field data for the reservoir studied, namely:

- characteristics of the geological structure;
- the geological and physical characteristics;
- the fluid dynamic characteristics (production rates, pressures, etc.);
- the approved calculation parameters – oil in place and recoverable reserves of oil, dissolved gas, free gas and condensate;
- the field history data, etc.

The information supporting the geological model consists of geological field data, the values directly measured in the wells (geological structure, geological and physical characteristics, field data etc.), details of field development and approved volume of recoverable reserves, cross section information near the wells drilled in the same area of the field.

The techniques to improve the accuracy of geological field data are highly location dependent. They give sufficiently reliable information on the formation parameters close to the well. Obtaining the necessary information for the whole of reservoir requires application of various methods of interpolation and statistical processing [61].

Obtaining the distribution of the geological and physical parameters of reservoir areally and with depth, modifying these parameters, using additional information on the field and processing the parameters using the statistical, engineering and fluid dynamic methods (inverse problems) results in the formulation of *a geological model* (GM) for the reservoir. Often the formulation of the GM for the reservoir is referred to as *the identification* of the reservoir's geological and physical parameters.

The book [61] discusses the means to obtain, store and process geological field data in sufficient detail.

Let us consider some specific problems of modelling geological field data using a simulation approach to model various units in the reservoir.

3.11.3.1 Selection of Calculation Domains

Any mathematical model of an oil field is designed to simulate its performance, to determine the causes of discrepancy between pre-development and actual indices, and to suggest corrective measures to eliminate any discrepancies. Geological differences in fields, their development history, substantial differences in field parameters etc., requires the modelling of each field individually.

Let us initially focus on the problems of planar (areal) flow of nonhomogeneous fluid in a formation. This case corresponds to the development of a specific formation consisting of several layers having similar permeability and located in a limited part of the field or covering the whole field. Let us refer to this complex system of strata as the oil bearing formations.

The practical analysis and the development plan for large fields is done using the zonal principle as described in [26, p. 20; 41; 86, p. 50; 197].

The reservoir to be studied (bed) is divided into calculation domains influenced by the reservoir arrangement (influencing the choice of the coordinate system), *geological properties or the field development plan* (they control the choice of domain for application of the mathematical model), and considerations of the symmetry or location of the section close to the field boundaries (the latter require selecting the special types of boundary conditions) (Fig. 3.35).

Fig. 3.35, *a* shows the split into the calculation domains A, B, C, corresponding to a different choice of the coordinate system. Fig. 3.35, *b* outlines the locations of the calculation domains for the various development schemes: A – row, B – transient, C – five-spot. The specific features for

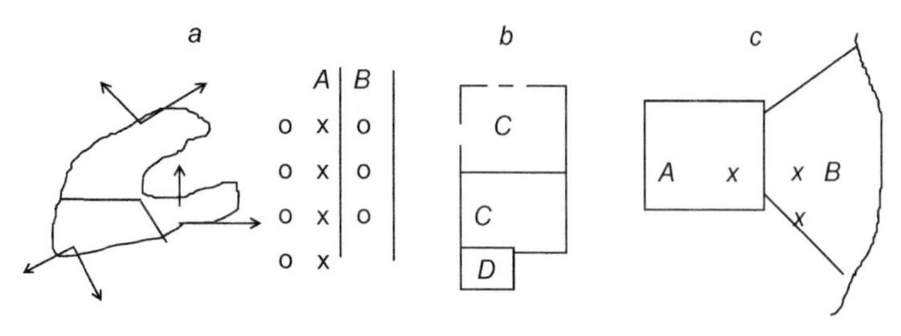

Figure 3.35 The reservoir to be studied (bed) is divided into calculation domains influenced by the reservoir arrangement.

division into calculation domains at the field boundaries are shown in Fig. 3.35, *c*.

Depending on the geometrical, geological and operational field characteristics the number of injection and production wells varies significantly in the calculation domain (from several to many hundreds of wells). The calculation domains are especially important as they feature a regular system of well spacing. They are referred to as the base symmetry elements. It is assumed that a closed system having symmetrical conditions at the boundaries can be considered as a base element. This element has the known boundary conditions for the wells (the production rate, pressure or fluid saturation) and homogeneous reservoir parameters within the boundaries of the element. The first stage of assessing the field development plant involves the calculation of development parameters within the base element followed by calculation of the parameters for the complete reservoir. For example, for the five-spot development scheme (Fig. 3.35, *b*, section *C*) the repeating base symmetry element of water flooding is provided by a rectangle having no flow boundaries to water with injection and production wells in the corners (see Fig. 3.35, *b*, section *D*).

The symmetry of the above base element allows calculation of field characteristics using one-dimensional mathematical models. However, with time any field develops asymmetry in well spacing and production rates which can be caused, for example, by drilling operations which differ from those planned. This process can only be taken into account with two-dimensional calculations using the base element and discarding the assumption of symmetrical boundary conditions which makes such a base element an integral calculation domain.

The next crucial moment in selecting the calculation domains deals with field parameters describing the oil production and the rate of water encroachment in the field. First of all, it is necessary to locate the areas in the field (if they exist) where production wells are water-free. For such areas the practical mathematical models are provided by homogeneous flow (not two-phase) models. These are the models proposed by Darcy, Navier-Stokes, and Zhukovsky.

The models describing the flow of homogeneous fluid can also be applied to domains featuring permanent water saturation (at least over a long period of time). Indeed, let $\bar{s}\in(0,1)$ be the constant water saturation in a domain, $k_i = k_i(\bar{s})\mu_i^{-1} = const, \bar{b}_1 = b_1(\bar{s}) = \bar{k}_1(\bar{k}_1 + \bar{k}_2)^{-1} = const, k = \bar{k}_1 + \bar{k}_2 = const.$ Then the equation of the ML- model (3.239) can be readily reduced to the following Darcy model: $\vec{v} = -k\nabla p, div\vec{v} = 0(p = p_2 - \bar{b}_1\bar{p}_c, \bar{p} = p_c(\bar{s}))$, where $\vec{v} = \vec{v}_1 + \vec{v}_2$. In the same manner we can obtain the Zhukovsky model with $\lambda = \frac{k}{\mu}, \vec{u} = \vec{u}_1 + \vec{u}_2$.

Now let us consider the case when process of water drive of oil takes place in the domain Ω having two poroperm features. That is throughout Ω the permeability equals k_1, and in the regions (inclusions) $\Omega_n \subset \Omega, n = \overline{1, N - k_2} \ll k_1$ [9, p 147; 23, p 92]. Then these areas are bypassed by water and oil can be recovered from them only due to capillary effects. If the domain $\Omega_n \subset \Omega$ is sufficiently large, then capillary flow can be ignored and we consider only streamlined flow around barriers Ω_n in Ω (Sections 3.4, 3.5 in this chapter).

When, according to the Irmey hypothesis (see Section 3.11.2.1), the flow of fluid is described by the Navier-Stokes model we obtain the description of the viscous flow in connected regions (Section 3.3 in this chapter).

3.11.3.2 Vertical Cross Section of the Reservoir

In the case of controlled flow the effects of sublayers with different porosity, permeability, and residual water saturation values can be accounted for in the mathematical model through averaging of the above parameters. To specify the flow scheme in a reservoir, depending on the geometrical (thickness, curvature) and fluid dynamic (porosity, permeability) characteristics of the sublayers, it is necessary to study flow in the vertical direction (in a cross section) of a reservoir. Therefore, depending on the nature of the water drive, the flow of the two-phase fluid is studied from injector to producer. Most often, the ML-model is used to describe this flow (see Section 3.11.2.1). In the case when for the given stage of field development the flow is steady then equations of the ML-model can be split to model the underlying physical processes:

$$\vec{v}_i^{n+1} = -\frac{k_i^n}{\mu_i} \cdot \nabla \varphi_i^{n+1}, div v_i^{n+1} = 0 \; i = 1, 2, n = 1, 2, \ldots \tag{3.242}$$

$$p_c(s^{n+1}, x) = p_2^n - p_1^n; \varphi_i^n = p_i^n + \rho_i g h, g \nabla h = \vec{g}. \tag{3.243}$$

We will note that such splitting also takes place in the steady state case.

With a fixed value $s = s^n(x)$ equation (3.242) describes the homogeneous flow of fluid in nonhomogeneous porous medium having the permeability coefficient $k_i^n = k_i(s^n) \equiv k_i^n(x)$, and the distribution of saturation at the $(n+1)$ iteration step is unambiguously given by the Laplace law (3.243) as $\left| \frac{\partial p_c}{\partial s} \right| \neq 0$ at $s \in (0, 1)$.

The Darcy models obtained for each of the components ($i = 1, 2$) may be substituted by Navier-Stokes equations (when the Irmey hypothesis is fulfilled) or by the Zhukovsky model.

In oil practice a field may feature multilayered connected beds [26, p. 167; 86, p. 106]. This leads to study of flow in several connected regions (Fig. 3.36). In the steady-state case as was shown above, the several connected regions may exhibit the types of flow described by the Navier-Stokes or Zhukovsky models (see Sections 3.3, 3.5 in this chapter).

An interesting class of problems described in this book (Chapter 1, 1.6) is encountered in the wellbore when there are two inflows: one slow rate and one fast (Fig. 3.37).

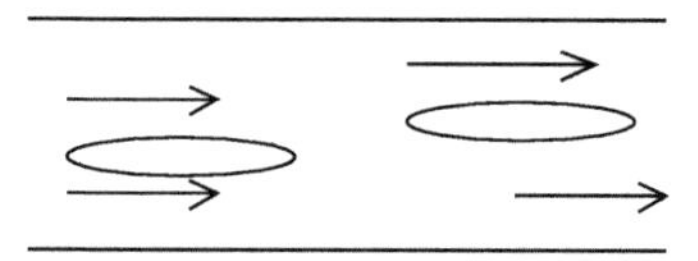

Figure 3.36 Several connected regions.

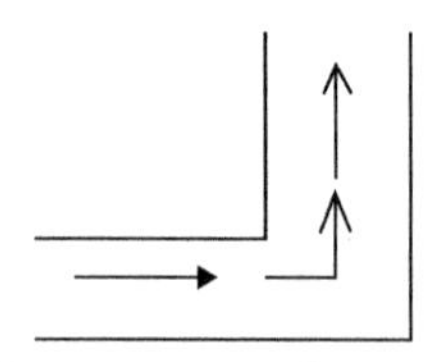

Figure 3.37 A wellbore when there are two inflows: one slow rate and one fast.

3.11.3.3 Combined Fluid Dynamic Characteristics

Combined fluid dynamic characteristics can be easily taken into account in a calculation domain (areal scheme) having multiple beds (horizons). In this case the total oil recovery and water flooding is obtained by summing the corresponding formation parameters.

A far more complicated problem occurs when cross flow of a two-phase fluid takes place through the common boundaries of various domains [86, p. 165]. In this case, if the calculation domain consists only of the base symmetry elements then the conditions given preclude fluid cross flow between these elements. Therefore, the main fluid dynamic characteristics of the domain are calculated as a sum of the characteristics of the base elements. The issue of combined fluid dynamic characteristics is solved in the same way as if it was possible to select the calculation domains as fluid dynamically isolated or to consider the whole field as one domain. Otherwise, the boundary conditions at the common boundaries must account for potential cross flow of the two- phase fluid. This leads to the evaluation of a complex interaction between the calculation domains (an example is provided by superposition of the volume and channel flows in a well – see Chapter 1, Section 1.6).

3.11.4 The Mathematical Model of a Reservoir

Any mathematical model of a complicated reservoir (a section having several wells or a section of a field containing several interlayers having different permeabilities) is informative when the relevant geological field data are processed to generate numerical values of the model parameters (coefficients of differential equations and initial- boundary conditions).

The convenience of using the geological and mathematical models to describe an oil formation depends on a degree of interaction between the automated system and the analysis to develop these reservoirs; the potential for direct access to geological field data and processing to derive the model parameters; the correspondence between the forecasted parameters of the formation developed and the additionally measured characteristics of flowing wells. In brief, the usefulness of the models depends on comparison between calculated and actual data and validity of the geological model.

The mathematical model of oil flow is a set of submodels developed for each calculation domain complying with the initial and boundary conditions as well as meeting the requirements for calculating the combined effect of adjacent domains. Operation of this complex system together with the geological and field data (GFD) provides a reliable control on the process of oil production. Solutions of these complex issues form the basis of an automated system of field development. The mathematical models and the geological field database are described in detail in paper [61].

3.11.4.1 Mathematical Models of Calculation Domains

Based on the above arguments we can make a short list of the mathematical models applied to various reservoir calculation domains.

a. The Muskat-Leverett model (see Section 3.11.2.1) is a universal instrument to describe the two- phase flow under almost all conditions in calculation domains including those featuring homogeneous fluid flow on reaching residual saturations ($s = 0$ or $s = 1$) when it becomes the Darcy model for a homogeneous fluid.
However, as was indicated in 1.4, if the water is flowing into a "dry" formation, that is the dynamic saturation is negative, $s = \frac{\sigma - \sigma_B}{1 - \sigma_B - \sigma_H}$, < 0, where [σ is the true water saturation, σ_α is the residual oil saturations ($\alpha = \mathrm{H}$) and water ($\alpha = \mathrm{B}$)], then it is impossible to describe them within terms of the Muskat-Leverett model.

b. The displacement model (see Section 3.11.2.4) is included into the group of mathematical models of an automated system to analyse field development (ASAFD) which is intended to describe the flow of two- and one-phase fluid into a "dry" formation (at $s < 0$ water flow, and at $s > 1$ oil flow in the absence of residual water).

c. The Darcy, Navier-Stokes (provided the Irmey hypothesis applies) and Zhukovsky models are mainly used to describe the flows of homogeneous fluid. However, as the preceding overview shows (see Section 3.11.3.1), they are also applicable in the following cases of flow of a two-phase fluid:
1. under conditions of constant water saturation;
2. under conditions of a steady two-phase flow;
3. under conditions of superposition of matrix and channel flows;
4. when the following iteration splitting two-phase flow into the underlying physical processes is made:

$$-\vec{v}^{n+1} = k^n(\nabla p^{n+1} + \vec{F}_n^0),\ div\vec{v}^{n+1} = 0 \tag{3.244}$$

$$m\frac{\partial s^{n+1}}{\partial t} = div(k_0^n a^n \nabla s^{n+1} - b^n \vec{v}^{n+1} + \vec{F}^n). \tag{3.245}$$

Here $\vec{F}_0 = k^{-1}\vec{f}$ ($\vec{f}$ is determined in [61, 8.0] and for any value of $\varphi(s,x) = (k, k_0, a, \vec{F}_0, \vec{F})$ it is assumed that $\varphi^n = \varphi(s^n, x)$.

For a given value $s^n = s^n(x,t)$ the system (3.244) is obviously a Darcy model describing a flow of a homogeneous incompressible fluid in a nonhomogeneous formation caused by force $\vec{F}_0$. Equation (3.245) describes the diffusion of water saturation $s^{n+1}(x,t)$ in a given distribution of flow velocity $\vec{v}^{n+1}$.

The convergence of this iteration scheme was proved in book [2] under given conditions of regular flows in which the function to be found – $s(x,t)$ satisfies the inequalities $0 < s^- \le s(x,t) \le s^+ < 1, s^\pm = const$. The authors found sufficient conditions for the existence of such regular flows.

We will note that when the ordinary conditions are met the Darcy model (3.244) can be replaced by the Navier-Stokes or Zhukovsky models.

d. The model of live oil flow (see Section 3.11.2.2) describes the flow of a homogeneous fluid in an oil formation when the pressure in some areas is below the saturation pressure ($p_{form} < p_{sat}$).

3.11.4.2 Interaction of Numerical Algorithms with the Data Base

When the numerical mathematical model algorithms are included into ASAFD it is necessary to ensure that they interact with the data bases. All initial information on oil and gas fields is contained in three sections of the data base: geological field, the operational and general field parameter data [61].

The standard scheme of interaction should include:

- access and readout of the necessary data to operate a specific algorithm from the general field parameter databank;
- access to the geological field databank and construction of a geological model of the part of field developed;
- access and readout from the operational parameter databank of the necessary data on injection and production wells, pressure, temperature etc.;
- calculation of the field development parameters using the algorithm for specific time intervals;
- storage of the calculation results in relevant parts of the data base.

A similar scheme is used to arrange for interaction with the data base to apply extrapolation and statistical analytical methods.

It should be noted, however, that the interaction of numerical algorithms with the data base is dependent on the specifics of the problem. With ASAFD running analysis in the "process engineer – computer" mode, for each specific case means that the interaction of the calculation algorithms with the data base depends on expert decisions, the process engineer in our case. This option is provided by ASAFD.

The option of interaction with the data base at various stages of implementation is ensured for all numerical algorithms included into the group of mathematical models to simulate oil reservoir.

3.11.4.3 Selecting, Applying and Matching the Mathematical Model

Let us consider the following example of applying the mathematical model. Let us take the problem of numerical calculations of two-phase fluid flow using the ML-model between the injection and production wells in a vertical cross section of the reservoir (a horizon, or group of sublayers).

It is assumed that the geological and physical characteristics are given for the wells and the method of determination (for example, the linear interpolation) is fixed for all points in the bed. In addition, we know the information on the wells: the flow rates for the phases (the fluid production rate and the water cut) and the average pressure profile.

Let us make numerical calculations for flow of the mixture with known production rates. We will determine the p_{calc} (calculated pressures 0 and compare them with the p_{field} 9 field pressures). Then, we will solve the opposite problem. Using the known distribution of pressure we calculate Q^k_{calc}, the phase production rates ($k = 1, 2$) and compare them with Q^k_{field}, the actual phase production rates.

Such a comparison of the calculated and actual characteristics enables one to improve the accuracy of the geological and physical parameters of the reservoir between wells applying the multivariant calculation technique. In this manner the formation geological model becomes more accurate, or, in other words, the selected mathematical model gets localised.

Let us provide another example of model localisation.

It is necessary to specify the pressure at the inlet (p_{inlet}) and at the outlet (p_{outlet}) to make a sufficiently accurate calculation of the flow of the two-phase liquid in a producing well. Whereas p_{outlet} can be easily measured, this is not the case with p_{inlet}. Without proof it is often assumed that $p_{inlet} = p_{form}$ (p_{form} is the formation pressure). In this case the best mathematical model to describe the formation not requiring the determination of p_{inlet}, is provided by the model of matched boundary layers (MMBL) or porous insert model (PIM) in which the NS and ZH models are matched (Chapter 1, Section 1.2).

For the last example we consider the situation when localisation of the initial mathematical model requires an unjustifiably large correction of the geological model. In this case one should consider new physical factors not accounted for in the initial mathematical models. Such factors, for example, may include fracturing of the porous medium which may lead to selecting another mathematical model.

We will dwell on the problem of matching mathematical models in various formations. This issue was studied in Section 3.10.2, where we mentioned the requirement to account for the combination at the boundaries of the formation domains (multiplication scheme). Some examples of matching various models have been considered for the vertical cross sections of reservoir (MMBL and PIM).

We will mention one more straightforward example of superposition of the flow without accounting for the capillary forces (BL-model) and in the case when the capillary effects are taken into account (ML-model) (Section 3.10 in this chapter). Obviously, it is not possible to superimpose these flows in the well and it is required to introduce a "transient" zone. This zone is the area which is distorted by flow into the well in which the flow is calculated using the radial flow scheme.

We note that ASAFD provides for variants of matching various mathematical models.

3.11.5 Numerical Reservoir Modelling

We will carry out the fluid dynamic analysis of the numerical simulation using various mathematical models developed for the reservoir as part of ASAFD (Sections 3.2–3.7, 3.10 in this chapter).

3.11.5.1 The Zhukovsky Numerical Model

Paper [50] (compare with 3.4) gives the results of numerical calculation of the mixed flow of viscous fluid in an L-shaped region having a porous section arbitrarily located at the bottom. The numerical algorithm is described and several fluid dynamic problems are solved for different Reynolds numbers and various locations of the porous section in the vertical flow (to model a well) as well as in case when there are no porous section at all.

Let us underline the main results of the fluid dynamic considerations and numerical calculations.

1. The hypothesis that forms the basis of the Zhukovsky model does not require of the unknown efficient viscosity $\mu_{ef} \neq \mu$ applied by other authors to describe the flow in porous media [43,114]. This circumstance leads to the fact that for superposition of the volume and channel flows considered in papers [50,59], the coefficients in front of higher derivatives in the equations remain continuous (when we have the same viscosity μ in $\Delta\vec{u}$ for the NS- and Zh-models). Therefore, on one hand, there is no need to develop "theories" further to determine μ_{ef}, and on the other, the stability of the difference approximations of equations is markedly improved.
2. The relations of the Kozeni type [72] are used to describe the drag coefficient λ, in flow models applied for porous media. Their experimental confirmation was only obtained for very specific porous media. So for the rest of physical parameters these relations have to be confirmed experimentally. In the Zhukovsky model (3.239) the coefficient $\lambda = \mu/k$ is linked only to the known geological parameters of the formation and flowing fluid.
3. The numerical calculations were carried out in regions having complex geometry (Sections 3.4—3.6 in this chapter), in which the flow of fluid sharply alters its direction. In order to overcome this difficulty it was required to introduce a special approximation of equations in the vicinity of the flow turning point that allowed the retention of approximation in equations along the flow lines (Section 3.5 in this chapter).

The type of difference equation to determine the pressure of the singularity point and its calculation algorithm are given in paper [50]. In addition we made calculations of controlled fluid flow in the porous medium having L-shaped structure that models a section of the oil formation (Section 3.4 in this chapter). Despite the complex geometry of the region the results of these calculations allowed comparing the Zhukovsky model and the Darcy model (see 2.1), proving that they are sufficiently close:

$$\rho|\vec{u}_t| \leq 10^{-1}\lambda|\vec{u}|, \rho|(\vec{u}\nabla)\vec{u}| \leq 3 \cdot 10^{-2}\lambda|\vec{u}|, \mu|\Delta\vec{u}| \leq 10^{-1}\lambda|\vec{u}|.$$

3.11.5.2 The Numerical Simulation of Formation Pressure

In paper [46] the pioneering results of numerical modelling of the flow of a viscous fluid in the L-shaped region having a known pressure $p = p_{outlet}$ at the outlet and the zero tangent velocity vector component at the inlet and outlet (uniform flow) were obtained. The problem considered models the formation pressure $p = p_{form}$ using the measured value $p = p_{outlet}$ at the producing well head which is important from a practical point of view (Section 3.6 in this Chapter).

The convergence of the difference approximation was proved for this problem implying that it has a solution in time. The numerical solution algorithm of a more complex problem was proposed and practically developed for computer calculation when instead of pressure p_{outlet} the full head is known, i.e. $q_{\text{outlet}} = p + 1/2|\vec{u}|^2$; in this case instead of the Navier-Stokes model the Zhukovsky model is used (Section 3.6 in this chapter).

A simpler problem for the Navier-Stokes equations in a rectilinear channel was studied theoretically by V.V. Ragulin [108], V.V. Ragulin and Sh.S. Smagulov *et al.* [109].

It should be noted that the problems involving the Zhukovsky model are also applicable to describe the flows of conducting fluid in a magnetic field. Numerical algorithms may be used to solve the applied problems in magnetic fluid dynamics.

3.11.5.3 The Muskat-Leverett Numerical Model

In paper [64] the principal basis of the ASAFD-Muskat-Leverett model system was developed numerically.

An example to calculate the problem of controlled two-phase flow using the base element of the nine-spot pattern was used to demonstrate the efficient and quick convergence of the proposed difference system. The latter provide the necessary criteria for the numerical algorithm to be used.

The solvability of the problem studied and convergence of method of splitting into physical process compiled in the proposed difference approximation were proven in book [61].

For the sake of comparison the numerical simulation of the same problem was made applying the scheme of Duglas, Pichman and Rechford [63] (see also [41]), in which the matrix of the steady elliptical system of the two equations describing water saturation s and phase pressure p has a non-diagonal character. It is natural that the calculation results were pretty close but from the application point of view the second scheme is not as good as the basic algorithm in which the matrix of the elliptical steady system has diagonal character (this is achieved by a special choice of the "average" pressure (Section 3.10 in this chapter).

3.11.5.4 The Numerical Model of the Two-Phase Temperature Flow in Analytical Variables

The theorem of existence and uniqueness of the analytical solutions was proved for the thermal flow model in a two-phase fluid proposed by O.B. Bocharov and V.N. Monakhov [16]. The smoothness of the solution was studied, the finite rate of perturbation propagation was established, and the algorithms to solve the boundary problems using numerical algorithms were proposed and implemented as computer programs (see Chapter 2, Section 2.2).

The comparative analysis of several difference schemes to develop analytical solutions was conducted. The method of Roter was used to solve the initial-boundary problems and the problems resulting in a one-dimensional equation (Chapter 2, Section 2.7).

It was established numerically that heating up of the formation brings about a dramatic improvement in flushing-out. The algorithms proposed to develop analytical solutions allow the prompt evaluation of the efficiency of the EOR methods to be applied to develop oil fields based on the thermodynamic properties of the fluid components.

3.11.5.5 The Numerical Model of the Boundary Layer

First of all we will point out that for two-phase fluid the flow is described by the Muskat-Leverett model. In this case the formation of the Prandtl boundary layer is impossible in a viscous fluid because the assumed averaging conditions at each point of the domain contain the porous medium and two fluid components.

The model of a two-phase fluid boundary layer proposed in paper [75] is a true mathematical approximation of the Muskat-Leverett equations in a thin formation.

We use the model of a two-phase fluid boundary layer to solve the "edge effect" problem [8; 142] in production wells ($x = x_e$), as for $x = x_e$ this model does not require stipulating the boundary condition for saturation.

The algorithm to obtain the boundary conditions proposed in paper [52] for $x = x_e$ consists of the following stages:

1. The solution of the boundary problem for the Muskat-Leverett model in the interval $x \in [0, x_e]$ ($x = 0$ corresponds to injection well) for fixed condition $x = x_e$ is used to determine water saturation value $s(x, y)\big|_{x=x_0} = s_0(y)$ for a known cross section $x = x_0 < x_e$, in proximity to the producing well.

2. Solving the corresponding problem for the flow boundary layer in the interval $x \in [x_0, x_e]$ having $s|_{x=x_0} = s_0(y)$ the value $s|_{x=x_e} = s_e(y)$ is restored.
3. The equations of the ML-model are solved for the new value of $s|_{x=x_e} = s_e(y)$. Then the cycle returns to stage 2 and so on.

Calculations made in paper [52] show a sufficiently quick convergence of the iteration process proposed to generate the water saturation value $s(x, y)$ in producing well. Together with solution of the relevant problem for the ML-model the value of $s(x, y)$ is converging over the interval $[0, x_e]$ between the injection and production wells.

3.11.5.6 Numerical Model of the Live Oil Flow

The physical aspects of the flow of the live oil are outlined in detail in paper [56], where its reduction to the Stefan type problem is proved.

Numerical simulation was carried out in the case when the pressure p of oil in interval $x \in [0, \xi_0]$ was less than the saturation pressure $p_{sat}(p < p_{sat})$ at the initial moment of time and $p = p_{sat}$when $x \geq \xi_0$.

Diagram 1 in paper [56] shows the distribution of the gas factor in a one-dimensional problem depending on the quantity $q = (p_{sat} - p)$. It was shown that the amount of gas released increases with an increase of q, which agrees with the experimental data. Diagrams 2–4 in paper [56] demonstrate the spatial scheme of propagation of gas front $p = p_{sat}$.

3.11.5.7 Numerical Modelling of Flows in Several Connected Regions

The numerical model bank ASAFD contains a number of algorithms to solve problems of the flow of homogeneous fluid in porous several connected media. These algorithms are based on a numerical approach proposed by one of the authors of this book to compile the Navier-Stokes model in several connected regions (Section 3.3 in this chapter).

The importance of studying the flow in vertical cross section for oil beds characterised by two different porosities and in multilayered systems was described above in detail.

REFERENCES

1. Alekseyev GV, Khusnutdinova NV. On solubility of the first boundary problem and Cauchy problem for equation of two-phase fluid one-dimensional flow. FEAS USSR 1972;202(2):310–2.
2. Antontsev SN, Kazhikhov AV, Monakhov VN. Boundary problems of nonhomogeneous fluid mechanics. Novosibirsk; 1983. 316 p.
3. Antontsev SN, Kazhikhov AV, Monakhov VN. Boundary value problems in mechanics of nonhomogeneous fluids. Studies in mathematics and its applications, vol. 22. North-Holland; 1990.
4. Antontsev SN, Monakhov VN. On some problems of two-phase incompressible fluid flow. Continuous medium dynamics. Novosibirsk; 1969. pp. 156–67.
5. Antontsev SN, Papin AA. Approximated methods of two-phase flow problems solution. Report, AS USSR 1979;247(3).
6. Antontsev SN, Papin AA. Approximated methods of regular and degenerate problems solution in two-phase flow. Continuous medium dynamics: Collection of sci. papers. Novosibirsk; 1982. Issue. 54. pp. 15–48.
7. Babakov AK, Gushchin VA, Davydov YM, Tolstykh AI. Numerical modelling of detached flows. V-th All-Union Congress of applied mechanics theory: Abstracts of Reports. Alma-Ata; 1981.
8. Barenblatt GI. The theory of two incompressible fluids flow in homogeneous porous medium. Numerical methods of continuous medium mechanics, 1971; 2(3): p. 103–117.
9. Barenblatt GI, Yentov VM, Ryzhik VM. Flow of fluids and gases in natural beds. Moscow: Nedra; 1984. 208 p.
10. Belotserkovsky OM. Numerical modelling in continuous media mechanics. Moscow: Nedra; 1984. 520 p.
11. Belotserkovsky OM, Belotserkovsky SO, Gushchin VA. Numerical modelling of non-stationary periodic flow of viscous fluid past a cylinder. ZHM & MPh 1984;24 (8):1207–16.
12. Belotserkovsky OM, Gushchin VA, Shchennikov VV. Application of splitting method to solve problems in viscous incompressible fluid dynamics. ZhVM and MF 1975;15 (1):197–207.
13. Belotserkovsky OM, Davydov Yu M. Investigation of schemes of coarse particles method with the help of difference approximations. Problems of applied mathematics and mechanics. Moscow: Nauka; 1971. p. 145–55.
14. Berkovsky BM, Nogotov EF. Difference methods to investigate thermal exchange problems. Minsk: Nauka i tekhnika; 1976.
15. Bocharov OB, Monakhov VN. Boundary-value problems of nonisothermal two-phase flow in porous media. Continuous medium dynamics. Collection of sci. papers, No. 86. Novosibirsk; 1988. p. 47–59.
16. Bocharov OB, Monakhov VN. Nonisothermal flow of incompressible fluids with variable residual saturations, No 88. Novosibirsk; 1988. p. 3–12.
17. Bocharov OB, Monakhov VN. Titles of reports of international conference "Mathematical models and numerical methods of continuous media mechanics". Novosibirsk; 1996. p. 398–399.
18. Bocharov OB, Monakhov VN. On solubility of boundary problems of two incompressible inhomogeneous fluids nonisothermal flow in porous media. Report of Academy of Sciences 1997;352(5):583–6.

19. Bocharov OB, Monakhov VN, Khusnutdinova NV. Proc. Congr. Intern. Modelisation Math, Ecoulements en MilieuxPoreux. Saint-Etiene, May, 1995.
20. Bocharov OB, Monakhov VN, Osokin AE. Numerical and analytical investigation methods of thermal two-phase flow. Mathematical models of flow and their applications, Novosibirsk; 1999. p. 46–59.
21. Bugrov AN. Iteration schemes to solve grid equations arising in method of virtual domains. Numerical analysis, Novosibirsk; 1978. p. 10–23.
22. Bugrov AN, Smagulov SS. Method of virtual domains in boundary problems for Navier-Stokes equations. Mathematical models of fluid flow, Novosibirsk; 1978. p. 79–90.
23. Bulygin VY. Formation hydromechanics. Moscow: Nedra; 1974. 230 p.
24. Ber Y, Zaslavski D, Irmey S. Physico-mathematical aspect of water flow. Moscow: Nedra; 1971. 446 p.
25. Vabishevich PN. Method of virtual domains for equations of mathematical physics. Moscow: 1992. 115 p.
26. Vakhitov GG. Efficient methods to solve problems of inhomogeneous oil- and water-bearing formations applying finite-difference scheme. Moscow: Gostechizdat; 1963. 216 p.
27. Vedernikov VV, Nikolayevsky VN. Equations of porous media mechanics saturated with two-phase fluid. Izv. AS USSR. Fluid and gas mechanics, No. 5. 1978. pp. 165–69.
28. Widlung OB. On the effects of scaling of the Peaceman-Rachford method. Math Comput January, 1971;25(113):33–41.
29. Vladimirova NN, Kuznetsov BG, Yanenko NN. Numerical calculations of symmetrical flow past a plate by viscous incompressible fluid flow. Some questions of computing and applied mathematics. Novosibirsk; 1966. p. 29–35.
30. Voevodin AF, Shugrin SM. Numerical methods of one-dimensional systems calculation. Novosibirsk: Nauka; 1981.
31. Voronovsky VR, Maksimov MM. System of information processing at development of oil deposits. Moscow: Nedra; 1975. 232 p.
32. Galkina EG, Papin AA. Analytical solution of equations of two fluids flow with account of dependence of their viscosities on velocity gradients. Applied mechanics and technical physics. Novosibirsk; 1999.
33. Gushchin VA. Numerical investigation of flow past a finite size body by incompressible viscous fluid flow. JHM and MPh 1980;20(5):1333–41.
34. Gushchin VA, Shchennikov VV. On one numerical method of Navier-Stokes equations solution. JHM and MPh 1974;14(2):512–20.
35. Gushchin VA, Shchennikov VV. Solution of problems of viscous incompressible fluid dynamics by splitting method. Direct numerical modelling of gas flows. Moscow: All-Union Centre of AS USSR; 1978. p. 114–33.
36. Daikovsky AG, Chudov LA. Influence of diagram factors at calculation of trace of poor flow past a body. Numerical methods of continuous medium mechanics, vol. 6, No. 5. Novosibirsk, 1975. p. 34–44.
37. Danayev NT, Zhumagulov BT, Smagulov SS. Numerical solutions of free convection equations. Titles of reports of the Ilndrep.conf. on probl. of сотр. math, and autom. of sci res Alma-Ata; 1988. p. 40.
38. Danayev NT, Zhumagulov BT, Kuznetsov VG, Smagulov Sh S. investigation of convergence of efficient finite-difference schemes for Navier-Stokes equation in variables (u,v,p). Mech model 1992;6(23)(2):25–57.
39. Fromm J. Non-stationary flow of incompressible viscous fluid. Computing methods in fluid dynamics. Moscow; 1967. p. 343–381.
40. Dzhaugashtin KE. Methodical development by course "Calculation of turbulent flows". Alma-Ata; 1986. 48 p.

41. Jim Jr D, Peaceman DW, Rachford Jr. A method for calculating multidimensional immiscible displacement. Trans AIME 1959;216:297–308.
42. Dulan E, Miller J, Shilderse U. Uniform numerical methods of investigation of problems with boundary layer. Moscow: Mir; 1983.
43. YershinSh A, Zhapbasbayev UK. Viscous incompressible fluid flow in contact apparatus with permeable partition. IPhZ 1986;50(4): (Dep in VINITI, No. 7245-B). 631 pp.
44. Ewing RE, Monakhov VN. Nonisothermal two-phase flow in porous media. Free boundary problems in continuum mechanics, International series of numerical mathematics, vol. 106. 1992. p. 121–30.
45. Zhumagulov BT. Numerical methods of solution of Navier-Stokes equations in multiply connected region: Dissert. Cand. of phys.-math. Sci., Alma-Ata; 1990. 153 p.
46. Zhumagulov BT. Numerical definition of the pressure in face zones of the wells. Report NAS RK 1995;(3):36–41.
47. Zhumagulov BT. Numerical definition of the pressure in face zones of the wells. Vestnik of Akmolinsk University; 1995;(2). p. 151–56.
48. Zhumagulov BT. Mathematical modelling of incompressible fluid flow in porous media. Mathematical and information technologies in education and science: Materials of scientific and methods seminar Part II. Almaty; 1996. p. 29–32.
49. Zhumagulov BT. Mathematical models of inhomogeneous fluid flow and their application in computer technologies for oil fields: Dissert. Of doctor of phys.- math. Sciences, Alma-Ata; 1997. 327 p.
50. Zhumagulov BT, Baigelov KZ, Temirbekov NM. Numerical modelling of formation development. Mechanics and modelling of technology processes. 1996. No. l. p. 16–20.
51. Zhumagulov BT, Baimirov KM. Numerical solution of one flow problem. Almaty; 1996. 23 pp. (Preprint EA RK No. 16).
52. Zhumagulov BT, Balakayeva GT. Numerical modelling of boundary layer flow motion. Almaty; 1996. 21 pp. (Preprint EA RK, No. 15).
53. Zhumagulov BT, Danayev NT. ε-approximation of one problem for Navier-Stokes equation. Alma-Ata; 1989. 14 pp. (Dep. in KazgosINTI, 25.12.98, No. 2955).
54. Zhumagulov BT, Danayev NT. Stability and convergence of one class solution of different schemes of incompressible fluid flow problem. Alma-Ata; 1989. 17 pp. (Dep. In KazGosINTI, 25.12.98, No. 2952).
55. Zhumagulov BT, Danayev NT, Smagulov SS, Orunkhanov MK. Numerical solutions of Navier-Stokes equations by coarse particles method in multiply connected regions. Materials of the IV conf. on difference equations and their application. Bulgaria; 1989. pp. 266.
56. Zhumagulov BT, Esekeyev KB, Tazhibekov ES. On one live oil flow problem. Almaty; 1996. 15 pp. (Preprint EA RK, No. 16).
57. Zhumagulov BT, Monakhov VN. Technologically efficient geological and mathematical models of formation. Oil and gas of Kazakhstan 1997;(3):41–56.
58. Zhumagulov BT, Monakhov VN. Theoretical analysis of thermal influence on oil production process. Oil and gas 1999;2(7):74–91.
59. Zhumagulov BT, Smagulov SS, Baimirov KM. Modelling of oil fields development. Materials of international conference "Actual problems of mathematics and mathematical modelling of ecological systems". Almaty: Gylym; 1996. p. 25.
60. Zhumagulov BT, Smagulov SS, Danayev NT, Temirbekov NM. Numerical methods of solution of Navier-Stokes equations in multiply connected region. Heads of III Int. Sem. on flame structure. Alma-Ata; 1992.
61. Zhumagulov BT, Smagulov SS, Monakhov VN, Zubov NV. New computer technologies in oil production. Almaty: Gylym; 1996. 167 p.

62. Zhumagulov BT, Smagulov SS, Orunkhanov MK. Numerical methods of solution of Navier-Stokes equations in multiply connected region. Izv AS KazSSR Ser Phys-math 1989;(3):23–7.
63. Zhumagulov BT, Temirbekov NM. Numerical method of natural convection equations solution in two-phase region. Oil and gas of Kazakhstan 1999;1(5):47–55.
64. Zhumagulov BT, Temirbekov NM. Methods of finding of oil production main operational parameters. Oil and gas of Kazakhstan 1997;(2):47–50.
65. Zolotukhin AB. Modelling of oil extraction processes from strata with use of methods of oil recovery increase: text-book. Moscow: MOGI; 1990. 267 p.
66. Kazhikhov AV. Some analytical problems of non-stationary flow and their numerical solution. Continuous medium dynamics, No. 3. 1969. pp. 33–49.
67. Kalashnikov AS. Construction of generalized solutions of quasi-linear equations of the 1st order. Far-Eastern Acad of Sciences USSR 1959;127(1):27–30.
68. Kalitkin NN. Numerical methods. Moscow: Nauka; 1978. 512 p.
69. Kantayeva TV, Osokin AE. Analytical problems of two-phase flow in case of variable residual saturations. Continuous media dynamics, No. 117. Novosibirsk; 2000. pp. 18–22.
70. Kobelkov GM. On methods of Navier-Stokes equations solution. Far- Eastern USSR Acad of Sciences 1978;123(4):843–6.
71. Kobelkov GM. On one iteration method of elasticity theory difference problems solution. Far Eastern USSR Acad Sci 1978;123(4):843–6.
72. Collins R. Fluid flows over porous materials. Moscow: Mir; 1964. 350 p.
73. Konovalov AN. Multiphase incompressible fluid flow problems. Novosibirsk; 1972. 128 p.
74. Konovalov AN. Multiphase incompressible fluid flow problems. Novosibirsk: Nauka; 1988. 166 p.
75. Konovalov AN, Monakhov VN. On some models of multiphase fluids flow. Continuous medium dynamics. Novosibirsk, No. 27. 1976. p. 51–65.
76. Kruzhkov SN. Generalized Cauchy problem solution as a whole for nonlinear equations of the first order. Far Eastern USSR Acad Sci 1969;187(1):29–32.
77. Kuznetsov BG, Smagulov S. Approximation of fluid dynamics equations. Numerical methods of continuous medium mechanics, vol. 6, No. 2. Novosibirsk; 1976. p. 158–75.
78. Kuznetsov BG, Smagulov S. On speed of solutions convergence of one system of equations with a small parameter to solutions of Navier-Stokes equations. Mathematical models of fluid flow. Novosibirsk; 1978. p. 158–75.
79. Kuznetsov BG, Smagulov S. On converging schemes of fractional steps for three-dimensional Navier-Stokes equations. Numerical methods of continuous medium mechanics, vol. 15, No. 2. Novosibirsk; 1984. p. 69–80.
80. Kurbanov AK. On two-phase fluids flow equations in porous medium. Theory and practice of oil production. Moscow: 1968. p. 281–86.
81. Kucher NA. On converging splitting scheme for multidimensional equations of viscous gas. Far Eastern USSR Acad of Sciences 1991;320(6):1315–8.
82. Ladyzhenskaya OA, Rivkind VY. On converging difference schemes for Navier-Stokes equations. Numerical method of continuous medium mechanics, vol. 2. No. 1. Novosibirsk; 1971. p. 55–73.
83. Ladyzhenskaya OA, Solonnikov VA, Uraltseva NN. Linear and quasi- linear equations of parabolic type. Moscow: Nauka; 1967. 736 p.
84. Leverett MC. Capillary Behaviour in porous solids trans. AIME 1941;142.
85. Lyugt H, Khosmich G. Laminar flow past flow past a flat plate under various angles of attack. Numerical methods in fluid mechanics. Moscow: 1973. p 269–76.
86. Maksimov MI, Rybitskaya LP. Mathematical modelling of oil field development processes. Moscow: Nedra; 1976. 264 p.

87. Marchuk GI. Numerical solution of atmosphere and ocean dynamics problems. L. Gidrometeroizdat; 1974. 304.
88. Meiramov AM. On solvability of Verigin problem in exact formulation. Far Eastern USSR Acad Sci 1980;253(3):588–91.
89. Monakhov VN. Boundary problems with free boundaries for elliptical systems of equations. Novosibirsk; 1977. 420 p.
90. Monakhov VN. Fluid flow with free boundary in nonperfect porous media. Mathematics and mechanics problem. Novosibirsk; 1983. p. 138–50.
91. Monakhov VN. Mathematical model of inhomogeneous fluid flow. Continuous medium dynamics, Novosibirsk; 1989. No. 94. p. 64–71.
92. Monakhov VN. Analytical solutions of thermal two-phase flow. Applied mechanics and technical physics, vol. 40, No. 3. Novosibirsk; 1999. p. 9–17.
93. Monakhov VN, Bocharov OB, Kantayeva TV, Osokin AE, Tlyusten SR. Analytical solutions of two-phase flow. Titles of reports of the Int. Conf. "Mathematical models and numerical methods of continuous media mechanics". Novosibirsk; 1997. p. 400–401.
94. Monakhov VN, Khusnutdinova NV. On superposition of channel and volume flows in viscous incompressible fluid. Applied mechanics and technical physics 1995;(1):95–9.
95. Nigmatulin RI. Multiphase media dynamics, vol. 1. M.: Nauka; 1987,464 p.
96. Nomofilov EV, Pirogov EP, Trevogda VM. Analysis of numerical methods of fluid dynamics problems modelling. Obninsk; 1983. 30p (Preprint/PhEI; PhEI-1400).
97. Orunkhanov MK. Use of virtual domains method for equations of viscous incompressible fluid in simply connected region. Novosibirsk; 1981. 18p (Preprint ITPM SD USSR AS; No. 16).
98. Oskolkov AP. On some converging difference schemes for Navier-Stokes equations. Proceedings of MI AS 1973;125:164–72.
99. Osokin AE. Numerical modelling of nonisothermal flow of two-phase fluid. Heads of Siberian conf. on mathematical physics nonclassical equations. Novosibirsk; 1995. p. 74.
100. Osokin AE. Substantiation of one approximated method in two-phase nonisothermal flow. Continuous medium dynamics, No. 113. Novosibirsk; 1998. p. 114–18.
101. Papin AA. Solvability "in small" in time of system of equations of one- dimensional flow of two interpenetrating viscous incompressible fluids. Continuous medium dynamics, No. 113 Novosibirsk; 1999. p. 64–70.
102. Paskonov VM, Polezhayev VI, Chudov AA. Numerical modelling of heat- and mass exchange processes. Moscow: Nauka; 1984. 288 p.
103. Pleshchinsky BI. Determination of phase permeability – flow speed dependence on quantitative relation of filtering fluids. Underground hydromechanics, No. 5. Kazan, 1968. p. 16–23.
104. Polezhayev VI. MAGS 1967;(2).
105. Polybarinov-Kochina PY. Theory of ground waters motion. Moscow: GITTL; 1952. 676 p.
106. Polybarinov-Kochina PY. Theory of ground waters motion. 2nd ed., revised and added. Moscow: Nauka; 1977. 664 p.
107. Polybarinov-Kochina PY. Theory of ground waters motion. Moscow: Nauka; 1987.
108. Ragulin VV. To the problem of viscous fluid flow through bounded region at given pressure of head drop. Continuous medium dynamics, No. 27 1975. p. 78–92.
109. Ragulin VV, Smagulov SS. On smoothness of solution of one boundary- value problem for Navier-Stokes equations. Numerical methods of continuous medium mechanics, vol. 11, No. 4. 1980. p. 113–121.
110. Development of investigations on flow theory in the USSR (1917–1967) /Under the editor ship of Polubarinova-Kochina P.Ya.: Nauka, 1969. 545 p.
111. Rappoport LA, Leas W. Properties of linear waterfloods. Trans AIME 1953;198.

112. Rakhmatulin KA. Gas dynamics basis of interpenetrating fluids of compressible media. PMM 1956;20(2):183–95.
113. Richardson E. Real fluid dynamics. Moscow: Mir; 1965. 328 pp.
114. Ryzhik VM. On capillary impregnation by water of oil-saturated hydrophilic stratum. Izv. USSR AS. Mechanics and machine building, No. 2. 1960. p. 149–51.
115. Samarsky AA. On monotonous difference schemes for elliptical equations in case of non self-conjugate elliptical operator. ZhVM and MF 1965;5(3).
116. Samarsky AA. Theory of difference schemes. Moscow: Nauka; 1983. 616p.
117. Samarsky AA. Theory of difference schemes. Moscow: Nauka; 1989.
118. Samarsky AA, Gulin AV. Stability of difference schemes. Moscow: Nauka; 1973. 415p.
119. Samarsky AA, Nikolayev ES. Methods of grid equations solutions. Moscow: Nauka; 1978. 590p.
120. Saulyev VK. On one method of automation of boundary problems solution on quick computers. Report USSR AS 1962;444(3):497–500.
121. Sirochenko VP. Numerical solution of one problem on viscous incompressible fluid flow in double connected region. Numerical methods of continuous medium mechanics, vol. 8. No. 1. Novosibirsk; 1977. p. 119–34.
122. Smagulov S. Method of virtual domains for boundary problem of Navier- Stokes equations: Preprint No. 68. Novosibirsk; 1979. p. 22.
123. Smagulov S, Orunkhanov MK. Approximated method of fluid dynamics equations solutions in multiply connected regions. Report. Far Eastern USSR AS 1981;(l):151–64.
124. Starovoytov VN. Measure-solution of the two phase flow problem. Continuous medium dynamics, No. 107. Novosibirsk, 1993. p. 124–33.
125. Elroid Jr S. Numerical solution of Navier-Stokes equations in double connected regions for incompressible fluid flow. Rocket Technol cosmonautics 1974;12 (5):76–82.
126. Surguchyov ML. Methods of control and regulation of oil fields development process. Moscow: Nedra; 1968. 300 p.
127. Tarunin EL. Calculation experiment in free convection problems. Irkutsk; 1990. p. 228.
128. Tarunin EL, Chernatynsky VI. Convection in closed cavity heated on the side at presence of viscosity temperature dependence. Transactions of Perm University Fluid Dynam 1972;(IV):71–85.
129. Telegin NG. Numerical realization of main two-phase fluid flow models. Continuous medium dynamics, No 117. Novosibirsk; 2000. p. 72–9.
130. Temam R. Unemethoded'approximation de la solution des equations de Navier-Stokes. Bull Soc Math France, 96:115–52.
131. Timukhin GI. On construction of some difference schemes and approximation of boundary thermal functions. Moscow: 1974. (Preprint /USSR AS, Institute of Problems of Mechanics; No. 40).
132. Timukhin GI, Timukhina MM. Fluid convection calculations in square region at side warm-up. Numerical methods of continuous medium mechanics, vol. 7, No. 7. Novosibirsk, 1976. p. 103–120.
133. Titchmarch EC. Expansions on eigenfunctions connected with difference equations of the second order, vol. 2. Moscow: IL; 1961. 555 p.
134. Turganbayev EM. Viscous-elastic fluid flow of Olroid type. Continuous medium dynamics, No 108. Novosibirsk; 1994.
135. Faizullayev DF, Umarov AI, Shakirova AA. Fluid dynamics of one- and two-phase media and its practical use. Tashkent: Fan; 1980.
136. Feorenko RP. Iteration methods of difference elliptical equations solutions. UMN 1973;28(2):121–81.

137. Khalimov EM, Levin BI, Dzyuba VI, Ponomaryev SA. Technology of formation oil recovery increase. Moscow: Nedra; 1984.
138. Khartman F. Ordinary difference equations. Moscow: Mir; 1970. 720 p.
139. Charny IA. Underground fluid dynamics. Moscow: Gostechizdat; 1963. 396p.
140. Chekalyuk EB. Formation thermodynamics. Moscow: Nedra; 1965 319 p.
141. Numerical methods of problem solution of incompressible fluid flow /Col. of scien. papers. Novosibirsk; 1975.
142. Shvidler MI, Levin BI. One-dimensional flow of immiscible fluids. Moscow: Nedra; 1970. 156 p.
143. Sheideger AE. Physics of fluid flow through porous media. M.: Gostoptechizdat; 1960. 250 p.
144. Shirko IV. Numerical investigation of flows granulated media. Moscow: 1986. p. 236–46.
145. Chia-Shun Y. Dynamics of nonhomogenious fluids. N.Y.L., Macmillan; 1963. 306 p.

INDEX

Note: Page numbers followed by "*f*" and "*t*" refer to figures and tables respectively.

G

H

I

K

L

M

T

Printed and bound by CPI Group (UK) Ltd, Croydon, CR0 4YY
08/05/2025
01864907-0002